Big Data-Driven Intelligent Fault Diagnosis and Prognosis for Mechanical Systems

Supported by the National Fund for Academic Publication in Science and Technology

大数据驱动的机械系统智能故障诊断与预测

Big Data-Driven Intelligent Fault Diagnosis and Prognosis for Mechanical Systems

雷亚国 李乃鹏 李 响 著

此版本仅限在中国大陆地区(不包括香港、澳门以及台湾地区)销售。
未经出版者预先书面许可,不得以任何方式复制或发行本书的任何部分。

图书在版编目(CIP)数据

大数据驱动的机械系统智能故障诊断与预测＝Big Data-Driven Intelligent Fault Diagnosis and Prognosis for Mechanical Systems：英文/雷亚国,李乃鹏,李响著. — 西安：西安交通大学出版社,2022.12(2025.3重印)
ISBN 978－7－5693－2802－8

Ⅰ. ①大… Ⅱ. ①雷… ②李… ③李… Ⅲ. ①机械系统-故障诊断-英文 Ⅳ. ①TH17

中国版本图书馆 CIP 数据核字(2022)第 183032 号

书　　名	大数据驱动的机械系统智能故障诊断与预测 DASHUJU QUDONG DE JIXIE XITONG ZHINENG GUZHANG ZHENDUAN YU YUCE
著　　者	雷亚国　李乃鹏　李　响
责任编辑	鲍　媛
责任校对	王　娜
出版发行	西安交通大学出版社 (西安市兴庆南路 1 号　邮政编码 710048)
网　　址	http://www.xjtupress.com
电　　话	(029)82668357　82667874(市场营销中心) (029)82668315(总编办)
传　　真	(029)82668280
印　　刷	西安五星印刷有限公司
开　　本	700mm×1000mm　1/16　印张 18.5　字数 567 千字
版次印次	2022 年 12 月第 1 版　2025 年 3 月第 3 次印刷
书　　号	ISBN 978－7－5693－2802－8
定　　价	228.00 元

如发现印装质量问题,请与本社市场营销中心联系、调换。
订购热线:(029)82665248　(029)82665249
投稿热线:(029)82665397
读者信箱:banquan1809@126.com

版权所有　侵权必究

Preface

With the increasing demands for high reliability, ensured operating safety and reduction of maintenance costs, fault diagnosis and prognosis for mechanical systems, which are the key components in prognostics and health management (PHM), are becoming more and more important in the modern industries. In the past decades, theories and methodologies of diagnosis and prognosis have been successfully established for mechanical systems, which further promote the developments of different industries, including manufacturing, aviation and aerospace industries, transportation, energy production, etc. Condition monitoring data are the foundation of diagnosis and prognosis. As the sensing, data storage, and computing technologies advance, the big data of mechanical systems have been available with significantly increased data volume, data types, data intensity, etc., propelling diagnosis and prognosis into the Big Data Era.

The industrial big data pose great opportunities as well as big challenges for diagnosis and prognosis of mechanical systems. It is promising to accurately identify and predict the machine health conditions through analyzing the big data. Meanwhile, how to effectively process the real industrial massive and messy data still remains difficult. In the current literature, some existing books have contributed much effort in the area of diagnosis and prognosis. However, the *Big Data-Driven Intelligent Fault Diagnosis and Prognosis for Mechanical Systems* is still a nascent and open research issue. This book aims at bridging this gap and offering the first attempt in presenting the latest advancements in this area from both the academic and engineering perspectives.

This book contains the descriptions of the fundamental theories in mechanical fault diagnosis and prognosis, as well as the cutting-edge research in the recent years. The presented theories and methodologies are all demonstrated by experimental validations or industrial applications in different fields. It is expected to offer a guideline for readers who are interested in the basic and advanced theories and methodologies of diagnosis and prognosis for mechanical systems. This book can be used as a textbook for graduate and senior undergraduate students in the related disciplines, and also an important reference for industrial engineers. Besides the mechanical engineering area, the presented works in this book can also benefit the other big data-related areas, such as computer science, electric engineering, automation science, power energy, etc.

The contents of this book were mostly developed at Xi'an Jiaotong University (XJTU), P. R. China. We would like to express sincere gratitude to our colleagues, Dr. Feng Jia, Dr. Liang Guo, Dr. Biao Wang, etc., who have provided invaluable contributions on the research works in this book. We also use this book to cherish the memory of Prof. Zhengjia He, who is the Ph.D. supervisor of the first author, Prof. Yaguo Lei. Even though he passed away in 2013 and has left us for 8 years, his insightful academic viewpoints on PHM are still the guidelines for our current and future research. Our memory of him will become more and more impressive as time goes.

Some research works in this book were conducted in cooperation with several world-renowned scholars and institutions in the area of diagnosis and prognosis. Here, we would like to give special thanks to Prof. Nagi Gebraeel from Georgia Institute of Technology in US, Prof. Steven X. Ding from the University of Duisburg-Essen in Germany, Prof. Ming J. Zuo from the University of Alberta in Canada, Prof. Jay Lee from the University of Cincinnati in US, and Prof. Chi-Guhn Lee from the University of Toronto in Canada. The communications with them further increase the depth of the theoretical research and the breadth of engineering cases in this book.

In the past years, the funding supports by Ministry of Science and Technology of the P. R. China, National Natural Science Foundation of China, Ministry of Education of the P. R. China, and Organization Department of the Chinese Communist Party, have largely helped form the extensive research materials in this book, as well as the grants by the industrial partners, including China Huadian Corporation Ltd., CRRC Corporation Ltd., China Huaneng Group Corporation Ltd., Sumyoung Technology Corporation Ltd., etc. The kind permissions from the industrial partners for reusing the research results are greatly appreciated. Moreover, the efforts of providing public machine condition monitoring datasets by the institutions including NASA in US, FEMTO-ST in France, and Case Western Reserve University in US, are very much acknowledged. The open-source datasets largely promote the research progress on diagnosis and prognosis, and further validate the methodologies addressed in this book.

We would also like to thank our Ph.D. students at XJTU, Bin Yang, Yuan Wang, Tao Yan, Saibo Xing, etc., without whom this book cannot be so comprehensive on the PHM theories, methodologies and applications.

At last, we would like to acknowledge the kind help and support from the editors of Springer and Xi'an Jiaotong University Press.

Xi'an, China	Yaguo Lei
November 2021	Naipeng Li
	Xiang Li

Contents

1 **Introduction and Background** 1
   1.1 Introduction 1
        1.1.1 AI Technologies for Data Processing 4
        1.1.2 Big Data-Driven Intelligent Predictive Maintenance 5
        1.1.3 Big Data Analytics Platform Practices 6
   1.2 Overview of Big Data-Driven PHM 9
        1.2.1 Data Acquisition 9
        1.2.2 Data Processing 11
        1.2.3 Diagnosis 12
        1.2.4 Prognosis 13
        1.2.5 Maintenance 15
   1.3 Preface to Book Chapters 16
   References 18

2 **Conventional Intelligent Fault Diagnosis** 21
   2.1 Introduction 21
   2.2 Typical Neural Network-Based Methods 23
        2.2.1 Introduction to Neural Networks 23
        2.2.2 Intelligent Diagnosis Using Radial Basis Function
             Network 27
        2.2.3 Intelligent Diagnosis Using Wavelet Neural Network 31
        2.2.4 Epilog 37
   2.3 Statistical Learning-Based Methods 37
        2.3.1 Introduction to Statistical Learning 38
        2.3.2 Intelligent Diagnosis Using Support Vector Machine 39
        2.3.3 Intelligent Diagnosis Using Relevant Vector Machine 49
        2.3.4 Epilog 57
   2.4 Conclusions 57
   References 58

3 Hybrid Intelligent Fault Diagnosis 61
3.1 Introduction 61
3.2 Multiple WKNN Fault Diagnosis 62
    3.2.1 Motivation 62
    3.2.2 Diagnosis Model Based on Combination of Multiple WKNN 63
    3.2.3 Intelligent Diagnosis Case Study of Rolling Element Bearings 67
    3.2.4 Epilog 69
3.3 Multiple ANFIS Hybrid Intelligent Fault Diagnosis 71
    3.3.1 Motivation 71
    3.3.2 Multiple ANFIS Combination with GA 72
    3.3.3 Fault Diagnosis Method Based on Multiple ANFIS Combination 73
    3.3.4 Intelligent Diagnosis Case of Rolling Element Bearings 75
    3.3.5 Epilog 80
3.4 A Multidimensional Hybrid Intelligent Method 81
    3.4.1 Motivation 81
    3.4.2 Multiple Classifier Combination 82
    3.4.3 Diagnosis Method Based on Multiple Classifier Combination 84
    3.4.4 Intelligent Diagnosis Case of Gearboxes 87
    3.4.5 Epilog 91
3.5 Conclusions 91
References 92

4 Deep Transfer Learning-Based Intelligent Fault Diagnosis 95
4.1 Introduction 95
4.2 Deep Belief Network for Few-Shot Fault Diagnosis 98
    4.2.1 Motivation 98
    4.2.2 Deep Belief Network-Based Diagnosis Model with Continual Learning 99
    4.2.3 Few-Shot Fault Diagnosis Case of Industrial Robots 106
    4.2.4 Epilog 110
4.3 Multi-Layer Adaptation Network for Fault Diagnosis with Unlabeled Data 111
    4.3.1 Motivation 111
    4.3.2 Multi-Layer Adaptation Network-Based Diagnosis Model 113
    4.3.3 Fault Diagnosis Case of Locomotive Bearings with Unlabeled Data 121
    4.3.4 Epilog 125
4.4 Deep Partial Adaptation Network for Domain-Asymmetric Fault Diagnosis 126

		4.4.1	Motivation	126
		4.4.2	Deep Partial Transfer Learning Net-Based Diagnosis Model	127
		4.4.3	Partial Transfer Diagnosis of Gearboxes with Domain Asymmetry	136
		4.4.4	Epilog	142
	4.5	Instance-Level Weighted Adversarial Learning for Open-Set Fault Diagnosis		144
		4.5.1	Motivation	144
		4.5.2	Instance-Level Weighted Adversarial Learning-Based Diagnosis Model	146
		4.5.3	Fault Diagnosis Case of Rolling Bearing Datasets	151
		4.5.4	Epilog	161
	4.6	Conclusions		163
	References			164
5	**Data-Driven RUL Prediction**			167
	5.1	Introduction		167
	5.2	Deep Separable Convolutional Neural Network-Based RUL Prediction		169
		5.2.1	Motivation	169
		5.2.2	Deep Separable Convolutional Network	169
		5.2.3	Architecture of DSCN	170
		5.2.4	RUL Prediction Case of Accelerated Degradation Experiments of Rolling Element Bearings	173
		5.2.5	Epilog	180
	5.3	Recurrent Convolutional Neural Network-Based RUL Prediction		181
		5.3.1	Motivation	181
		5.3.2	Recurrent Convolutional Neural Network	181
		5.3.3	Architecture of RCNN	182
		5.3.4	RUL Prediction Case Study of FEMTO-ST Accelerated Degradation Tests of Rolling Element Bearings	188
		5.3.5	Epilog	194
	5.4	Multi-scale Convolutional Attention Network-Based RUL Prediction		195
		5.4.1	Motivation	195
		5.4.2	Multi-scale Convolutional Attention Network	195
		5.4.3	Architecture of MSCAN	196
		5.4.4	RUL Prediction Case of a Life Testing of Milling Cutters	202
		5.4.5	Epilog	207
	5.5	Conclusions		208
	References			209

6 Data-Model Fusion RUL Prediction ... 213
- 6.1 Introduction ... 213
- 6.2 RUL Prediction with Random Fluctuation Variability ... 215
  - 6.2.1 Motivation ... 215
  - 6.2.2 RUL Prediction Considering Random Fluctuation Variability ... 216
  - 6.2.3 RUL Prediction Case of FEMTO-ST Accelerated Degradation Tests of Rolling Element Bearings ... 222
  - 6.2.4 Epilog ... 227
- 6.3 RUL Prediction with Unit-to-Unit Variability ... 227
  - 6.3.1 Motivation ... 227
  - 6.3.2 RUL Prediction Model Considering Unit-to-Unit Variability ... 229
  - 6.3.3 RUL Prediction Case of Turbofan Engine Degradation Dataset ... 237
  - 6.3.4 Epilog ... 239
- 6.4 RUL Prediction with Time-Varying Operational Conditions ... 241
  - 6.4.1 Motivation ... 241
  - 6.4.2 RUL Prediction Model Considering Time-Varying Operational Conditions ... 243
  - 6.4.3 RUL Prediction Case of Accelerated Degradation Experiments of Thrusting Bearings ... 252
  - 6.4.4 Epilog ... 255
- 6.5 RUL Prediction with Dependent Competing Failure Processes ... 256
  - 6.5.1 Motivation ... 256
  - 6.5.2 RUL Prediction Model Considering Dependent Competing Failure Processes ... 258
  - 6.5.3 RUL Prediction Case of Accelerated Degradation Experiments of Rolling Element Bearings ... 270
  - 6.5.4 Epilog ... 275
- 6.6 Conclusions ... 275
- References ... 276

Glossary ... 279

About the Authors

Prof. Yaguo Lei is a full professor of the School of Mechanical Engineering at Xi'an Jiaotong University, P. R. China. Before that, he was an Alexander von Humboldt fellow at University of Duisburg-Essen, Germany and a postdoctoral research fellow at University of Alberta, Canada. He received the B.S. degree and the Ph.D. degree both in mechanical engineering from Xi'an Jiaotong University, P. R. China, in 2002 and 2007, respectively. He is a Fellow of ASME, IET and ISEAM, a senior member of IEEE, CMES, ORSC and CAA, respectively, and a senior editor, associate editor or editorial board member of more than ten journals such as IEEE Transactions on Industrial Electronics, Mechanical Systems and Signal Processing, etc. His research interests include condition monitoring and operation maintenance large model, big data-driven intelligent fault diagnostics and prognostics, reliability evaluation and remaining useful life prediction, mechanical signal analysis and processing, and robotic system dynamic modeling. He has published four monographies, more than 100 peer-reviewed papers, and over 50 invention patents on fault diagnostics and prognostics. He has developed AI-based condition monitoring and diagnostic systems for real industrial applications, such as wind turbines, new-energy vehicles, high-speed trains, industry robots, etc. Moreover, he has received the National Award for Technological Invention, the Natural Science Award and Young Scientist Award from the Ministry of Education, and the Xplorer Prize from the New Cornerstone Science Foundation. He has been recognized as a Global Highly Cited Researcher by Clarivate Analytics and a Chinese Most Cited Researcher by Elsevier. He is also listed in the Stanford/Elsevier Global Top 2% Scientists.

Assoc. Prof. Naipeng Li is an associate professor in School of Mechanical Engineering of Xi'an Jiaotong University, P. R. China. Before that, he was a visiting scholar at Georgia Institute of Technology, US. He received the B.S. degree in mechanical engineering from Shandong Agricultural University, P. R. China, in 2012, and the Ph.D. degree in mechanical engineering from Xi'an Jiaotong University, Xi'an, P. R. China, in 2019. His research interests include condition monitoring, intelligent fault diagnostics and RUL prediction of mechanical systems. He has published more than 40 peer-reviewed papers in fault diagnosis and prognosis including 4 ESI hot papers and 7 ESI highly cited papers. He has participated in the development of 4 national standards, and has been granted more than 30 invention patents. He has developed many fault diagnosis and prognosis platforms for mechanical systems, such as wind turbines and industrial robots. His doctoral dissertation was awarded the Outstanding Doctoral Dissertation in Shaanxi Province. He was awarded the First Prize in Natural Science in Shaanxi Province, the First Prize in Science and Technology in Higher Education Institutions in Shaanxi Province, and the Young Elite Scientists Sponsorship Program by CAST.

Prof. Xiang Li is a full professor in School of Mechanical Engineering of Xi'an Jiaotong University, P. R. China. Before that, he was a postdoctoral fellow at University of Cincinnati, US, and a visiting scholar at University of California at Merced, US. He received the B.S. and Ph.D. degrees both in mechanics from Tianjin University, P. R. China, in 2012 and 2017, respectively. His research interests include industrial AI, machine vision, large multimodal model, neuromorphic computing, etc. He is an IET Fellow. He has published over 80 high-level research papers, including 24 ESI highly cited papers and 8 ESI hot papers. He is serving as the Associate Editors for IEEE Transactions on Neural Networks and Learning Systems, Expert Systems with Applications, and Pattern Recognition. He is on the editorial boards of IEEE/CAA Journal of Automatica Sinica, Measurement, Measurement Science and Technology, etc. His citations on Google Scholar are over 10000 with H-index of 46. He is honored to be selected as Highly Cited Researcher by Clarivate, and National Young Talent of China. His research works have been applied in different industries, including aviation, intelligent manufacturing, transportation, etc.

Chapter 1
Introduction and Background

1.1 Introduction

With the rapid development of modern industries, effective fault diagnosis and prognosis of mechanical systems are becoming more and more important, which are the key components of prognostics and health management (PHM). They can enhance mechanical system operational safety, prevent unexpected failures, reduce maintenance costs and increase production profits (Zio 2022). In general, three types of maintenance strategies are used in the industries, i.e., reactive maintenance, preventive maintenance, and predictive maintenance. The first two strategies follow the conventional patterns in the past decades, and additional costs are generated, from both mechanical system fatal break-downs and unnecessary mechanical system shut-downs in operations. The predictive maintenance strategy has been attracting more and more attention in the past years, which can effectively avoid mechanical system fatal failure while preventing unnecessary shut-downs. The three maintenance strategies are illustrated as follows (Randall 2021).

(1) Reactive maintenance

This is a classic maintenance strategy in which the mechanical system is left running until failure. This strategy, in theory, provides the longest operational time before breakdown. However, after mechanical system failure, it can only give a passive reaction, which might be disastrous and result in significant damage or accidents. This strategy is appropriate in some cases where there are generally a large number of mechanical systems to select from even if one of the mechanical systems fails. For mechanical systems like steam turbine generator sets, heavy oil catalytic cracking units, etc., however, the failure of any mechanical system may result in significant costs, which makes this strategy ineffective in many practical cases. This strategy is also unsuitable in the cases where the mechanical system breakdown may threaten human safety, such as aircrafts, trains, and automobiles.

(2) Preventative maintenance

The mechanical systems are periodically re-checked in the process of preventative maintenance, which is more conservative than reactive maintenance in order to prevent breakdowns. To avoid catastrophic effects, if an early fault is recognized, the system is generally shut down promptly and the defective component is replaced. In this strategy, the right time interval for inspections is critical. If the time period is exceedingly lengthy, there will be some unanticipated failures. On the other hand, if the time interval is too short, there will be too much maintenance and too many replacement components to be used. This strategy tries to predict faults and arrange maintenance solutions ahead of time to avoid catastrophic breakdowns. However, frequent shutdowns are necessary, which eventually reduce production efficiency and increases maintenance costs.

Preventative maintenance is acceptable if the failure time of the mechanical system can be correctly predicted based on the statistical analytics of the lifetimes of a large number of comparable mechanical systems. Nonetheless, due to the variability of different mechanical systems and operating conditions, the lifetimes of different mechanical systems are generally different from each other, resulting in a wide statistical distribution of lifetimes. The statistical results from a large number of mechanical systems may remarkably depart from the real failure time of the mechanical system of interest.

(3) Predictive maintenance

Predictive maintenance, often known as condition-based maintenance (CBM), is a maintenance strategy that makes maintenance recommendations based on data collected during the condition monitoring process. Through the examination of condition monitoring data, this strategy reveals the deterioration trend of the mechanical system. Then, the future deterioration pattern can be predicted by using certain prediction models or methodologies. The remaining useful life (RUL) of the mechanical system can be estimated using pre-defined failure thresholds. Before the inevitable breakdown happens, the optimal maintenance approach is planned based on the expected RUL. Predictive maintenance provides a number of distinct advantages compared with reactive and preventive maintenance. It may predict the mechanical system's possible failure by real-time condition monitoring, avoiding tragedies from occurring. Furthermore, depending on the predicted RUL, it may schedule maintenance action at the optimal moment. As a result, the mechanical system will have the longest uptime while requiring the least amount of maintenance. Predictive maintenance has obtained a lot of attention in recent years because of the aforementioned advantages (Dimaio et al. 2021).

Since condition monitoring data are one of the key components in condition-based maintenance, the quality and quantity of the collected data have large impacts on the predictive maintenance performance. In the past years, with the rapid development of sensing and storage technologies, condition monitoring systems are used to collect real-time data from the mechanical system to evaluate the health conditions. Huge amounts of data are generated over a lengthy period of operation, propelling mechanical system health monitoring into the Big Data Era. The major causes of big data generation are listed as follows.

1.1 Introduction

- The modern industrial mechanical systems generally include a large number of sub-systems and components, which are mostly supposed to be accurately monitored to ensure smooth operations. That leads to big data at the system level.
- The structure of the mechanical system is usually complex, especially for the large-scale ones. Therefore, multiple sensors should be considered to achieve effective monitoring of the whole mechanical system. Figure 1.1 illustrates the multiple sensors for different types of mechanical systems, such as industrial robots, wind turbines, etc.
- The mechanical systems are generally monitored in long-term operations, which directly leads to large amounts of condition monitoring data in the life cycle.
- The sampling frequency has largely increased with the advancement of sensing technology in the past decades. That further enlarges the data size in condition monitoring.

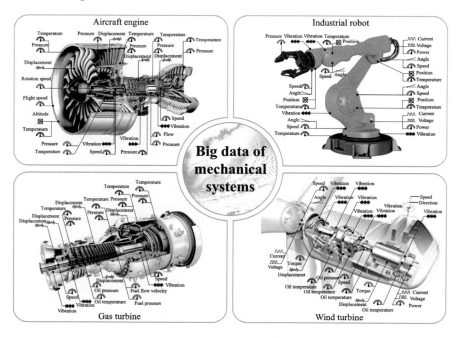

Fig. 1.1 Illustration of the multiple sensors for different mechanical systems

In general, the effective exploration of industrial big data has potential benefits for economic transformation and a new wave of production growth (Manyika et al. 2011). The use of valuable industrial big data analytics will become the basic competition for industrial enterprises, and intelligent predictive maintenance will surely benefit from the utilization of industrial big data (Ding and Li 2021). Industrial big data analytics has aroused widespread research interests in both academia and industry (Lei et al. 2020). In the following, as the key tools for processing big data, the artificial intelligence (AI) technologies are discussed, which are followed by the big data-driven predictive maintenance approaches using the AI methods. Afterwards, the existing big data-driven analytics platforms on PHM for mechanical systems are briefly reviewed.

1.1.1 AI Technologies for Data Processing

AI technologies are the key methodologies for big data processing at present. AI refers to the simulation of human intelligence in machines, which is programmed to think like humans and imitate the behaviors of the brain. This term can also be applied to any method that exhibits characteristics related to human thinking, such as learning, analyzing, understanding, and problem-solving (Minsky 1961).

In general, modern AI technologies are developed from statistics into machine learning. In the current literature, the artificial neural network (ANN)-based machine learning methods have been one of the most effective AI technologies in many cases, compared with the conventional machine learning algorithms such as support vector machine (SVM), etc. The ANN models have become more and more complex with increasing capabilities. Following this technical direction, the deep neural network (DNN) has been well developed, which is also known as deep learning. A large number of state-of-the-art methods have been achieved using deep learning, compared with the conventional ANN methods. For instance, the deep learning-based agent AlphaGo beat the best human player in the go game in 2016, which is a significant milestone in the development of AI technologies, since the go game strategy is extremely high-dimensional and complex in the searching space. The success of AlphaGo demonstrates the strong abilities and prospects of AI (Silver et al. 2016).

In order to further increase model generalization abilities in different application scenarios, deep transfer learning has been further well developed (Pandhare et al. 2021), which has also benefitted many industrial tasks such as cross-operating condition fault diagnosis for mechanical systems, etc. The model generalization ability has been largely increased with transfer learning, and the model robustness is also enhanced. More details of transfer learning on PHM will be discussed in Chap. 4 in this book.

Moreover, a large number of advanced machine learning algorithms have been further proposed in the recent literature to address more challenging industrial problems, including unsupervised learning, few-shot learning, meta-learning (Li et al. 2021a), reinforcement learning (Dai et al. 2020), etc. From the algorithmic perspective, a lot of cutting-edge research has been carried out by researchers all over the world. One of the latest advancements in AI is the data-model fusion method, which integrates human expert knowledge and data science to achieve better performance. This novel idea has also been introduced in PHM, and that will be presented in Chap. 6 in this book.

Meanwhile, the great merits of AI largely lie in the ability to solve real-world issues. In the past decades, AI methods have been successfully applied to different practical areas and achieved great success in engineering (Barr and Feigenbaum 2014). Besides PHM of mechanical systems, the aforementioned AlphaGo, and other daily-life applications, different areas of mechanical systems have also benefitted a lot from AI. For instance, researchers used AI for analyzing mechanical dynamics (Izzo et al. 2019), designing mechanical systems (Noor 2017), optimizing manufacturing processes (Kinkel et al. 2021), etc., all of which have achieved great success in engineering. In the future, AI will continue serving as a powerful tool for big data analytics.

1.1.2 Big Data-Driven Intelligent Predictive Maintenance

As stated in the previous sections, industrial big data can be well explored for intelligent predictive maintenance for mechanical systems. AI technologies can help find the underlying patterns and rules from the massive and messy practical data collected from the operating mechanical systems. The integration of industrial big data and AI technologies promotes the development of intelligent predictive maintenance of mechanical systems. Big data-driven intelligent predictive maintenance has become an important part of Industry 4.0 (Yan et al. 2017).

During the mechanical system operating periods, the health condition of the mechanical system can be well reflected by the measured data from multiple sensors. The features of the collected data are different with respect to different health levels. For instance, when the mechanical system is healthy, the data are generally stable with limited fluctuations. However, when the mechanical system is largely degraded, fault impact components or significant fluctuations can be observed in the data. In the common cases, the small anomalies found during the automatic evaluation of data from the sensors can indicate the components and equipment that require maintenance. These components can then be replaced or repaired appropriately before a complete failure occurs. In this way, by analyzing the big data, we can correctly identify the mechanical system health conditions and achieve predictive maintenance (Ainapure et al. 2020).

The new problems in predictive maintenance thus include effectively extracted features from such big data, properly diagnosing mechanical system health conditions using advanced methods, and further predicting the RUL of mechanical systems before the end of life. This process is not easy in industrial scenarios, since the collected data from the practical operating mechanical systems are usually noisy with the disturbances of environments and working modes. Furthermore, the completeness of the data cannot be guaranteed, since it is always difficult to keep accurately monitoring the mechanical system in a long term. These factors pose great challenges for actual big data analytics. How to effectively extract useful features for predictive maintenance is thus a quite difficult problem.

For processing industrial big data, AI technologies have been powerful for effective feature extraction, pattern recognition, health identification, etc. The whole implementation procedure can be automatic and accurate, which leads to intelligent predictive maintenance. However, there are two major challenging aspects in the current big data-driven intelligent predictive maintenance, i.e., weak model generalization ability, and low RUL prediction performance.

(1) Weak model generalization ability

In the current literature, the data-driven predictive maintenance models have been rapidly developed with the advancements of the AI technologies. However, the existing methods generally assume the training and testing data are from the same distribution, that indicates the data are supposed to be collected from the same mechanical system in the same working condition (Yang et al. 2021a). In practice, the mechanical system can be operated in different conditions, and multiple types of the mechanical system can be used. That results in performance degradation of the data-driven predictive maintenance methods in the real cases (Zhang et al. 2021). In order to address this challenging issue, this book presents the deep transfer learning-based intelligent fault diagnosis methods for mechanical systems in Chap. 4. By using the transfer learning strategies, the knowledge learned from the training data can be generalized to the testing data collected in different scenarios. That largely enhances the model applicability in the pratical engineering cases (Yang et al. 2022).

(2) Low RUL prediction performance

RUL prediction is always a challenging problem since the uncertainty of the mechanical system is very strong in the long-term operations, due to the complex mechanical structure, environmental disturbance, operating mode, etc. Traditionally, the physics-based prognosis models can capture the general degradation patterns of the mechanical systems. However, as the mechanical systems are becoming more and more sophisticated, it is quite difficult to establish an accurate physics-based model for RUL prediction. In the recent years, the availability of the run-to-failure big data for mechanical systems brings new opportunities for prognosis (Li et al. 2021b). The big data-driven RUL prediction methods have been well developed, which are presented in Chap. 5 in this book. Furthermore, relying on the data alone may lead to overfitting in prognosis. The latest research shows that the data-model fusion RUL prediction methods can well integrate the advantanges of both data-driven and model-based methods (Arias Chao et al. 2022; Lei et al. 2018), and are thus more promising. They are described in Chap. 6 in this book.

1.1.3 Big Data Analytics Platform Practices

In recent years, some commercial companies have already been taking advantages of both AI technologies and big data for improvements of mechanical systems, including General Electric, Rolls Royce, Sany Group, etc. Many cloud platforms have been developed for big data acquisition and analytics. Internet of Things (IoT) technologies

1.1 Introduction

have been embedded in the mechanical systems and production lines within the data cloud platforms. Predictive maintenance has been a critical domain that leverages the advanced technologies. Since industrial stream big data have extremely high volume, high velocity, and complex data types, many challenges occur in the practical scenarios. Addressing the practical big data problems has been quite difficult for practitioners.

Figure 1.2 presents a general structure for the big data analytics platform for mechanical systems. First, with respect to the multiple distributed mechanical systems, the condition monitoring data are collected using sensors, and stored using the distributed file systems. The locally collected data are then transmitted to the distributed databases. Next, the distributed data are uploaded to the big data analytics platform for central processing, which is in charge of the whole process. Diagnosis and prognosis reports can be generated, as well as the maintenance recommendations, etc. Afterwards, the analytics reports are sent to the managers, engineers, maintainers, customers, etc., of the mechanical systems. In this way, the information of the mechanical systems can be effectively delivered to all the people involved, and the best maintenance plan can be scheduled accordingly. In the following, some representative examples of the industrial big data analytics platforms are presented.

(1) General Electric

According to a report by General Electric (GE), a contemporary General Electric wind farm with 200 turbines offers a promising method for data processing in an industrial big data platform. Each turbine is embedded with around 50 sensors and control loops, and the data are sampled at various speeds based on the sort of analytics and process.

At the edge, the first level of analytics and interaction is at the wind turbines. The turbine controller's real-time analytics collects and saves sensor data to create an on-board data history record, optimizes the transmission of turbine blade rotation energy every 40 ms, converts it to electrical energy, and determines whether electrical energy should be converted into electrical energy. On the farm, the second stage of analytics and interaction takes place. At 160 ms intervals, the farm controller gets more than 30 signals from each turbine, and its real-time processing guarantees that the proper combination of turbines provides predictable electricity to the utility.

The turbine data are then sent to a distant monitoring center a few miles away every minute from this wind farm and other wind farms. Individual turbines and whole wind farms are analyzed by teams of data scientists and engineers to fine-tune asset algorithms and procedures, which are then applied to mechanical systems. Finally, in the form of energy production reports and predictions, an analytics report that combines operational data from the farm with financial and other data from the operator's business system is sent to the CFO's office. All of these provide a contemporary user interface for the big data platform.

(2) Rolls Royce

Facing the increasing technological changes, Rolls-Royce is focusing on innovation strategies in the aerospace industry (Rolls-Royce). Through the utilization of big data

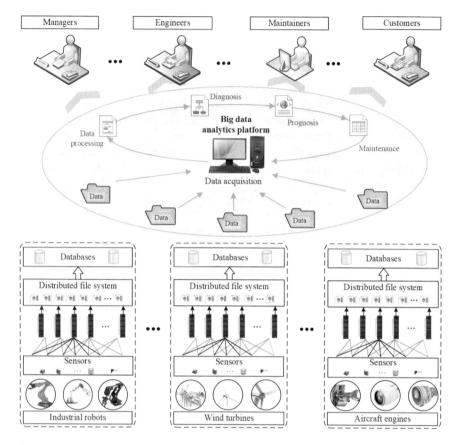

Fig. 1.2 Illustration of big data analytics platform for mechanical systems

analytics, data engineering, AI, machine learning, IoT, and the other technologies in the entire value chain, Rolls Royce can identify areas where data insights can help improve efficiency, so as to achieve strategic execution. Process innovation and optimization span the end-to-end customer experience, from engine design to providing customer-centric after-sales service and the use of robots for maintenance. Their goal is to help their customers achieve benefits and satisfy them by providing fast customer service, reducing customer costs, improving customer profitability, and ultimately increasing customer satisfaction. Rolls-Royce has accelerated digital transformation by establishing new business operating models under the Big Data Era.

(3) Sany Group

Excavators are powerful and complex systems. At present, they are showing growing trends of high efficiency, low energy consumption, intelligence, precision, and better user experience. As a traditional manufacturing company, Sany group makes manufacturing, service, and production digitized. High-performance computers are used for processing big data of excavators (SanyGroup).

Sany Group developed a big data storage and analytics platform, which is the "ECC Customer Service Platform". The platform includes the hardware and software parts for the underlying control of all the excavators, which can realize two-way interaction and remote control of excavators. It is able to connect more than 200000 customer excavators. The real-time operating status data can be transmitted to the server through sensors for analytics and optimization. The daily real-time monitoring of the mechanical system operation information is implemented, including location, working hours, speed, main pressure, fuel consumption, etc. The system is oriented to four types of main users: agents, operators, excavator owners, and R&D personnel. Users can capture the status of all the aspects of the mechanical system anytime and anywhere through the web or mobile APP. Based on big data analytics, precise control of the commonly used parts is achieved. The fuel consumption can be also effectively reduced by 10%.

The aforementioned industrial big data analytics platforms provide basic frameworks for the conduction of intelligent predictive maintenance. Lots of advanced technologies involving databases, cloud computing, web design, etc., are integrated. Nevertheless, the key point hidden in the industrial big data platform is the PHM module which is mostly composed of diagnosis and prognosis methodologies. The task of the PHM module is to transform industrial big data into valuable information for maintenance decisions. This is the major concern of industrial engineers as well as academic researchers. A powerful platform should be equipped with a strong PHM module. Otherwise, it is unable to create values for customers. The following part of this book will focus on the key points of the industrial big data platform, i.e., the hidden PHM methodologies. The following section will first give an overview of the big data-driven PHM.

1.2 Overview of Big Data-Driven PHM

In this section, the concept of big data-driven PHM is described, and the flow chart of the typical PHM for mechanical systems is presented in Fig. 1.3. Big data-driven PHM basically includes five major processes, i.e., data acquisition, data processing, diagnosis, prognosis, and maintenance (Hamadache et al. 2019).

1.2.1 Data Acquisition

Data acquisition is the process of capturing measurement signals from a monitored mechanical system using various types of sensors and storing the data on a computer.

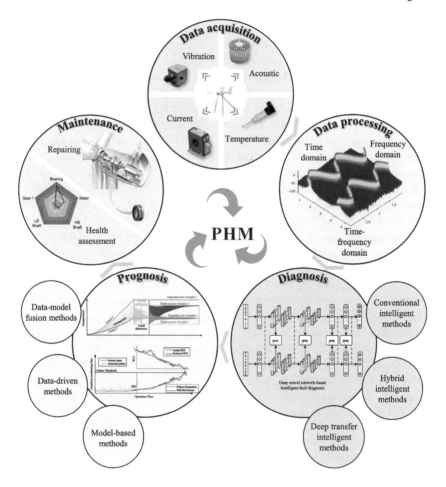

Fig. 1.3 The flowchart of the typical big data-driven PHM strategy

The measurement signal should be related to the health condition of the mechanical system that is being monitored. In other words, the measuring signal provides information that may be used to determine the mechanical system's health condition. Vibration signals, sound signals, temperature signals, and current signals are a few examples of commonly used monitoring data. The signal is collected and sent to a computer through a data acquisition (DAQ) device, where it is stored for future analytics. Under the Big Data Era, different kinds of sensors for mechanical systems have been created to gather different types of signals. Many new data acquisition equipment and technologies have been created and implemented in the industries as a result of the fast developments of computing and sensing technologies. The invention of additional sensors has also benefited real-world companies and factories in a variety of ways. Big data collection for PHM has been convenient and practicable with such devices.

1.2.2 Data Processing

Data processing is to examine the captured data stored in the data collecting process using signal analytics techniques and algorithms. The goal of data processing is to extract useful information from the original measurement signal that can disclose the health condition of mechanical systems. Many signal processing approaches and algorithms have been proposed and well developed in the literature (Wang et al. 2017). Time-domain analytics, frequency domain analytics, and time–frequency domain analytics are the three categories that they fall within.

The initial measurement signals are in the form of time domain and are typically sampled frequently between pre-specified time intervals. As a result, the raw measurement signals are directly used in time-domain analytics. Conventional time-domain analytics uses statistical features such as peak, root mean square (RMS), kurtosis, and skewness to describe the health conditions of mechanical systems. Time synchronous average (TSA), the autoregressive (AR) model, the autoregressive moving average (ARMA) model, principal component analysis (PCA), and other time-domain analytic approaches are widely employed.

Frequency-domain analytics is based on frequency-domain transformed signals. The capacity to break down the original signals into a sequence of frequency components is a benefit of frequency-domain analytics over time-domain analytics. The fast Fourier transform (FFT) is the most commonly utilized frequency-domain analytics method. The primary goal of spectrum analytics is to extract and discover certain frequency components related to mechanical system failure features. One popular FFT-based approach is the power spectrum. Cepstrum is also widely used because it can identify harmonics and sideband patterns in the power spectrum. Frequency filters, envelope analytics, sideband structure analytics, and other auxiliary tools for spectrum analytics are generally quite useful. The Hilbert transform, which is important in envelope analytics, has also been adopted to recognize and diagnose mechanical system faults.

Frequency-domain analytics has a restriction that it can only deal with stationary measurement signals and cannot deal with non-stationary measurement signals. Mechanical system measurement signals, however, are basically non-stationary. Thus, time-frequency-domain analytics has been proposed to deal with non-stationary measurement signals, which analyzes measurement signals in both time and frequency domains. To better identify mechanical system failure patterns, time-frequency-domain analytics evaluates the properties of measurement signals as two-dimensional functions of both time and frequency.

Short-time Fourier transform (STFT) is a typical time-frequency-domain analytic technique that splits measurement signals into various segments with short-time windows and then applies Fourier transform to each segment. The frequency-domain resolution is unavoidably reduced as a result of signal segmentation. Furthermore, each signal segment is roughly regarded as a stationary process. Therefore, the STFT can only be used on non-stationary signals in limited cases.

Wigner-Ville distribution is another prominent time-frequency-domain analytic technique. Instead of signal segmentation, it uses a bilinear transform. As a result, it can overcome STFT's frequency resolution restriction. The wavelet transform is another technique for time-frequency analytics. The wavelet transform, unlike the STFT, may be utilized for multi-scale analytics of the signal via dilation and translation, allowing it to well extract time-frequency features. The wavelet transform can generate high-frequency resolution at low-frequency areas, and high time resolution at high-frequency areas for signals with lengthy low frequencies and short high frequencies, due to its multi-scale analytics capability. The implementation of the wavelet transform necessitates the use of a pre-determined wavelet function, which limits its applications.

As a result, empirical mode decomposition (EMD), which is a new time-frequency domain analytic tool, is developed. It is based on the measurement signals' local characteristic time scales and may decompose signals into a collection of full and nearly orthogonal components, that are known as intrinsic mode functions (IMFs) (Lei et al. 2013). The signals in each IMF contain a natural oscillatory mode. Correspondingly, the EMD technique may adaptively divide the measurement signals into various modes based on their local characteristic time scales. Although the EMD technique performs well in the processing of nonlinear and non-stationary signals, it also has significant flaws, including a lack of theoretical proof, end effects, filtering stop criterion, and excessive interpolation.

It should be noted that each of the signal analytics tools described above has its own set of benefits and drawbacks. In real situations, with respect to the practical big data of the mechanical system, the case-specific knowledge and investigation are necessary to choose the best signal processing approach from a variety of options (Yang et al. 2021b). In this book, the more advanced methodologies in intelligent fault diagnosis and prognosis are focuse on, and the data processing part is less described since that is more well-established methodologies. Readers can refer to (Lei 2017) for more details on the specific data processing methodologies.

1.2.3 Diagnosis

Diagnosis is defined as the process of discovering and determining the relationship between the condition monitoring data collected in the measurement space and the mechanical system fault patterns in the health condition space. In general, diagnosis consists of three key steps: anomaly detection, fault isolation, and fault identification (Chen et al. 2021). Anomaly detection is the process of determining if a fault has already occurred in the mechanical system being monitored (Cheng et al. 2021). The goal of fault isolation is to identify the fault component and its location (Lei et al. 2020). The next stage in diagnosis is fault identification, which aims to identify the pattern and severity of the fault. It should be noted that the three processes are interconnected. Since the later phase is dependent on the output of the former one, they cannot be completed independently.

Conventional diagnosis approaches include comparing the signal processing outputs of mechanical systems in unknown health conditions to those in healthy conditions, and identifying their discrepancies. As a result, these manual diagnosis procedures need much diagnosis knowledge in the relevant disciplines. In the big data scenario, the conventional methods have been less capable of effectively diagnosing faults, since the data processing difficulty significantly increases. Compared with the conventional manual methods, the modern intelligent fault diagnosis methods, with the help of AI technologies, are developed for automatically and effectively replacing diagnosis experts in order to offer correct fault diagnosis results. Machine learning is a representative AI technique and has been popularly adopted. As a result, in the Big Data Era, it is expected to be the best tool for dealing with massive data for mechanical system failure diagnosis (Ma et al. 2020). The deep learning-based fault diagnosis method, which is the advanced algorithm in machine learning, has further been one of the most representative methods for intelligent diagnosis (Lei et al. 2020). In this book, Chaps. 2 and 3 present the well-established intelligent fault diagnosis methods using the AI technologies, which are the conventional and hybrid intelligent fault diagnosis methods, respectively.

In recent years, deep transfer learning-based fault diagnosis methods have also been well developed. The practical problems in the industries are focused on, where the training and testing data are from different distributions (Azamfar et al. 2020). Specifically, the training data and testing data are usually collected from different scenarios in practice, such as different operating conditions, different environments, different mechanical systems, etc. With respect to this challenge, the deep transfer learning-based fault diagnosis methods have been proposed, where the deep learning method is used as the framework and the transfer learning strategy is integrated. The knowledge of fault diagnosis is expected to be transferred from the source domain, which is the training data, to the target domain, which is the testing data. The deep transfer learning-based fault diagnosis methods have been the emerging technologies in the intelligent fault diagnosis and prognosis field, and they are elaborated in Chap. 4 in this book. Many successful deep transfer fault diagnosis methods and their applications in practical scenarios are also presented.

1.2.4 Prognosis

Diagnosis is the process of detecting the existence of defects in mechanical systems as early as possible, as well as identifying the types, locations, and degrees of faults. However, it can only be carried out after a defect has occurred, which is insufficient to prevent disasters from occurring. Prognosis should be performed during the PHM process to avoid unexpected hazards and reduce maintenance costs (She et al. 2020). Prognosis is the process of predicting the future performance of mechanical system using the prediction methods and calculating the RUL before the mechanical system loses its capacity to operate. Predictive maintenance or replacement is possible to be implemented with an accurate RUL prediction of mechanical systems using big data, saving expensive unexpected maintenance costs (Wang et al. 2021). As a result, RUL prediction becomes a hot topic that has attracted a lot of interests in recent years (Zang et al. 2021).

Prognosis is far more effective than diagnosis in achieving zero-downtime performance because we can schedule maintenance strategies before faults occur (Fink et al. 2014). Prognosis, on the other hand, cannot entirely replace diagnosis. The following are the main reasons:

- Prognosis is dependent on the diagnosis outputs and hence cannot be performed in isolation.
- There are always some unexpected failures that are difficult to anticipate.
- No prognosis approach can guarantee that the RUL will be precisely the same as the actual one.

As a result, diagnosis is still required to assist in maintenance decision-making. The prognosis process of mechanical systems has three key phases: state estimation, state prediction, and RUL prediction (Lei et al. 2018). The output of fault identification in diagnosis, which has inferred the pattern and severity of the defect, is used in state estimation. The goal of state estimation is to quantify the severity of the problem and estimate the state of deterioration of the mechanical system. The goal of state prediction is to forecast the state's deterioration trend and rate based on data from previous degradation curves. The temporal duration of the deterioration curve from the present condition to eventual failure is calculated using a pre-determined failure threshold in the RUL prediction stage.

Model-based, data-driven and data-model fusion methods are the three primary categories of mechanical system prognosis methods (Lei et al. 2018). Model-based methods aim to create mathematical or physical models that characterize mechanical system degradation processes (Cubillo et al. 2016), and then update model parameters with the observed data. Expert knowledge and real-time data from mechanical systems can both be included in the model-based methods. As a result, they can be useful in the RUL prediction of mechanical system. These methods also require less information than the data-driven methods. Markov-based models, Winner process models (Lei et al. 2016), and inverse Gaussian process models (Pan et al. 2016) are the widely-used models. Model-based methods, on the other hand, may not be appropriate for dealing with mechanical system prognosis in the following scenarios:

- Mechanical system failure patterns vary largely depending on the operations.
- The mechanical system is so complicated that it is difficult to understand the principles of operation and construct an accurate model.

Instead of creating models based on complete system physics and human knowledge, the data-driven methods seek to deduce mechanical system degradation mechanisms from measurement signals. These methods are based on the idea that, unless a fault arises, the statistical features of data are generally constant. Based on the past measurement signals, the approaches can generate the RUL prediction results. Therefore, the precision of the data-driven methods is determined by the quantity and quality of the previous data collection. However, in most situations, collecting enough quality data are quite challenging.

For the RUL prediction of the mechanical systems, both data-driven and model-based methods offer certain advantages and disadvantages. In recent years, the data-model fusion methods have been developed to supplement their merits (Liu et al. 2012). For condition monitoring, relying only on data or physics is insufficient to adequately characterize system behavior. Especially, for a more reliable mechanical system RUL prediction, we would like to combine the capabilities of both data-driven and model-based techniques. In this book, we also discuss the promising data-model fusion prognosis and remaining useful life prediction methods. The benefits of data-driven methods and model-based methods may be combined to further improve the performance of the prognosis models.

Chapters 5 and 6 of this book will focus on the topic of prognosis. The key prognosis theories are first explained. Then the methodologies and application cases in mechanical system RUL prediction are introduced by using several prognosis methods as examples. The more advanced data-driven and data-model fusion RUL prediction methods are illustrated in detail, which are presented in Chaps. 5 and 6, respectively. The experimental validations and application scenarios serve as a valuable resource for readers who want to learn how to apply the prognosis methods.

1.2.5 Maintenance

Diagnosis and prognosis are important in the development of the actual maintenance strategies. In big data-driven PHM, maintenance is the final step. Maintenance decision-making is the process of assessing diagnosis and prognosis outputs, and devising maintenance plans, in order to make an optimal replacement or repairing decisions for the failing or failed components in the mechanical system (Jardine et al. 2006; Alaswad and Xiang 2017). The assessment of the RUL and the optimization of the maintenance operations are used to make the best maintenance decisions. As a result, diagnosis and prognosis must be developed to correctly track the changes in mechanical system health conditions following maintenance and predict the changes in mechanical system reliability (Xu et al. 2021).

Many intelligent maintenance solutions have lately been created based on the PHM strategy with big data. Early detection, isolation, and identification of the incipient faults can be anticipated from a good PHM system, that can also well estimate and predict the mechanical system degradation levels. In this way, an appropriate maintenance plan and asset management decision can be accordingly made. Incipient faults should be recognized and tracked as they develop from minor to bigger faults, until the severity of the fault exceeds a pre-specified failure threshold or the degree of dependability decreases to an acceptable level. The health of the mechanical system or essential components should be monitored at any time with such a system, and the eventual breakdown can be predicted and avoided, allowing for near-zero downtime performance. Meanwhile, needless and costly preventative maintenance can be removed, maintenance schedules can be streamlined, and spare parts can be minimized, all of which can effectively save costs.

Since the early 2000s, a new term which is called e-maintenance has developed as a result of the advancement of e-technology, and it has become highly popular in maintenance decision making. It is described as the maintenance assistance that encompasses the resources, services, and management required to execute proactive decision processes. Not only e-technology (ICT, web-based, tether-free, wireless, informatics, etc.) but also e-maintenance activities such as e-monitoring, e-diagnosis, and e-prognosis are included. The origin of e-maintenance may be traced back to two primary causes.

- The advancement of sophisticated e-technologies allows for increased maintenance efficiency, velocity, and proactivity, resulting in an optimized maintenance workflow.
- The requirement to integrate business performance also raises maintenance needs, such as openness, integration, and collaboration with the other enterprise's services.

The introduction of e-maintenance has resulted in a number of practical improvements in maintenance types and techniques. First, with the help of wireless and Internet technology, any operator, management, or specialist may connect to a factory's mechanical system remotely through the Internet, allowing them to do remote maintenance. Second, an e-maintenance system connects customers, managers, engineers, and mechanical systems via an information architecture that relates geographically distant subsystems, allowing collaborative maintenance amongst different human operators, enterprise areas, and organizations. Third, real-time remote monitoring and high-frequency communications enable maintenance operators to check the health of remote mechanical system, obtain expert advice and ideas rapidly, and respond to any mechanical system monitoring situations.

All the five procedures described above form the typical big data-driven PHM methodology, which are also the key components in the intelligent predictive maintenance for mechanical systems. As the core parts in the big data-driven PHM, the intelligent fault diagnosis and prognosis methods for mechanical systems are the main contents in this book.

1.3 Preface to Book Chapters

In this book, a comprehensive introduction of the big data-driven intelligent fault diagnosis and prognosis strategies for mechanical systems is presented, including both the well-established conventional methods and the latest cutting-edge research results. The reported methodologies are all validated using practical data. The

1.3 Preface to Book Chapters

contents start with the typical and hybrid intelligent fault diagnosis methods with respect to the mechanical condition monitoring data. Afterwards, more focuses are placed on the recent developments of the advanced algorithms in big data-driven PHM, including the deep transfer learning-based fault diagnosis, big data-driven RUL prediction, and data-model fusion prognosis, etc. Both the theoretical research and the applications in the actual mechanical system settings are elaborated.

It should be pointed out that in the predictive maintenance of the mechanical systems, the rotating machines and the associated components are the most vulnerable to failures. Rotating machine usually operates in harsh working environments and therefore often fails (Lee et al. 2014). Any failure of the rotating machine may cause the failure of the entire mechanical system, thereby reducing the reliability, safety, and availability of the mechanical system. With the rapid development of science and technology, the rotating machine in modern industry is becoming larger and more sophisticated. The structure of the rotating machine is becoming more and more complex. As a result, its latent failures become more difficult to be detected (Li et al. 2020). Therefore, in this book, while the presented methods can be applied to different kinds of mechanical systems, the fault diagnosis and prognosis of the rotating machines are attached with more attention.

The remainder of this book starts with the conventional intelligent fault diagnosis methods in Chap. 2, which includes the typical artificial neural networks and statistical learning theories. Chapter 3 reviews the hybrid intelligent fault diagnosis approaches, that enhance the PHM methodologies from different perspectives. In Chap. 4, the deep transfer learning methods are presented, which discusses deep learning-based transfer learning strategies in mechanical system fault diagnosis. The transfer case studies across different operating conditions and different mechanical systems are provided as the validations of the presented methodologies. The generalization ability of the intelligent fault diagnosis models is largely enhanced with the deep transfer learning algorithms. Chapter 5 presents the big data-driven RUL prediction methods following the fault detection of the mechanical systems. The health assessment, health indicator establishment, and RUL estimation algorithms are provided. In Chap. 6, the data-model fusion methods for prognosis are illustrated, which integrates the advantages of both data-driven and model-based methods. That is currently at the frontier of the PHM-related research, and also validated using practical mechanical system data.

This book offers a guideline for the readers in the areas of intelligent fault diagnosis and prognosis of mechanical systems. The fundamental concepts are expected to be delivered, as well as the cutting-edge research and state-of-the-art algorithms. While the methodologies presented in this book are mostly focused on mechanical systems, the algorithms can be generally applied in different kinds of scenarios, such as image processing, audio recognition, control engineering, etc. We expect the readers from different research areas on big data could benefit from the contents in this book.

References

Ainapure A, Li X, Singh J, Yang Q, Lee J (2020) Deep learning-based cross-machine health identification method for vacuum pumps with domain adaptation. Proced Manuf 48:1088–1093. https://doi.org/10.1016/j.promfg.2020.05.149

Alaswad S, Xiang Y (2017) A review on condition-based maintenance optimization models for stochastically deteriorating system. Reliab Eng Syst Saf 157:54–63

Arias Chao M, Kulkarni C, Goebel K, Fink O (2022) Fusing physics-based and deep learning models for prognostics. Reliab Eng Sys Saf 217:107961. https://doi.org/10.1016/j.ress.2021.107961

Azamfar M, Li X, Lee J (2020) Deep learning-based domain adaptation method for fault diagnosis in semiconductor manufacturing. IEEE Trans Semicond Manuf 33(3):445–453. https://doi.org/10.1109/TSM.2020.2995548

Barr A, Feigenbaum EA (2014) The handbook of artificial intelligence, vol 2. Butterworth-Heinemann

Chen Z, Liang K, Ding SX, Yang C, Peng T, Yuan X (2021) A comparative study of deep neural network-aided canonical correlation analysis-based process monitoring and fault detection methods. IEEE Trans Neural Netw Learn Syst 1–15. https://doi.org/10.1109/TNNLS.2021.3072491

Cheng C, Zou W, Wang W, Pecht M (2021) Construction of a deep sparse filtering network for rotating machinery fault diagnosis. Proc Inst Mech Eng Part D J Automob Eng 09544070211014852. https://doi.org/10.1177/09544070211014852

Cubillo A, Perinpanayagam S, Esperon-Miguez M (2016) A review of physics-based models in prognostics: application to gears and bearings of rotating machinery. Adv Mech Eng 8(8):1687814016664660

Dai W, Mo Z, Luo C, Jiang J, Zhang H, Miao Q (2020) Fault diagnosis of rotating machinery based on deep reinforcement learning and reciprocal of smoothness index. IEEE Sens J 20(15):8307–8315. https://doi.org/10.1109/JSEN.2020.2970747

Dimaio F, Scapinello O, Zio E, Ciarapica C, Cincotta S, Crivellari A, Decarli L, Larosa L (2021) Accounting for safety barriers degradation in the risk assessment of oil and gas systems by multistate bayesian networks. Reliab Eng Syst Saf 216:107943. https://doi.org/10.1016/j.ress.2021.107943

Ding SX, Li L (2021) Control performance monitoring and degradation recovery in automatic control systems: a review, some new results, and future perspectives. Control Eng Pract 111:104790. https://doi.org/10.1016/j.conengprac.2021.104790

Fink O, Zio E, Weidmann U (2014) Predicting component reliability and level of degradation with complex-valued neural networks. Reliab Eng Syst Saf 121:198–206. https://doi.org/10.1016/j.ress.2013.08.004

GE. https://www.ge.com/digital/blog/case-industrial-big-data

Hamadache M, Jung JH, Park J, Youn BD (2019) A comprehensive review of artificial intelligence-based approaches for rolling element bearing PHM: shallow and deep learning. JMST Adv 1(1):125–151

Izzo D, Märtens M, Pan B (2019) A survey on artificial intelligence trends in spacecraft guidance dynamics and control. Astrodynamics 3(4):287–299. https://doi.org/10.1007/s42064-018-0053-6

Jardine AK, Lin D, Banjevic D (2006) A review on machinery diagnostics and prognostics implementing condition-based maintenance. Mech Syst Signal Process 20(7):1483–1510

Kinkel S, Baumgartner M, Cherubini E (2021) Prerequisites for the adoption of AI technologies in manufacturing—evidence from a worldwide sample of manufacturing companies. Technovation 102375. https://doi.org/10.1016/j.technovation.2021.102375

Lee J, Wu F, Zhao W, Ghaffari M, Liao L, Siegel D (2014) Prognostics and health management design for rotary machinery systems—reviews, methodology and applications. Mech Syst Signal Process 42(1):314–334. https://doi.org/10.1016/j.ymssp.2013.06.004

Lei Y, Lin J, He Z, Zuo MJ (2013) A review on empirical mode decomposition in fault diagnosis of rotating machinery. Mech Syst Signal Process 35(1–2):108–126

References

Lei Y, Li N, Lin J (2016) A new method based on stochastic process models for machine remaining useful life prediction. IEEE Trans Instrum Meas 65(12):2671–2684

Lei Y, Li N, Guo L, Li N, Yan T, Lin J (2018) Machinery health prognostics: a systematic review from data acquisition to RUL prediction. Mech Syst Signal Process 104:799–834

Lei Y, Yang B, Jiang X, Jia F, Li N, Nandi AK (2020) Applications of machine learning to machine fault diagnosis: a review and roadmap. Mech Syst Signal Process 138:106587

Lei Y (2017) 2—Signal processing and feature extraction. In: Lei Y (ed) Intelligent fault diagnosis and remaining useful life prediction of rotating machinery. Butterworth-Heinemann, pp 17–66. https://doi.org/10.1016/B978-0-12-811534-3.00002-0

Li C, Li S, Zhang A, He Q, Liao Z, Hu J (2021) Meta-learning for few-shot bearing fault diagnosis under complex working conditions. Neurocomputing 439:197–211. https://doi.org/10.1016/j.neucom.2021.01.099

Li X, Li X, Ma H (2020) Deep representation clustering-based fault diagnosis method with unsupervised data applied to rotating machinery. Mech Syst Signal Process 143:106825. https://doi.org/10.1016/j.ymssp.2020.106825

Li X, Zhang W, Ma H, Luo Z, Li X (2021b) Degradation alignment in remaining useful life prediction using deep cycle-consistent learning. IEEE Trans Neural Netw Learn Syst 1–12. https://doi.org/10.1109/TNNLS.2021.3070840

Liu J, Wang W, Ma F, Yang Y, Yang C (2012) A data-model-fusion prognostic framework for dynamic system state forecasting. Eng Appl Artif Intell 25(4):814–823

Ma Z-S, Li X, He M-X, Jia S, Yin Q, Ding Q (2020) Recent advances in data-driven dynamics and control. Int J Dyn Control 8(4):1200–1221. https://doi.org/10.1007/s40435-020-00675-2

Manyika J, Chui M, Brown B, Bughin J, Dobbs R, Roxburgh C, Hung Byers A (2011) Big data: the next frontier for innovation, competition, and productivity. McKinsey Global Institute

Minsky M (1961) Steps toward artificial intelligence. Proc IRE 49(1):8–30

Noor AK (2017) AI and the future of the machine design. Mech Eng 139(10):38–43. https://doi.org/10.1115/1.2017-Oct-2

Pan D, Liu J-B, Cao J (2016) Remaining useful life estimation using an inverse Gaussian degradation model. Neurocomputing 185:64–72

Pandhare V, Li X, Miller M, Jia X, Lee J (2021) Intelligent diagnostics for ball screw fault through indirect sensing using deep domain adaptation. IEEE Trans Instrum Meas 70:1–11. https://doi.org/10.1109/TIM.2020.3043512

Randall RB (2021) Vibration-based condition monitoring: industrial, automotive and aerospace applications. Wiley

Rolls-Royce. https://www.rolls-royce.com/media/our-stories/insights

SanyGroup. https://www.sanygroup.com/socialMedia/3748.html

She D, Jia M, Pecht MG (2020) Sparse auto-encoder with regularization method for health indicator construction and remaining useful life prediction of rolling bearing. Meas Sci Technol 31(10):105005. https://doi.org/10.1088/1361-6501/ab8c0f

Silver D, Huang A, Maddison CJ, Guez A, Sifre L, van den Driessche G, Schrittwieser J, Antonoglou I, Panneershelvam V, Lanctot M, Dieleman S, Grewe D, Nham J, Kalchbrenner N, Sutskever I, Lillicrap T, Leach M, Kavukcuoglu K, Graepel T, Hassabis D (2016) Mastering the game of Go with deep neural networks and tree search. Nature 529(7587):484–489. https://doi.org/10.1038/nature16961

Wang D, Tsui K-L, Miao Q (2017) Prognostics and health management: a review of vibration based bearing and gear health indicators. IEEE Access 6:665–676

Wang X, Bin J, Wu S, Lu N, Ding S (2021) Multivariate relevance vector regression based degradation modeling and remaining useful life prediction. IEEE Trans Ind Electron 1–1. https://doi.org/10.1109/TIE.2021.3114724

Xu Y, Pi D, Wu Z, Chen J, Zio E (2021) Hybrid discrete differential evolution and deep Q-network for multimission selective maintenance. IEEE Trans Reliab 1–12. https://doi.org/10.1109/TR.2021.3111737

Yan J, Meng Y, Lu L, Li L (2017) Industrial big data in an industry 4.0 environment: challenges, schemes, and applications for predictive maintenance. IEEE Access 5:23484–23491

Yang Q, Jia X, Li X, Feng J, Li W, Lee J (2021) Evaluating feature selection and anomaly detection methods of hard drive failure prediction. IEEE Trans Reliab 70(2):749–760. https://doi.org/10.1109/TR.2020.2995724

Yang B, Lei Y, Xu S, Lee CG (2021a) An optimal transport-embedded similarity measure for diagnostic knowledge transferability analytics across machines. IEEE Trans Ind Electron 1–1. https://doi.org/10.1109/TIE.2021.3095804

Yang B, Xu S, Lei Y, Lee C-G, Stewart E, Roberts C (2022) Multi-source transfer learning network to complement knowledge for intelligent diagnosis of machines with unseen faults. Mech Syst Signal Process 162:108095. https://doi.org/10.1016/j.ymssp.2021.108095

Zang Y, Shangguan W, Cai B, Wang H, Pecht MG (2021) Hybrid remaining useful life prediction method. A case study on railway D-cables. Reliab Eng Syst Saf 213:107746. https://doi.org/10.1016/j.ress.2021.107746

Zhang W, Li X, Ma H, Luo Z, Li X (2021) Universal domain adaptation in fault diagnostics with hybrid weighted deep adversarial learning. IEEE Trans Industr Inf 17(12):7957–7967. https://doi.org/10.1109/TII.2021.3064377

Zio E (2022) Prognostics and health management (PHM): where are we and where do we (need to) go in theory and practice. Reliab Eng Syst Saf 218:108119. https://doi.org/10.1016/j.ress.2021.108119

Chapter 2
Conventional Intelligent Fault Diagnosis

2.1 Introduction

After a long operation time of mechanical systems, the condition monitoring system which is used on the machines usually collects a large amount of real-time data, which leads to data explosion in the engineering cases (Liu et al. 2018). The data explosion has pushed mechanical fault diagnosis into the era of big data. Mechanical big data not only challenge the conventional methods of fault diagnosis, but also provide a new perspective for condition-based maintenance. Effectively extracting useful features from the big data and using advanced technologies to accurately identify the health condition of the mechanical systems have been attracting increasing attention from researchers all over the world (Jia et al. 2016; Lei et al. 2016).

Fault diagnosis plays an important role in the maintenance of the mechanical system, because it helps determine which part of the mechanical system needs maintenance. Generally, fault diagnosis is to determine the invisible situation of the internal components by analyzing the external information, such as the vibration signal generated by the mechanical system (Zhang et al. 2021). By vibration analysis, the important diagnosis information can be obtained. Traditionally, the fault diagnosis methods filter the signal and obtain the frequency spectrum for analytics by using the signal processing technologies. Then, the diagnosis result has to be visually inspected by comparing the signal processing result of the data of interest, with that of the data of the mechanical system in the healthy condition. With the above process, the fault features can be found. It can be seen that the conventional fault diagnosis methods significantly depend on experts' professional knowledge. In addition, the fault diagnosis experts have to analyze the collected signals one by one. It is clear that the conventional fault diagnosis methods may fail in dealing with massive condition monitoring data in the Big Data Era (Yang et al. 2022).

Therefore, it is essential to introduce the AI technologies into the field of fault diagnosis and investigate the intelligent fault diagnosis methods. Different from the conventional fault diagnosis methods, the intelligent fault diagnosis methods can quickly and effectively analyze signals and give accurate diagnosis results based on the AI technologies (Lei et al. 2015). Since the intelligent fault diagnosis methods do not need the experts to check data manually, they are promising tools to process the mechanical big data (Wu et al. 2019).

The framework of conventional intelligent fault diagnosis includes four main steps: data acquisition, feature extraction, feature selection, and fault classification (Lei et al. 2020), as shown in Fig. 2.1. In data acquisition, the vibration data are widely used because they provide the most intrinsic information about mechanical failures. Feature extraction aims at extracting representative features from the collected data by the signal processing techniques, such as time-domain statistical analysis, Fourier spectrum analysis, and demodulation analysis. Feature selection is used to select sensitive features through dimensionality reduction strategies, such as PCA and distance evaluation techniques. In fault classification, the selected features are used to train ANNs, support vector machines, and other AI technologies (Wang et al. 2020). Those technologies can determine the health condition of the mechanical systems.

This chapter introduces two representative AI methods, i.e., typical neural network-based methods and statistical learning-based methods, and develops the intelligent fault diagnosis methods for mechanical systems based on them.

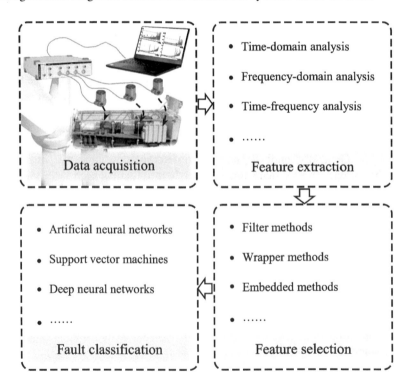

Fig. 2.1 Flow chart of the conventional intelligent fault diagnosis methods

2.2 Typical Neural Network-Based Methods

2.2.1 Introduction to Neural Networks

The ANN is an information processing paradigm inspired by the way the human brain processes information. They have been powerful intelligent technologies in the fault diagnosis of mechanical systems (Azamfar et al. 2020). Using the ANN as classifiers for fault diagnosis can greatly improve the reliability of fault diagnosis. In this section, we will introduce the basic concepts, symbols, and basic learning algorithms of ANNs. Additionally, we briefly reviewed the basic concepts of neural networks and back-propagation algorithms that are usually used to train the neural networks.

The basic ANN is the multilayer perceptron (MLP). It is a feed-forward neural network consisting of units, i.e., neurons, that are connected to each other using directed edges. Note that the neurons are placed in multiple layers. Data flow through the network in a forward direction, and the final layer is the output layer of the network. Before MLP can be used to diagnose mechanical faults, we need to train MLP by the back-propagation algorithm.

The original back-propagation algorithm was first described in the paper of Paul Werbos in 1974 (Werbos 1974). However, this paper had not been widely spread. Until the 1980s, the back-propagation algorithm was independently rediscovered by David Rumelhart, Geoffery Hinton and Ronald Williams (Rumelhart et al. 1986), Yann LeCun (LeCun et al. 1989), and David Parker (Parker 1985). After the publication of the book Parallel Distributed Processing by David Rumelhart and James McClelland, the back-propagation algorithm became popular (Rumelhart et al. 1988). That book has inspired a lot of research about neural networks. After that, the back-propagation algorithm has become the most commonly used algorithm to train networks. The training process is divided into the following stages.

- In the forward phase, the parameters are fixed. The input data are propagated layer by layer through MLP until they reach the output layer.
- In the backward phase, the error signal is calculated by comparing the output of MLP with the expected label. The error signal is first generated in the output layer. Then the error signal is back-propagated through the layers of the network. In this stage, the weights and deviations of MLP are continuously adjusted. Calculating these adjustments is simple in the output layer, but very challenging for the middle layer.

Assume that MLP has L layers and the layer l ($l = 1, \cdots, L$) contains s^l nodes. We use superscripts to indicate the layer number. The output of the lth layer is the input of the $(l+1)$th layer. The input layer is considered as the first layer, which may be slightly different from other references. The elements of the weight matrix $\boldsymbol{W}^l$ are represented by $w_{i,j}^l$ ($i = 1, \cdots, s^l; j = 1, \cdots, s^{l-1}$) and elements of the b_i as matrix $\boldsymbol{b}^l$ are a vector denoted by b_i^l. The nonlinear function of the unit i in the lth layer is denoted by f^l.

Assume that we have a training dataset $\{x_q, y_q\}_{q=1}^{Q}$ where $x_q \in \Re^{s^1 \times 1}$ is the qth feature vector, i.e., the qth training sample. Use $x_q \in \Re^{s^1 \times 1}$ as the input and $y_q \in \Re^{s^L \times 1}$ as the corresponding labels, i.e., the output data. With respect to x_q, the output data of MLP are:

$$z_i^l = \sum_{j=1}^{s^{l-1}} w_{i,j}^l \cdot a_j^{l-1} + b_i^l \tag{2.1}$$

$$a_i^l = f^l(z_i^l) \tag{2.2}$$

where $2 \leqslant l \leqslant L$ and $a_j^1 = x_j$.

With respect to the qth input x_q, we define the sum of squared errors as the difference between the output a_q^L and the desired labels y_q:

$$E_q = \frac{1}{2}(y_q - a_q^L)^2 \tag{2.3}$$

where $a_q^L = [a_1^L, \cdots a_{s^L}^L]$. The back-propagation algorithm tries to find the parameters to minimize the error. In the back-propagation algorithm, the gradient descent method is used. One iteration of gradient descent to the update of $w_{i,j}^l$ and b_i^l can be shown as follows:

$$w_{i,j}^l = w_{i,j}^l - \alpha \frac{\partial E(w,b)}{\partial w_{i,j}^l} \tag{2.4}$$

$$b_i^l = b_i^l - \alpha \frac{\partial E(w,b)}{\partial b_i^l} \tag{2.5}$$

where α denotes the learning rate. The partial derivatives can be written as follows:

$$\frac{\partial E(w,b)}{\partial w_{i,j}^l} = \frac{\partial E(w,b)}{\partial z_i^l} \cdot \frac{\partial z_i^l}{\partial w_{i,j}^l} \tag{2.6}$$

$$\frac{\partial E(w,b)}{\partial b_i^l} = \frac{\partial E(w,b)}{\partial z_i^l} \cdot \frac{\partial z_i^l}{\partial b_i^l} \tag{2.7}$$

The second term in each equation can be calculated easily, because the input of the lth layer is a function of weights and bias of the layer.

Hence, the second term can be expressed as:

$$\frac{\partial z_i^l}{\partial w_{i,j}^l} = a_j^{l-1}, \quad \frac{\partial z_i^l}{\partial b_i^l} = 1 \tag{2.8}$$

2.2 Typical Neural Network-Based Methods

However, the first terms have to be calculated. Define:

$$s_i^l = \frac{\partial E(w,b)}{\partial z_i^l} \tag{2.9}$$

where s_i^l is the sensitivity of $E(w,b)$ to the changes of the ith units in the lth layer. Equations (2.6) and (2.7) can be simplified as follows:

$$\frac{\partial E(w,b)}{\partial w_{i,j}^l} = s_i^l \cdot a_j^{l-1} \tag{2.10}$$

$$\frac{\partial E(w,b)}{\partial b_i^l} = s_i^l \tag{2.11}$$

Next, the sensitivities s_i^l can be obtained. This computation gives the term backpropagation, since it describes a recurrence relationship where the sensitivity in the lth layer can be computed from the sensitivity in the lth layer.

In order to deduce the recurrence relationship with respect to the sensitivities, the following Jacobian matrix is used:

$$\frac{\partial z^{l+1}}{\partial z^l} = \begin{bmatrix} \partial z_1^{l+1}/\partial z_1^l & \partial z_1^{l+1}/\partial z_2^l & \cdots & \partial z_1^{l+1}/\partial z_{s^l}^l \\ \partial z_2^{l+1}/\partial z_1^l & \partial z_2^{l+1}/\partial z_2^l & \cdots & \partial z_2^{l+1}/\partial z_{s^l}^l \\ \vdots & \vdots & & \vdots \\ \partial z_{s^{l+1}}^{l+1}/\partial z_1^l & \partial z_{s^{l+1}}^{l+1}/\partial z_2^l & \cdots & \partial z_{s^{l+1}}^{l+1}/\partial z_{s^l}^l \end{bmatrix} \tag{2.12}$$

For each element in this Jacobian matrix:

$$\frac{\partial z_i^{l+1}}{\partial z_j^l} = \frac{\partial \left(\sum_{j=1}^{s^l} w_{i,j}^{l+1} \cdot a_j^l + b_i^{l+1} \right)}{\partial z_j^l} = w_{i,j}^{l+1} \cdot \frac{\partial a_j^l}{\partial z_j^l} \tag{2.13}$$

$$= w_{i,j}^{l+1} \frac{\partial f^l(z_j^l)}{\partial z_j^l} = w_{i,j}^{l+1} \dot{f}^l(z_j^l)$$

where $\dot{f}^l(z_j^l) = \frac{\partial f^l(z_j^l)}{\partial z_j^l}$.

According to Eq. (2.13), the Jacobian matrix can be expressed as:

$$\frac{\partial z^{l+1}}{\partial z^l} W^{l+1} \dot{F}^l(z^l) \tag{2.14}$$

where $\dot{\boldsymbol{F}}^l(z^l) = \begin{bmatrix} \dot{f}^l(z_1^l) & 0 & \cdots & 0 \\ 0 & \dot{f}^l(z_2^l) & \cdots & 0 \\ \vdots & \vdots & & \vdots \\ 0 & 0 & \cdots & \dot{f}^l(z_{s^l}^l) \end{bmatrix}.$

The recurrence relation for the sensitivities is written in the matrix form:

$$s^l = \frac{\partial E(w,b)}{\partial z^l} = \left(\frac{\partial z^{l+1}}{\partial z^l}\right)^T \frac{\partial E(w,b)}{\partial z^{l+1}} = \dot{F}^m(z^l)(W^{l+1})^T \frac{\partial E(w,b)}{\partial z^{l+1}} \quad (2.15)$$

Due to the fact $s^{l+1} = \frac{\partial E(w,b)}{\partial z^{l+1}}$, Eq. (2.15) is rewritten as:

$$s^l = \dot{F}^l(z^l)(W^{l+1})^T s^{l+1} \quad (2.16)$$

The sensitivities propagate through the network from the last layer to the first layer:

$$s^L \to s^{L-1} \to \cdots \to s^l \to \cdots \to s^2 \quad (2.17)$$

We still have one more step to complete the back-propagation algorithm. The starting sensitivity s^L needs to be calculated and obtained at the output layer by Eq. (2.3).

$$s_i^M = \frac{\partial E(w,b)}{\partial z^L} = -(y_i - a_i^L)\frac{\partial a_i^L}{\partial z_i^L} = -\sum_{q=1}^{Q}(y_i - a_i^L)\dot{f}^L(z_{s^L}^L) \quad (2.18)$$

In this part, the back-propagation algorithm is summarized in a matrix form. Firstly, propagate the input data forward through the network:

$$a^1 = x \quad (2.19)$$

$$a^l = f^l(W^l a^{l-1} + b^l) \text{ for } l = 2, \cdots, L \quad (2.20)$$

Next, propagate the sensitivities backward through the network:

$$s^L = -(y - a^L)\dot{F}^L(z^L) \quad (2.21)$$

$$s^l = \dot{F}^l(z^l)(W^{l+1})^T s^{l+1} \quad (2.22)$$

Finally, the weights and biases are updated by the following equations:

$$W^l = W^l - \alpha s^l (a^{m-1})^T \quad (2.23)$$

2.2 Typical Neural Network-Based Methods

$$b^l = b^l - \alpha s^l \tag{2.24}$$

MLP is popularly applied because of its strong nonlinear ability, which actually originates from f^l. This function is usually a differentiable non-linear activation function. The most widely used activation functions are the sigmoid function and tanh function.

2.2.2 Intelligent Diagnosis Using Radial Basis Function Network

Micchelli and Powell used the radial basis function (RBF) method to independently study the interpolation problem in high-dimensional data spaces (Micchelli 1984; Powell 1987). Inspired by this work, Broomhead used RBF to improve the performance of ANN, i.e., the RBF network. The RBF network is a feedforward network containing one hidden layer, i.e., the radial basis layer. The activation function of the hidden layer is the Gaussian radial basis function. It is a good approximation for the nonlinear input, and can also automatically determine the number of neurons in the radial base layer during the training process. RBF networks have attracted more and more interest in intelligent fault diagnosis of mechanical systems.

Traditionally, to obtain accurate fault features from vibration signals, signal processing techniques are required to preprocess the vibration signals, especially for early fault signals. Among signal processing techniques, wavelet packet transform (WPT) and EMD are very powerful for mechanical fault diagnosis in many applications. Therefore, WPT and EMD are usually used to preprocess the vibration signal to highlight the fault features. Firstly, time-domain features are extracted from both the vibration signal and the signals preprocessed by WPT and EMD. Secondly, we can select some features that clearly characterize the health state of mechanical systems. The commonly used feature selection methods are principal component analysis, conditional entropy, distance evaluation technology, and so on. Among these methods, distance evaluation technology is widely used in intelligent fault diagnosis.

According to the above analysis, an intelligent fault diagnosis method using the RBF network is presented. This method first uses WPT and EMD to preprocess the vibration signals to obtain more fault features. Second, the time domain features are extracted from the original signal and the preprocessed signal. Third, the distance evaluation technology is performed on the feature set to select sensitive features. Finally, the RBF network is used to identify the health condition of mechanical systems.

1) RBF network

For a training set $\{x_i, y_i\}_{i=1}^{M}$, the detailed structure of RBF is described as follows:

(1) Input layer: This layer consists of N source nodes, where N is the dimension of $\boldsymbol{x}_i$.
(2) Radial basis layer, i.e., the hidden layer: This layer includes the neurons using radial basis functions as the activation function. The number of neurons is identical to the training sample number M. Generally, the Gaussian function is applied as the radial basis function. The Gaussian function holds the property that each basis function is dependent on the Euclidean distance from a center $\boldsymbol{c}_j$. The Gaussian function can be expressed as:

$$\varphi_j(\boldsymbol{x}_i) = \exp(-\frac{(\boldsymbol{x}_i - \boldsymbol{c}_j)^2}{2b_j^2}) \tag{2.25}$$

where b_j denotes the width of the jth radial basis function. Because Gaussian is centered, this function only responds to a small area of the input space. The Gaussian function reaches its peak at zero distance, and its value becomes smaller as the distance to the center increases. Therefore, the output of the neuron in the radial base layer is 1 when the input is the center. The output value decreases as the distance between the input and the center increases.

(3) Output layer: This layer is connected by the radial basis layer. Generally, the size of the output layer is much smaller than the hidden layer. The size of the output layer is selected by the number of health conditions, which is the dimension of $\boldsymbol{y}_i$. The value of the output layer can be a linear combination of radial basis functions:

$$f(\boldsymbol{x}_i) = \sum_{j=1}^{M} w_j \varphi_j(\boldsymbol{x}_i) \tag{2.26}$$

Practically, in the situation of processing large training samples, it can be a waste of computation resource when the RBF network holds a hidden layer of the same size with the number of training samples. The radial basis layer is presented in Eq. (2.25). It can be observed that any correlation between adjacent data of the training sample can be transplanted into the adjacent neurons. In other words, there may be redundant neurons. Furthermore, the training set $\{\boldsymbol{x}_i, \boldsymbol{y}_i\}_{i=1}^{M}$ can be noisy, especially in studies about intelligent fault diagnosis of mechanical systems. If the noisy samples are selected as the centers, misleading results will be obtained. Therefore, to successfully implement the RBF network, a suitable center for the Gaussian function needs to be found. An easy way is randomly selecting a subset of the training samples. A more systematic method is called orthogonal least squares (Chen et al. 2014). In this method, the next data point selected as the center of the basis function in each step corresponds to the one with the largest reduction in the sum of square errors. The value of the expansion coefficient is selected as part of the algorithm. We can also use clustering algorithms which may give centers that no longer coincide with the training samples.

2.2 Typical Neural Network-Based Methods

2) RBF-Based fault diagnosis

Six time-domain features are used including skewness, kurtosis, crest indicator, gap indicator, shape indicator, and pulse indicator. First, the features of the original vibration signal are extracted. Second, we use WPT and EMD to preprocess each vibration signal to obtain additional fault features. Six time-domain features are extracted from WPT frequency band signal and EMD intrinsic mode function.

Suppose that the decomposition level of WPT is I and the IMF number of EMD is L. The feature set contains $6 + 6 \times 2^I + 6L$ features for each vibration signal. Then we select sensitive features from the above feature set. The distance evaluation technique is used for the feature selection (Lei et al. 2008). After calculating the distance evaluation factors of all features, these features are sorted according to their evaluation factors. From the sorted feature set, we select the first feature and apply it to the diagnosis of faults in mechanical systems.

To sum up the process, the intelligent fault diagnosis method based on the RBF network can be displayed in Fig. 2.2 as follows:

(1) Obtain vibration signals and preprocess them by WPT of db10 and EMD, respectively.
(2) Extract the 6 time-domain features from both the signals and preprocessed signals.
(3) Examine these features by the distance evaluation method. For the evaluation factors, the features can be rearranged to obtain the ranked feature set.
(4) Feed the selected features to the RBF network for training. If the training accuracy increases to 100% or the training accuracy hardly increases as the number of sensitive features increases to F, the training process stops.

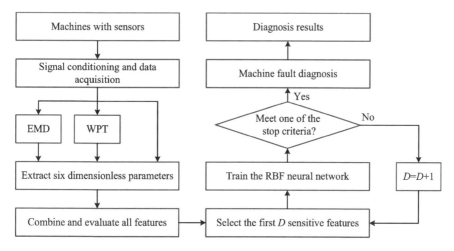

Fig. 2.2 Flow chart of the intelligent fault diagnosis method based on RBF network

3) Intelligent fault diagnosis case of a heavy oil catalytic cracking unit

The heavy oil catalytic cracking unit is one of the key units used in the refinery and is a typical mechanical system. The presented method is applied to the minor friction fault diagnosis of heavy oil catalytic cracking unit in the refinery. The unit consists of a gas turbine, compressor, gearbox, and drive motor (Lei et al. 2009). The gas turbine converts thermal energy into mechanical energy. Bearings #1 and #2 are used to support the gas turbine shaft. Bearings #3 and #4 are used to support the compressor shaft. Bearing #5 is used to support the gearbox shaft. The hub and blades on the shaft (the left part of the gas turbine) are cantilevered structures. The speed of the unit is 5745 r/min (95.75 Hz). The Bently 3300 system is used to monitor its operating conditions. In order to collect vibration signals, eddy current sensors are fixed on each bearing box in the vertical and horizontal directions. A computer online monitoring system is used for data collection and recording.

After the overhaul, the unit was running again. However, the unit vibration exceeded the alarm limit and must be shut down. When it was disassembled, there was a large area of friction between the hub (rotating with its shaft) and the air seal (stationary). After the friction area was polished, the friction became slight and the vibrations were reduced. However, friction failure was not completely eliminated. When the unit was running, the vibration monitored in the vertical direction of bearing box #1 was larger and more complicated than other monitoring points. Therefore, we analyzed to identify the different situations of the mechanical system group. Under the three working conditions of pre-overhaul (normal state), post-overhaul (large abrasion) and after grinding (slight abrasion), 82 data samples were collected respectively, the sampling frequency was 2000 Hz, and the data were sampled with 512 sampling points. Therefore, a total of 246 samples were collected from this mechanical system.

The presented intelligent fault diagnosis method is applied to unit fault diagnosis. In order to identify the severity of the fault, we solved the three-level classification problem. The 246 data samples are divided into 123 training samples and 123 testing samples. Each raw vibration signal is decomposed by db10 WPT at the level 3 and EMD. We then choose the top six IMFs that carry the main information. Six dimensionless parameters are obtained from each original signal, band signal, and IMF. In this way, 90 eigenvalues are obtained to represent a data sample.

When the training accuracy is 100%, the first stop criterion of the presented method is met, and the training process is terminated. The three features selected in the friction fault diagnosis are the skewness of the signal, the impulse index of the first frequency band signal of WPT, and the shape index of the first IMF of EMD. Due to the use of WPT and EMD, the fault diagnosis accuracy reaches 86.18%. After evaluating the combined feature set extracted from the original signal and the preprocessed signal and selecting sensitive features, the computational burden of the RBF network can be reduced. In addition, the fault diagnosis accuracy of the RBF network can increase to 100%. Therefore, the presented intelligent fault diagnosis method can not only distinguish large-area friction faults from normal conditions, but also identify minor friction faults from the normal state of the unit. The above

2.2 Typical Neural Network-Based Methods

results verify the effectiveness of the presented fault diagnosis method integrating WPT, EMD, distance evaluation technology, and RBF network.

2.2.3 Intelligent Diagnosis Using Wavelet Neural Network

In the previous section, the RBF network is used to overcome the shortcomings of MLP that limit the representations of local features. However, the activation function of the RBF network is a Gaussian function, which is non-adaptive. On the contrary, the wavelet base has adaptive space and frequency localization features, so wavelet transform is effective for the characterization and detection of local features of mechanical systems. A neural network model was presented, called wavelet neural network (WNN) (Yang and Chen 2019). Due to the combination of the time–frequency positioning features of wavelet transform and the self-learning ability of the ANN, WNN combines the advantages of wavelet transform and ANN.

This section presents an intelligent fault diagnosis method based on WNN. This method first decomposes the collected vibration signal by ensemble empirical mode decomposition (EEMD) (Wu and Huang 2009). Then, the most sensitive IMF is selected based on the kurtosis index. Third, we extract the time domain and frequency domain features from the most sensitive IMF, its spectrum, and envelope spectrum. Finally, these extracted features are put into WNN to identify the health of the mechanical system.

1) Wavelet neural network

The structure of the wavelet neural network is similar to that of a multi-layer perceptron. Generally, WNN consists of the input layer, one hidden layer, and the output layer. The hidden layer includes neurons, whose activation functions replace the ordinary basis function networks with a wavelet basis and neurons. The inputs of the wavelet neuron contain the sum of weighted inputs and parameters of the wavelet function. In the whole training process, both weights and wavelet parameters can be updated using the back-propagation algorithm.

The output of the wavelet neuron can be expressed as

$$c = \psi\left(\frac{u_k - b}{a}\right) \qquad (2.27)$$

where a and b denote the dilation and translation parameters, and the sum of weighted inputs u_k is expressed by

$$u_k = \sum_{i=1}^{p} w_{ki} x_i \qquad (2.28)$$

The commonly used wavelet functions are Morlet, Haar, Mexican Hat, Shannon, and Meyer. Since the Morlet wavelet has simple explicit expression and its partial derivative of the corresponding error function can be easily calculated, the Morlet wavelet is used as the activation function of hidden layer neurons here. The Morlet wavelet can be presented as follows:

$$\psi(t) = \cos(5t)\exp(-\frac{t^2}{2}) \qquad (2.29)$$

Its derivative can be given by

$$\dot{\psi}(t) = -5\sin(5t)\exp(-\frac{t^2}{2}) - t\cos(5t)\exp(-\frac{t^2}{2}) \qquad (2.30)$$

The training of WNN is implemented through the back-propagation algorithm. The training continues until a specified value of the error function is satisfied or the maximum number of iterations is reached.

2) Fault diagnosis using WNN

As previously stated, the WNN-based fault diagnosis method first decomposes the collected vibration signal by EEMD. Compared with EMD, EEMD not only retains the advantages of EMD, but also alleviates the mode mixing problem of EMD. Among the IMFs obtained by EEMD, some are sensitive and related to failures, while some are not. In order to avoid the interference of components that are not related to the fault, we can choose a sensitive IMF for further processing. Here we present a new method based on kurtosis to select sensitive IMFs.

(1) Compute the kurtosis values of I IMFs of each data sample. Let $k_{i,s}$ denote the kurtosis of the ith IMF of the sth data sample, and it is presented as

$$k_{i,s} = \frac{N\sum_{n=1}^{N}(a_{n,i}-\overline{a_i})^4}{\left(\sum_{n=1}^{N}(a_{n,i}-\overline{a_i})^2\right)^2}, n=1,2,\cdots,N;\ i=1,2,\cdots,I;\ s=1,2,\cdots,S$$

(2.31)

where $a_{n,i}$ denotes the nth data point in the ith IMF a_i, $\overline{a_i}$ represents the average of a_i, N denotes the number of data points of a_i, i denotes the number of IMFs, and S denotes the number of the data samples.

(2) Calculate the mean and standard deviation of the kurtosis values of each IMF:

$$m_i = \frac{1}{S}\sum_{s=1}^{S}k_{i,s},\ i=1,2,\cdots,I \qquad (2.32)$$

2.2 Typical Neural Network-Based Methods

where m_i denotes the mean of the kurtosis values.

$$\text{std}_i = \sqrt{\frac{1}{(S-1)} \sum_{s=1}^{S} (k_{i,s} - m_i)^2}, \quad i = 1, 2, \cdots, I \quad (2.33)$$

where std_i denotes the standard deviation of the kurtosis values.

(3) Construct the criterion of the sensitive IMF selection as follows:

$$f_i = \begin{cases} m_i \cdot \text{std}_i & \text{for normal bearings} \\ \frac{\text{std}_i}{m_i} & \text{for faulty bearings} \end{cases} \quad i = 1, 2, \cdots, I \quad (2.34)$$

(4) Choose the sensitive IMF of the data sample under each health condition. According to the definition of the selection criterion, the smallest f_i in Eq. (2.34) suggests the sensitive IMF.

For fault diagnosis of mechanical systems, time and frequency-domain features can be extracted from the sensitive IMF. As presented in Table 2.1, they are standard deviation (x_{std}), kurtosis (x_{kur}), shape factor (SF) and impulse factor (IF) in the time domain, and mean frequency (x_{mf}), root mean square frequency (x_{rmsf}), standard deviation frequency (x_{stdf}) and spectrum peak ratio of bearing outer race (SPRO), inner race (SPRI) and roller (SPRR) in the frequency domain. The features will be fed into the WNN to automatically identify the mechanical health conditions.

SPRO, SPRI and SPRR can be defined as

$$\text{SPRO} = \frac{K \sum_{i=1}^{H} p_O(h)}{\sum_{k=1}^{K} s(k)}, \quad \text{SPRI} = \frac{K \sum_{i=1}^{H} p_I(h)}{\sum_{k=1}^{K} s(k)}, \quad \text{SPRR} = \frac{K \sum_{i=1}^{H} p_R(h)}{\sum_{k=1}^{K} s(k)} \quad (2.35)$$

where $s(k)$ denotes a spectrum for $k = 1, 2, \cdots, K$, and K denotes the number of spectrum lines; f_k represents the frequency value of the kth spectrum line; $p_O(h)$, $p_I(h)$ and $p_R(h)$ are respectively the peak values of the hth ($h = 1, 2, \cdots, H$, and H is

Table 2.1 The extracted 10 features

1	Standard deviation	x_{std}	6	Root mean square frequency	x_{rmsf}
2	Kurtosis	x_{kur}	7	Standard deviation frequency	x_{stdf}
3	Shape factor	SF	8	Spectrum peak ratio of bearing outer race	SPRO
4	Impulse factor	IF	9	Spectrum peak ratio of bearing inner race	SPRI
5	Mean frequency	x_{mf}	10	Spectrum peak ratio of bearing roller	SPRR

the number of harmonics.) harmonics of the characteristic frequencies for bearing outer race (f_O), inner race (f_I) and roller (f_R), that can be computed based on the following equations:

$$f_O = \frac{f_r}{2} N_R (1 - \frac{B}{C} \cos\alpha), \quad f_I = \frac{f_r}{2} N_R (1 + \frac{B}{C} \cos\alpha), \quad f_R = \frac{f_r C}{2B}[1 - (\frac{B}{C} \cos\alpha)^2] \quad (2.36)$$

where f_r denotes the shaft rotational frequency; N_R denotes the roller number; α represents the contact angle; B and C denote the roller and pitch diameters, respectively.

The intelligent fault diagnosis method based on EEMD and WNN is shown in Fig. 2.3. It contains the following four processes. First, we can use EEMD to decompose each vibration signal obtained from mechanical systems into a series of IMFs. Second, select sensitive IMFs from these IMFs through the kurtosis-based method. Third, extract 10 features from the most sensitive IMF, its spectrum and envelope spectrum. Finally, WNN is used to identify the health of the mechanical systems and get the fault diagnosis result.

3) Intelligent fault diagnosis case of roller bearings

(1) Experimental setup and data acquisition

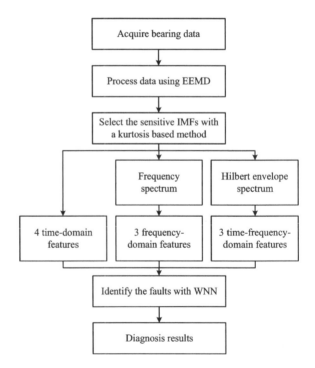

Fig. 2.3 Flow chart of the WNN-based fault diagnosis method

2.2 Typical Neural Network-Based Methods

The locomotive roller bearing test bench consists of a hydraulic motor, two supporting pillow blocks, a bearing under test (552732QT) loaded on the outer ring by a hydraulic cylinder, a hydraulic radial load applying system, and a tachometer for shaft speed measurement (Lei et al. 2011). The bearing is fixed in a mechanical system driven by a hydraulic motor. The 608A11 ICP accelerometer is installed on the load module next to the outer ring of the tested bearing to measure its vibration. Use Sony EX's advanced data acquisition and analysis system to collect data. Some parameters in the experiment are shown in Table 2.2. Nine different health conditions of the bearing are shown in Table 2.3. In the nine bearing working conditions, the serious failure of the outer ring (working condition 2) is a serious peeling failure, and the other failures are all minor scratches (Lei et al. 2011). Each data subset contains 50 samples corresponding to one health condition. The sample length is 8192.

In order to verify the effectiveness of the presented method, three experiments were performed on three different datasets selected from the entire dataset of locomotive roller bearings. Three-layer MLP and RBF neural networks are used for comparison. The MLP neural network, RBF network and WNN have the same architecture.

Table 2.2 Parameters in the experiment

Parameter	Value
Bearing specs	552732QT
Load	9800 N
Inner race diameter	160 mm
Outer race diameter	290 mm
Roller diameter	34 mm
Roller number	17
Contact angle	0°
Sampling frequency	12.8 kHz

Table 2.3 Description of the faulty bearings

Condition	Rotating speed/ (r/min)	Label
Normal condition	490	1
Slight rub fault in the outer race	490	2
Serious flakingfault in the outer race	480	3
Slight rub fault in the inner race	500	4
Roller rub fault	530	5
Compound faults in the outer and inner races	520	6
Compound faults in the outer race and rollers	520	7
Compound faults in the inner race and rollers	640	8
Compound faults in the outer and inner races and rollers	550	9

(2) Experiment 1: effectiveness in diagnosing different faults

The effectiveness of WNN in identifying different types of bearing faults is evaluated using a bearing dataset composed of four data subsets under normal conditions, serious outer race faults, inner race faults, and roller faults. Each of the four data subsets contains 50 samples and the entire dataset contains a total of 200 samples. 100 samples are used for training, and the remaining 100 samples are for testing.

We take the 10 features in Table 2.1 as the inputs, and train the MLP neural network, RBF neural network, and WNN. The testing results are shown in Table 2.4. In this experiment, the MLP neural network, RBF neural network, and WNN achieved classification accuracies of 89%, 90% and 100% respectively. The comparison results show that the presented method performs better than the compared methods in identifying bearing faults with higher accuracy.

(3) Experiment 2: effectiveness in diagnosing different fault severities

In this experiment, the dataset applied contains 150 samples. Among them, 50 samples are normal, 50 samples have minor faults in the outer ring, and the remaining 50 samples have serious faults in the outer ring. 75 samples are used as the training samples, and the remaining 75 samples are used for testing.

The testing results of the three kinds of neural networks to identify different fault severity are shown in Table 2.4. Since this experiment is a three-class classification problem and is relatively simple, the testing accuracy of all neural networks has reached 100%.

(4) Experiment 3: effectiveness in diagnosing compound faults

In this experiment, we analyzed a dataset that contains not only different fault classes and severities, but also compound faults. This dataset includes nine cases of locomotive roller bearings (as described in Table 2.3) and consists of 450 samples. Half of the 450 data samples are used for training, and the remaining 225 samples are used for testing.

The testing results of MLP neural network, RBF neural network, and WNN in this experiment are shown in Table 2.4. It can be seen that the testing accuracies of the three methods are 67.56%, 78.67% and 91.56% respectively. The fault diagnosis accuracy drops because this experiment has ten classes and is relatively difficult. However, the presented method still achieves the highest testing accuracy.

Table 2.4 The testing accuracies (%) of MLP neural network, RBF neural network, and WNN

Methods	Experiment 1	Experiment 2	Experiment 3
MLP	89	100	67.56
RBF	90	100	78.67
WNN	100	100	91.56

2.2.4 Epilog

This section presents two intelligent fault diagnosis methods based on the RBF network and WNN, respectively. In the former method, WPT and EMD are used to obtain abundant fault information from the original vibration signal, firstly. Then, six time-domain features are extracted from the original signal, WPT frequency band signal and EMD IMFs. Thirdly, all features are used to generate a combined feature set, where sensitive features are selected through distance evaluation technology. Finally, these sensitive features are input to the RBF network to improve the accuracy of fault diagnosis. The fault diagnosis results of rolling bearings and heavy oil catalytic cracking units show that this method is an effective method for fault diagnosis of mechanical systems. The latter method decomposes the collected vibration signal by EEMD, firstly. Then, use the kurtosis-based method to select the most sensitive IMF. Third, extract the time domain and frequency domain features from the most sensitive IMF, its spectrum and Hilbert envelope spectrum. Finally, these features are input into WNN to diagnose the health of mechanical systems. The results show that the presented method can not only reliably distinguish a single fault and its different fault severity, but also a compound fault.

2.3 Statistical Learning-Based Methods

At present, most AI technologies, including ANNs, are to find out the subject of the sample when the number of training samples is close to infinite. This is called empirical risk minimization (ERM), which aims to minimize errors on training samples. Neural networks based on ERM are often applied in a large number of engineering cases. The training process is to establish a neural network model of the observed system, that is to use the observed data to derive the response of the system. Therefore, the quantity and quality of the observed data may dominate the performance of the neural network model. Because the obtained data are limited and non-uniform sampling is traditionally used, the neural network method has generalization problems, which may cause the model to overfit. In addition, the limited number of data samples usually leads to insufficient performance in practical applications. Therefore, the SVM based on structural risk minimization (SRM) is used to deal with the problem of neural networks trained by ERM. Different from the idea of ERM, SRM is achieved by minimizing the upper limit of expected risk and providing better generalization capabilities, especially when the number of samples is limited. SRM is rooted in statistical learning theory (SLT) and has received increasing attention since the 1990s. SLT is based on a solid foundation, provides a unified framework for limited sample learning tasks, and integrates a few existing methods. Therefore, it is beneficial to address the challenges.

Section 2.3 first briefly introduces the statistical learning theory, and then focuses on the application of support vector machines in intelligent fault diagnosis. Finally, we present a fault diagnosis method using a new technology relevant vector machine based on the statistical learning theory.

2.3.1 Introduction to Statistical Learning

Statistical learning theory is a machine learning framework that draws on the fields of statistics and functional analysis. It deals with the problem of finding functions that can describe data patterns. The basic concepts of statistical learning theory include VC (Vapnik–Chervonenkis) dimensionality and structural risk minimization (Burges 1998).

1) VC dimension

The VC dimension is a scalar value that measures the capacity of a set of functions. The VC dimension of a set of functions is p if and only if there exists a set of points $\{x^i\}_{i=1}^{p}$ such that these points can be isolated in all the 2^p possible configurations, and that no set $\{x^i\}_{i=1}^{p}$ exists where $q > p$ satisfying this property.

2) Structural risk minimization

For a structure, S_h is a hypothesis space of VC dimension h and

$$S_1 \subset S_2 \subset \cdots \subset S_\infty \tag{2.37}$$

SRM is expected to solve the following problem:

$$\min_{S_h} R_{\text{emp}}[f] + \sqrt{\frac{h \ln(\frac{2l}{h} + 1) - \ln(\frac{\delta}{4})}{l}} \tag{2.38}$$

If the underlying process of modeling is not deterministic, the modeling problem will become more stringent, so this chapter is limited to the deterministic process. Multi-output problems can usually be transformed into a set of single-output problems that can be considered independently. Therefore, it is appropriate to consider a process with multiple inputs that requires a single output.

Now we can summarize the principles of SRM. Please note that the VC confidence term depends on the selected function category, while the empirical risk and actual risk depend on a specific function selected during the training process. We hope to find a subset of the selected function set to minimize the risk limit.

2.3.2 Intelligent Diagnosis Using Support Vector Machine

SVM is a supervised learning technique using the statistical learning theory. SVM is widely used in classification problems in various fields (Li et al. 2018). As we all know, SVM has a simple structure, global optimization, good generalization performance, and fast learning speed. SVM uses a nonlinear transformation to map the original sample space to the high-dimensional feature space, and find the optimal classification plane to classify the samples in the high-dimensional feature space. Even when the sample size is small, SVM is able to obtain high accuracy. Since the classification problem in the intelligent fault diagnosis of mechanical systems is often limited in samples, it is a good choice to use SVM as a fault identification technology (Shi and Zhang 2021).

Since the vibration signal of mechanical systems has nonlinear and non-stationary features, the SVM input are expected to characterize these features. Compared with the features extracted by the other fault diagnosis methods, nonlinear dynamic parameter estimation technology may provide a promising method for extracting fault-related features hidden in complex nonlinear fault signals. At present, many nonlinear parameter extraction techniques have been studied and introduced into fault diagnosis, such as fractal dimension, Lempel–Ziv complexity, and approximate entropy. Among these feature extraction techniques, permutation entropy (PE) is a powerful tool in analyzing mechanical signals (Bandt and Pompe 2002). Furthermore, multi-scale displacement entropy is introduced to quantify the complexity of mechanical signals on multiple scales (Aziz and Arif 2005).

Based on MPE and SVM, we develop a new intelligent method for mechanical fault diagnosis. This method uses MPE to extract features that measure the complexity of mechanical vibration signals. Then, SVM acts as a fault classifier (Yan and Jia 2018).

1) Support vector machine

Basically, given a set of training samples that are labeled by two classes, a binary SVM is trained to classify the new samples into one category or the other. For the training set that treats the samples as p-dimensional vectors, it is hoped to find a $(p-1)$-dimensional hyperplane, and divide the sets into two categories. As displayed in Fig. 2.4, many hyperplanes can divide datasets. A reasonable hyperplane represents the maximum separation or margin between two categories. In other words, the distance from the selected hyperplane to the closest data point in each category is maximized. This hyperplane is called the maximum margin hyperplane. SVM aims at finding the hyperplane to classify the dataset.

Traditionally, the classification of datasets is a non-linear problem. Therefore, the linear SVM can be extended to the non-linear version, which uses the kernel technique to effectively implement the non-linear classification. The kernel function is used to map the original samples of the dataset to the high-dimensional space.

Fig. 2.4 The hyperplanes: H1 does not separate the classes. H2 does, but only with a small margin. H3 separates them with the maximum margin

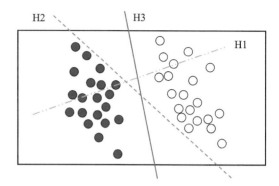

Then the SVM is to find the optimal hyperplane to classify the samples in the high-dimensional space. Here, the linear and non-linear SVMs are described first. Then SVM for multi-class classification is introduced next.

(1) Linear SVM

Given a set of samples x_i ($i = 1, 2, \cdots, M$), where M denotes the number of samples. The set includes two classes, i.e., the positive class and the negative class. $y_i = 1$ denotes the positive class and $y_i = -1$ denotes the negative class, respectively. It is possible to find a hyperplane $f(x) = 0$ that classifies the given dataset:

$$f(x) = \mathbf{w}^T \mathbf{x} + b = \sum_{j=1}^{M} w_j x_j + b = 0 \qquad (2.39)$$

where $\mathbf{w}$ denotes a M-dimensional vector, and b represents a scalar. The above two parameters can be adopted to determine the hyperplane.

If the samples can be linearly separated, two hyperplanes can be found to separate the samples. These hyperplanes are named as separating hyperplanes, and the area enclosed by them is named as the "edge". The purpose of SVM is to maximize this margin and then improve its generalization capability. Two separated hyperplanes can be described by equations:

$$f(x_i) = 1 \text{ if } y_i = 1 \qquad (2.40)$$

$$f(x_i) = -1 \text{ if } y_i = -1 \qquad (2.41)$$

or the equation in the following form:

$$y_i f(x_i) = y_i(\mathbf{w}^T \mathbf{x}_i + b) \geq 1 \text{ for } i = 1, 2, \cdots, M \qquad (2.42)$$

The separating hyperplane that produces the largest margin is called the optimal hyperplane. SVM tries to find a hyperplane between two different classes and locates it in such a way that the margins of the two hyperplanes are maximized. The samples used to determine the hyperplanes are named as the support vectors and are marked as gray squares and circles. Because the support vector contains the information needed to define the classifier, no remaining samples are needed after the support vector is selected.

In geometry, the margin is represented by $\|w\|^{-2}$ based on Eq. (2.42). As a result, the optimal separating hyperplane can be obtained through solving the following optimization problem:

$$\text{minimize} \quad \frac{1}{2}\|w\|^2 \tag{2.43}$$

$$\text{subject to} \quad y_i(w^T x_i + b) \geq 1, \quad i = 1, \cdots, M \tag{2.44}$$

This problem can be considered as addressing the equivalent Lagrangian dual problem under the assistance of the Karush–Kuhn–Tucker (KKT) condition. The dual quadratic optimization problem can be shown as

$$\text{maximize} \quad L(a) = \sum_{i=1}^{M} a_i - \frac{1}{2}\sum_{i,j=0}^{M} a_i a_j y_i y_j x_i^T \cdot x_j \tag{2.45}$$

$$\text{subject to } a_i \geq 0, \ i = 1, \cdots, M, \ \sum_{i=1}^{M} a_i y_i = 0 \tag{2.46}$$

Through solving the above optimization problem, the coefficients a_i can be calculated, and the decision function can be computed:

$$f(x) = \text{sign}(\sum_{i,j=1}^{M} a_i y_i x_i^T \cdot x_j + b) \tag{2.47}$$

(2) Soft margin

If there is no hyperplane that is able to separate the samples of two due to the existence of the noisy samples, the linear SVM will not be able to find a feasible solution. Therefore, to address this problem, a soft margin method is presented. A further optimization problem can be introduced in this algorithm to relax the constraints in Eqs. (2.40) and (2.41). Through introducing positive slack variables ξ_i, the new optimization problem is described by

$$\text{minimize} \quad \frac{1}{2}\|w\|^2 + C\sum_{i=1}^{M}\xi_i \tag{2.48}$$

$$\text{subject to} \quad \begin{cases} y_i(w^T x_i + b) \geq 1 - \xi_i, & i = 1, \cdots, M \\ \xi_i \geq 0, & i = 1, \cdots, M \end{cases} \tag{2.49}$$

In those equations, ξ_i measures the distance between the margin and $\boldsymbol{x}_i$ lying on the wrong side of the margin. As the soft margin method is able to weaken the impact of the outliers on SVM, it is popularly adopted in different classification problems.

(3) Nonlinear classification

The SVM method can be generalized to address the non-linear classification problem by applying kernel functions. By kernel functions, the samples can be mapped into a high-dimensional feature space. In the high-dimensional feature space, the linear classification can be handled. A kernel function $k(\boldsymbol{x}, \boldsymbol{y})$ represents an inner product between the samples where $k(\boldsymbol{x}, \boldsymbol{y}) = \langle \varphi(\boldsymbol{x}), \varphi(\boldsymbol{y}) \rangle$. Valid kernel functions are supposed to satisfy Mercer's conditions, which indicates $k(\boldsymbol{x}, \boldsymbol{y})$ must equal $k(\boldsymbol{y}, \boldsymbol{x})$.

By adopting the kernel function, a nonlinear version of SVM is able to be developed. The optimization problem of the nonlinear SVM can be presented (in the dual form) as

$$\text{maximize} \quad L(\boldsymbol{a}) = \sum_{i=1}^{M} a_i - \frac{1}{2} \sum_{i,j=0}^{M} a_i a_j y_i y_j k(\boldsymbol{x}_j, \boldsymbol{x}_j) \tag{2.50}$$

$$\text{subject to} \quad \begin{cases} 0 \leq a_i \leq C, \quad i = 1, \cdots, M \\ \sum_{i=1}^{M} a_i y_i = 0 \end{cases} \tag{2.51}$$

The decision function of the nonlinear SVM can be expressed by

$$f(\boldsymbol{x}) = \text{sign}(\sum_{i,j=1}^{M} a_i y_i k(\boldsymbol{x}_i, \boldsymbol{x}_j) + b) \tag{2.52}$$

Note that the samples closest to the hyperplane, i.e., support vectors, are those with non-zero coefficients. The support vector includes the necessary information to construct the optimal hyperplane, while the other samples have no influence on it. Therefore, SVM can work in the classification tasks with limited samples.

Different kernel functions can be used in SVM, such as linear, polynomial, and Gaussian radial basis functions, as shown in Table 2.5. Choosing an appropriate kernel function is critical to the performance of SVM, because the kernel determines the high-dimensional space in which the sample will be classified. Among the kernel functions, the Gaussian radial basis function is the most commonly used in intelligent fault diagnosis.

2) Multi-class support vector machine

The SVM discussed above works in binary classification problems. However, in the fault diagnosis of mechanical systems, there are traditionally more than two classes.

2.3 Statistical Learning-Based Methods

Table 2.5 Formulation of kernel function

Kernel	$k(x, x_j)$
Linear	$x^T x_j$
Polynomial	$(\gamma x^T x_j + r)^d, \gamma > 0$
Gaussian radial basis function	$\exp(-\|x - x_j\|^2 / 2\gamma^2)$
Hyperbolic tangent	$\tanh(\kappa x^T x_j + c)\kappa > 0 \text{ and } c < 0$

For example, the fault categories of gears include pitting, cracks, missing teeth, missing teeth, and so on. Therefore, the multi-class classification strategy of SVM should be studied.

(1) One-against-all (OAA)

The commonly adopted approach to the multi-class classification problem of SVM is the OAA method. It builds k SVM models where k denotes the class number. The ith SVM ($i = 1, \cdots k$) can be trained using all samples in the ith class with positive labels, and remaining samples with negative labels. Hence, given M training instances $(x_1, y_1), \cdots, (x_M, y_M)$, where $x_j \in \Re^n$, $j = 1, \cdots M$ and $y_j \in \{1, \cdots, k\}$ denotes the label of x_j, the ith SVM addresses the problem as follows:

$$\text{minimize} \quad \frac{1}{2}\|w^i\|^2 + C \sum_{i=1}^{M} \xi_j^i (w^i)^T$$

$$\text{subject to} \quad (w^i)^T \varphi(x_j) + b^i \geq 1 - \xi_j^i \quad \text{if } y = i \quad (2.53)$$

$$(w^i)^T \varphi(x_j) + b^i \leq 1 - \xi_j^i \quad \text{if } y \neq i$$

$$\xi_j^i \geq 0, \quad j = 1, \cdots, M$$

where C denotes the penalty parameter.

(2) One-against-one (OAO)

Another popular approach is the OAO method. This approach establishes $[k(k-1)]/2$ SVMs. Each of the SVMs can be trained on samples from the two classes. For training instances of the ith and the jth classes, the following classification problem can be solved:

$$\text{minimize} \quad \frac{1}{2}\|w^{ij}\|^2 + C \sum_{t} \xi_t^{ij} (w^{ij})^T$$

$$\text{subject to} \quad (w^{ij})^T \varphi(x_t) + b^{ij} \geq 1 - \xi_t^{ij} \quad \text{if } y_t = i \quad (2.54)$$

$$(w^{ij})^T \varphi(x_t) + b^{ij} \leq 1 - \xi_t^{ij} \quad \text{if } y_t \neq i$$

$$\xi_t^{ij} \geq 0, \quad t = 1, \cdots, M$$

The decision of the training samples with the OAO method is made by the following method. If $\text{sign}\left[(w^{ij})^T\varphi(x) + b^{ij}\right]$ indicates a sample x in the ith class, the votes of x for the ith class increase by one. In the other situation, the ith is added by one. Afterward, the label of x can be predicted as the largest vote in the class.

3) Fault diagnosis using SVM

In this section, MPE is introduced in detail first. Then the intelligent fault diagnosis method based on MPE and SVM is described.

(1) Multi-scale permutation entropy

MPE is a coarse-grained program that obtains multi-scale time series from raw signals. As a measure of signal space complexity, MPE is a simple approach that can measure the complexity of nonlinear and non-stationary signals. The details of MPE are described below.

For a vibration data signal $\{x_i, i = 1, 2, \cdots, n\}$, multiple coarse-grained time series can be generated by averaging the data within non-overlapping windows of multiple lengths s. The coarse-grained process can be presented by the following equation:

$$y_j^s = \frac{1}{s} \sum_{i=(j-1)s+1}^{js} x_i, \ 1 \leqslant j \leqslant n/s \tag{2.55}$$

where s denotes the scale factor and y_j^s represents the coarse-grained time series with multiple scales. It should be noted that when the scale is 1, indicating $s = 1$, the coarse-grained time series are the original time series.

After the coarse-grained process is completed, we can calculate the PE of the coarse-grained time series y_j^s with multiple scale factors. This procedure is MPE.

First, y_j^s can be embedded in the phase space:

$$Y_t^s = \left[y_t^s, y_{t+\tau}^s, \cdots, y_{t+(m-1)\tau}^s\right], \ t \in [1, n/s - m + 1] \tag{2.56}$$

where m denotes the embedding dimension and τ represents the delay time.

With respect to each t, the m components of the vector Y_t^s are arranged in increasing order:

$$\left[y_{t+(k_1-1)\tau}^s \leqslant y_{t+(k_2-1)\tau}^s \leqslant \cdots \leqslant y_{t+(k_m-1)\tau}^s\right] \tag{2.57}$$

Afterward, we are able to find a vector π_t in order to identify the permutation pattern of Y_t^s:

$$\pi_t = [k_1, k_2, \cdots, k_m] \tag{2.58}$$

2.3 Statistical Learning-Based Methods

In the phase space of embedding dimension m, the total number of the permutations of Y_t^s is $m!$. N_l is adopted to denote the occurrence times of lth permutation for each t, where $1 \leq l \leq m!$. The occurrence probability of lth permutation is expressed as follows:

$$p_l^s = \frac{N_l}{n/s - m + 1} \tag{2.59}$$

where n/s denotes the length of coarse-grained time series using the scale factor s. Hence, the permutation entropy of coarse-grained time series with scale s can be defined as

$$H_p^s = -\sum_{l=1}^{m!} p_l^s \ln p_l^s \tag{2.60}$$

When all permutations of the time series have the same probability, the permutation entropy of the time series obtains the maximum value. In addition, its features are similar to noise features. Conversely, the more regular the time series are, the lower the permutation entropy is. Therefore, when a time series is collected from the linear prediction system, the permutation entropy of the time series will be minimum.

(2) Phase space reconstruction

In the approach of MPE, the mechanical vibration signals can be embedded into phase space with the embedding dimension m and the delay time τ. The above process can be regarded as phase space reconstruction, which can be applied to get the potential complexity information of the vibration signals. The phase space reconstruction aims at selecting m and τ. In this work, the mutual information approach is introduced to select τ and the Cao method is applied to select m(Cao 1997).

Fraser and Swinney first introduced the concept of mutual information to measure the inherent dependence of one-dimensional signals (Fraser and Swinney 1986). For vibration signals, mutual information can be defined as

$$I_\tau = \sum_{i=1}^{n} P(x_i, x_{i+\tau}) \log \left\{ \frac{P(x_i, x_{i+\tau})}{P(x_i) P(x_{i+\tau})} \right\} \tag{2.61}$$

where $x_{i+\tau}$ denotes the delayed vibration signal, $P(x_i)$ represents the probability of the measure of x_i, $P(x_{i+\tau})$ denotes the probability of the measure of $x_{i+\tau}$, and $P(x_i, x_{i+\tau})$ represents the joint probability distribution. When I_τ equals 0, it means that x_i and $x_{i+\tau}$ are totally uncorrelated. When I_τ reaches the minimal value x_i and $x_{i+\tau}$ are as uncorrelated as possible and the redundant information of the two sequences is thus minimal. Hence, τ can be determined when I_τ reaches the first local minimal value.

The Cao method was proposed by improving false neighbors (Cao 1997). In the Cao method, the embedding dimension of stochastic signals can be successfully calculated using a little amount of data after τ has been computed. Its main motivation is to compare the change value of the distance between two adjacent points from dimension m to dimension $m+1$. There are two judgment criterions, one for deterministic signals and the other one for stochastic signals. The signals of mechanical systems can be stochastic. Hence, we select the judgment criterion for stochastic signals:

$$E'_m = \frac{1}{n-m\tau} \sum_{i=1}^{n-m\tau} \left| x_{i+m\tau} - x^N_{i+m\tau} \right| \tag{2.62}$$

where $x^N_{i+m\tau}$ denotes the nearest neighbor of $x_{i+m\tau}$. The judgment criterion can be computed as

$$E(m) = E'_{m+1} / E'_m \tag{2.63}$$

With respect to the vibration signal containing noise, the correlation between the data changes with the change of embedding dimension. Then, an appropriate embedding dimension is able to be selected when the value of $E(m)$ does not change.

(3) The intelligent fault diagnosis method based on SVM

Based on SVM, an intelligent fault diagnosis strategy is presented as shown in Fig. 2.5. The scale factor of MPE is selected as 1 ~ 5, and the delay time and embedding dimension of MPE are determined by the mutual information method and Cao method. The RBF kernel is selected as the kernel function of SVM, and the optimal values C and γ are obtained by the grid search method as the optimal parameters of SVM.

The feature vector of the signal computed by MPE can be defined as

$$\boldsymbol{H}_{\text{MPE}} = [H_p^{s_1} \ H_p^{s_2} \cdots H_p^{s_n}] \tag{2.64}$$

Based on the feature vector, the fault diagnosis approach can be presented as follows:

- Obtain the vibration signals.
- Choose the delay time and embedding dimension with the phase space reconstruction.
- Extract the feature vectors $\boldsymbol{H}_{\text{MPE}}$ of the signal with MPE.
- Split the feature vectors $\boldsymbol{H}_{\text{MPE}}$ into a training set and a testing set.
- Treat the training set as the input of SVM with the RBF kernel and calculate the optimal parameters of SVM using the grid search method.
- Predict the health conditions of the testing set through the utilization of the trained SVM.

2.3 Statistical Learning-Based Methods

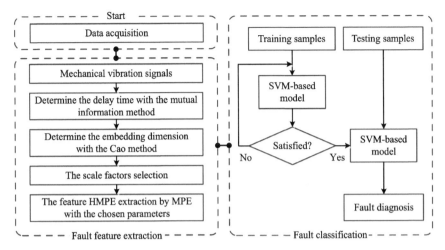

Fig. 2.5 Flow chart of the method based on SVM

4) Intelligent fault diagnosis case of motor bearings

(1) Data description

The data are from Case Western Reverse University (CWRU), and collected from a motor-driven mechanical system (Smith and Randall 2015). The system contains a 2-horsepower three-phase induction motor (left), a torque sensor (middle), a dynamometer (right) and several control electronics. The vibration data are obtained by an accelerometer placed on the drive end of the motor. Fault types are the inner race failure (IF), outer race failure (OF), and rolling element failure (RF) of the bearing. The fault diameter is 0.18, 0.36, 0.53 mm, and the point fault depth is 0.28 mm. The type of bearing in the drive end is 6205-2RS JME SKF. To collect vibration data, the sampling frequency is 48 kHz.

The detailed information of the dataset is in Table 2.6. The experiment was conducted under a load of 1 hp. Each health condition contains 48000 data points. The data points of each condition are divided into 100 samples. 20% of the samples are used as training samples and the remaining samples are used as testing samples.

(2) Diagnosis results

First, we investigate the determination of parameters of MPE. In the presented method, the mutual information method can be adopted to select the delay time τ. The Cao method is adopted to select the embedding dimension m.

The mutual information can be adopted to predict the joint probability distribution $P(x_i, x_{i+\tau})$ of the samples with many partitions where the size of each element is tailored to the local situation. We used 128 partitions to compute the mutual information I_τ of each health condition in the dataset. In Table 2.7, the dependence of I_τ on τ is provided. The selected τ for each health condition with the first minimum I_τ is presented in bold. When I_τ reaches its first minimum, it

Table 2.6 Description of the bearing dataset

Datasets	Fault type	Fault diameter /mm	Samples	Training samples	Testing samples	SVM label
Bearing dataset	RF	0.18	100	20%	80%	1
	RF	0.36	100	20%	80%	2
	RF	0.53	100	20%	80%	3
	IF	0.18	100	20%	80%	4
	IF	0.36	100	20%	80%	5
	IF	0.53	100	20%	80%	6
	OF	0.18	100	20%	80%	7
	OF	0.36	100	20%	80%	8
	OF	0.53	100	20%	80%	9
	Normal	0	100	20%	80%	10

can be observed that most of the delay time for the samples is four. Therefore, the delay time for the bearing dataset is selected as four.

Then, the Cao method is adopted to numerically select the embedding dimension. In Fig. 2.6, the value $E(m)$ is plotted as a function of m for the samples of each health condition. It indicates that $E(m)$ is stable if m is equal to four. After $m = 6$, the $E(m)$ of samples for all health conditions change in a stable interval. In other words, the values are stable enough. Hence, $m = 6$ for the bearing dataset.

After determining the MPE parameters, the fault diagnosis accuracy of the presented method is 99.50%. To verify the effectiveness of the presented method, we compare it with other methods. In the first method, the feature is the EMD energy ratio and it is a type of feature that can characterize well the health

Table 2.7 The dependence of I_τ on τ in the bearing dataset

Health condition	τ					
	1	2	3	4	5	6
RF 0.18	1.4252	0.5554	0.1659	**0.0285**	0.0765	0.2886
RF 0.36	1.6858	0.7983	0.3718	0.1531	**0.0932**	0.1247
RF 0.53	1.5506	0.6624	0.2446	**0.0682**	0.0705	0.2382
IF 0.18	1.4530	0.6055	0.2442	**0.1142**	0.1212	0.2076
IF 0.36	1.1112	0.3748	0.0953	**0.0446**	0.0813	0.1113
IF 0.53	1.6008	0.7028	0.2691	0.0717	**0.0484**	0.1860
OF 0.18	1.4743	0.6181	0.2667	**0.2019**	0.3864	0.8599
OF 0.36	1.4359	0.5643	0.1720	**0.0306**	0.0743	0.2862
OF 0.53	1.7595	0.8770	0.4651	0.2852	**0.2587**	0.3467
Normal	0.8513	0.2027	0.0576	**0.0466**	0.0709	0.0704

2.3 Statistical Learning-Based Methods

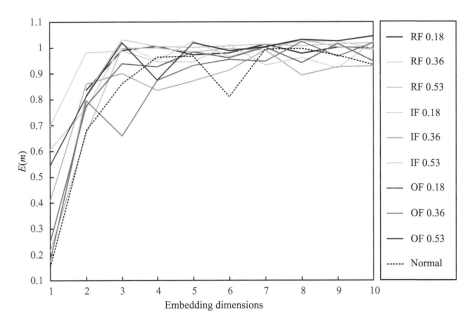

Fig. 2.6 Values of $E(m)$ for the bearing dataset

conditions of the bearings (Chen et al. 2014). The second method uses EEMD and singular value decomposition to extract features (Zhang and Zhou 2013). Both the compared methods use optimized SVM for classification, and their accuracies are shown in Table 2.8. It can be seen that the two compared methods provide the testing accuracies of 94.28% and 97.07% respectively, which are lower than the accuracy of the presented method.

Table 2.8 Compared results using different methods

Method	Features	Testing accuracy
The presented method	MPE	99.50%
The first compared method	EMD energy features	94.28%
The second compared method	EEMD-based features	97.07%

2.3.3 Intelligent Diagnosis Using Relevant Vector Machine

Relevance vector machine (RVM) (He et al. 2017) is a new classification kernel technology based on the Bayesian framework. The advantages of RVM in solving classification problems can be summarized as follows. (1) RVM can build a sparser model than SVM. Therefore, RVM can obtain faster performance on the testing samples and maintain comparable generalization capabilities. (2) RVM provides the member probability of the output. (3) The kernel function of RVM algorithm is no longer restricted by Mercer's theorem and can be constructed freely. (4) The parameters of RVM are automatically allocated during the training process, rather than subjectively allocated. Due to the above advantages, Widodo et al. (Widodo et al. 2009) used RVM to diagnose the failure of rolling bearings and achieved good results. Wang et al. (Wang et al. 2014) developed an RVM-based monitoring system to classify tool wear during milling.

Here, a new method based on the multiclass relevance vector machine (MRVM) is developed to classify the health conditions of planetary gearboxes. In this method, two features, i.e., accumulative amplitudes of carrier orders (AACO) and energy ratio based on difference spectra (ERDS) are used to characterize the health conditions of planetary gearboxes. Then MRVM is adopted as the classifier.

1) Relevance vector machine

For a set of samples $\{x_i, t_i\}_{i=1}^{N}$, $x_i \in R^D$ denotes the ith D-dimensional feature and $t \in \{1, 2, 3, \cdots, C\}$ represents the class labels. The functional relationship between x_i and t_i can be modeled using RVM as

$$t_i = \sum_{i=1}^{N} w_i K(x, x_i) + w_0 \tag{2.65}$$

The scalar-valued output functions are assumed to be RVM class labels from the samples with additive noise, as shown in the following standard probabilistic formulation:

$$t_i' = t_i + \varepsilon_i = \sum_{i=1}^{N} w_i K(x, x_i) + w_0 + \varepsilon_i \tag{2.66}$$

where ε_i denotes the ith independent sample from a noise process which subjects to zero-mean Gaussian distribution with variance σ^2. $K(x, x_i)$ denotes a kernel function, and w_i represents the ith weight of the linear combination of the kernel functions.

The target of RVM can only take the value 0 or 1 with respect to the binary classification problems. Therefore, the logistic sigmoid function $\sigma(y) = 1/(1 + e^{-y})$ can be used to generalize the model of RVM. Its likelihood can be presented as

$$P(t|w) = \prod_{i=1}^{N} \sigma[y(x_i; w)]^{t_i} \{1 - y(x_i; w)\}^{1-t_i} \tag{2.67}$$

2.3 Statistical Learning-Based Methods

As $P(t|w)$ is a non-normal distribution, the closed-form solutions of the weight posterior $p(w|t, \alpha)$ and the marginal likelihood $P(t|w)$ cannot be computed analytically. Additionally, the weight vector w can also be computed analytically. Therefore, an approximation procedure with the Laplace method is applied for addressing RVM.

(1) Due to the fact that $p(w|t, \alpha) \propto P(t|w) p(w|\alpha)$, the maximum posterior weight with a fixed value of α could be calculated:

$$\log\{P(t|w) p(w|\alpha)\} = \sum_{i=1}^{n} \left[t_i \log y_i + (1 - t_i) \log(1 - y_i) \right] - \frac{1}{2} w^T A w \tag{2.68}$$

where $A = \text{diag}(\alpha_0, \alpha_1, \cdots, \alpha_N)$ and $y_i = \sigma\{y(x_i; w)\}$.

(2) The Laplace method adopted in this section is a quadratic approximation approach for the log-posterior around its mode. We differentiate the maximum posterior weight in Eq. (2.68) and present the results as

$$\nabla_w \nabla_w \log p(w|t, \alpha)|_{w_{\text{MP}}} = -(\Phi^T B \Phi + A) \tag{2.69}$$

where ∇ represents the gradient operator and $B = \text{diag}(\beta_0, \beta_1, \cdots, \beta_N)$ is a diagonal matrix with $\beta_n = \sigma\{y(x_n)\}[1 - \sigma(x_n)]$.

(3) With the statistics $\Sigma = (\Phi^T B \Phi + A)^{-1}$ and $w_{\text{MP}} = \Sigma \Phi^T B t$ of the Gaussian approximation, the parameters α_i can be correspondingly updated by

$$\alpha_i^{\text{new}} = \frac{1 - \alpha_i \sum_{ii}}{w_{\text{MP}i}^2} \tag{2.70}$$

where $w_{\text{MP}i}$ denotes the ith element of the most possible maximum posterior weight w_{MP}, and $\sum_{ii}$ represents the ith diagonal element of the posterior weight covariance.

2) Multiclass relevance vector machine

The regression targets of MRVM, $Y \in R^{C \times N}$, are the auxiliary variables, which are subject to a standardized noise model:

$$y_{cn}|w_c, k_n \sim N_{y_{cn}}(w_c^T k_n, 1) \tag{2.71}$$

Using the multinomial probit link, the regression targets of MRVM are transformed into the labels:

$$t_n = i, y_{ni} > y_{nj}, \forall i \neq j \tag{2.72}$$

Therefore, the probabilistic output of a sample can be calculated by the resultant multinomial probit likelihood function:

$$P(t_n = i|\boldsymbol{w}, \boldsymbol{k}_n) = \varepsilon_{p(u)}\left\{\prod_{j \neq i} \Phi(u + (w_i - w_j)^{\mathrm{T}} \boldsymbol{k}_n)\right\} \quad (2.73)$$

where $u \sim N(0, 1)$, and Φ is the Gaussian cumulative distribution function.

In order to ensure the sparsity of MRVM, the weight w_{nc} follows a standard normal distribution with zero-mean and variance α_{nc}^{-1}. α_{nc} denotes the element of the scale matrix $A \in R^{N \times C}$ and follows a Gamma distribution. When the parameters a and b are small enough, most elements in $\boldsymbol{w}$ are limited to zero, resulting in a sparse solution.

The closed-form solution of weight $\boldsymbol{w}$ can be inferred as

$$P(\boldsymbol{w}|Y) \propto P(Y|\boldsymbol{w})P(\boldsymbol{w}|A) \propto \prod_{c=1}^{C} N((\boldsymbol{KK}^{\mathrm{T}} + A_c)^{-1}\boldsymbol{Ky}_c^{\mathrm{T}}, (\boldsymbol{KK}^{\mathrm{T}} + A_c))^{-1} \quad (2.74)$$

where A_c is derived from the cth column of A. Therefore, the maximum of a posterior estimator can be presented as

$$\hat{\boldsymbol{w}} = \arg\max_{\boldsymbol{w}} P(\boldsymbol{w}|Y, A, \boldsymbol{K}) \quad (2.75)$$

Next, the parameters can be optimized by maximizing the maximum a posterior estimator value

$$\hat{\boldsymbol{w}}_c = (\boldsymbol{KK}^{\mathrm{T}} + A_c)^{-1}\boldsymbol{Ky}_c^{\mathrm{T}} \quad (2.76)$$

In the last step, the posterior probability distribution of priori parameters of weight vector can be illustrated as

$$p(A|\boldsymbol{w}) \propto p(\boldsymbol{w}|A)p(A|a,b) \propto p\prod_{c=1}^{C}\prod_{n=1}^{N} G(a + \frac{1}{2}, \frac{w_{nc}^2 + 2b}{2}) \quad (2.77)$$

3) Fault diagnosis using MRVM

The diagram of the intelligent fault diagnosis method using MRVM is presented in Fig. 2.7. The details are as follows.

- Collect the vibration signals using sensors fixed on different stages of planetary gearboxes.

2.3 Statistical Learning-Based Methods

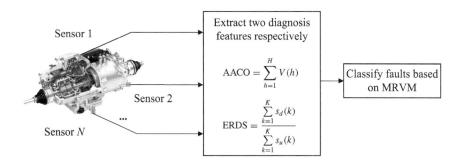

Fig. 2.7 Flow chart of the presented method based on MRVM

- Obtain AACO and ERDS from the collected signals. The extracted features together with their labels can be prepared into a dataset. Then split the dataset into a training set and a testing set.
- Utilize the training set to train the MRVM-based classifier.
- Adopt the trained MRVM to separate the testing samples.

In the above steps, details of the two features, i.e., AACO and ERDS are illustrated as follows:

(1) AACO

The characteristic frequency of the components in the planetary gearbox is usually a multiple of the carrier rotation frequency. Therefore, when these components fail, the magnitude of the order of the carrier rotation frequency will increase. In order to identify the change in order amplitude, the order spectrum of the carrier rotation frequency is used to calculate AACO. In other words,

$$\text{AACO} = \sum_{h=1}^{H} V(h) \tag{2.78}$$

where $V(h)(h = 1, 2, \cdots, H)$ indicates the amplitude at the hth order of the carrier rotating frequency in the order spectrum. H denotes the number of orders in the order spectrum.

(2) ERDS

Generally, if the health state of the planetary gearbox changes, its vibration signal and the amplitude of the frequency spectrum will also change. Therefore, in order to characterize the change of the frequency spectrum, the ERDS was developed using the signal difference spectrum of the planetary gearbox under healthy conditions and unknown conditions. ERDS is calculated by the sum of the absolute amplitudes of the normalized difference spectrum:

$$\text{ERDS} = \frac{\sum_{k=1}^{K} s_d(k)}{\sum_{k=1}^{K} s_u(k)} \tag{2.79}$$

where $s_d(k) = |s_u(k) - s_h(k)|(k = 1, \cdots, K)$ denotes the absolute amplitude at the kth spectrum line in the difference spectrum. K represents the number of spectrum lines. $s_u(k)$ and $s_h(k)$ indicate the amplitudes at the kth spectrum line of the frequency spectra with the unknown health condition and that under the healthy condition.

4) Intelligent fault diagnosis case of a planetary gearbox

(1) Introduction of the experimental system and data collection

The test bench of a planetary gearbox is displayed in Fig. 2.8. In the test bench, the inner sun gear of each stage of the planetary gearbox is surrounded by a stationary outer ring gear and three or four rotating planetary gears. The torque is transmitted to the planet gears fixed on the planet carrier through the sun gear, and the planet carrier transmits the torque to the output shaft. Six faults were simulated in the test bench, i.e., the first-stage sun gear tooth crack, the first-stage sun gear dent tooth, the first-stage planet gear tooth crack, the planet bearing needle falling off in the first stage, and the missing teeth of the second stage sun gear as shown in Table 2.9 (Lei et al. 2015).

In order to collect the vibration signals, two accelerometers, namely accelerometer 1 and accelerometer 2 are installed on the first and second stage housings of the planetary gearbox. The sampling frequency is 5120 Hz. The motor speeds are set as 2100, 2400, 2700, and 3000 r/min, respectively. The load conditions include no load and a load equaling 13.5 Nm. In the experiment, a vibration signal containing 20,480 data points constitutes a data sample. For each

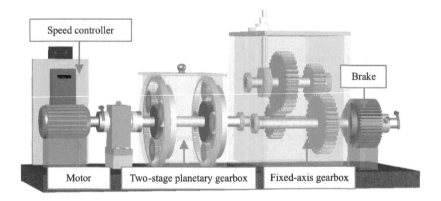

Fig. 2.8 Schematic model of the planetary gearbox test bench

2.3 Statistical Learning-Based Methods

Table 2.9 Seven health conditions of the planetary gearbox

Label	Health conditions	
	First stage	Second stage
1	Normal	Normal
2	The sun gear having a cracked tooth	Normal
3	The sun gear having a pitted tooth	Normal
4	A planet gear having a cracked tooth	Normal
5	A planet bearing having a flaked needle	Normal
6	Normal	The sun gear having a chipped tooth
7	Normal	The sun gear having a missing tooth

health state at each motor speed and under each load condition, each accelerometer collects 30 data samples. The same health conditions under different speeds and different load conditions are considered as one category. Therefore, for each health condition, two accelerometers are used to obtain 480 data samples.

In order to thoroughly verify the presented method, three different data subsets (data subset 1, data subset 2, and data subset 3) are used to compose three tests (test 1, test 2, and test 3).

(2) Test 1: fault diagnosis with the data collected from the first-stage housing

Data subset 1 consists of 1680 data samples collected by accelerometer 1 mounted on the first-stage housing of the planetary gearbox. This subset involves seven different health states at four motor speeds and under two load conditions. 50% of the samples are applied as training samples, and the remaining 50% are used for testing. The features extracted from the vibration signals are shown in Fig. 2.9(a) and (b). Figure 2.9(c) shows the clustering results of the two features. It can be seen that the samples of the same label cluster well, while the samples of different labels are separated. We train MRVM with these extracted featuresand the corresponding health state labels. After training, we use the trained MRVM to classify features of each testing sample to obtain the health state of the testing sample. As shown in Table 2.10, the accuracy obtained by MRVM on data subset 1 is 81.79%.

(3) Test 2: fault diagnosis with the data obtained on the second-stage housing

Different from data subset 1, data subset 2 is collected by accelerometer 2 installed on the second-stage housing. Data subset 2 also includes 1680 data samples. 50% of the samples are used for training and the rest are applied as testing samples. The classification accuracy on this data subset is 81.43%, which is very close to the accuracy in Test 1. Therefore, the classification accuracy of the 7 health conditions is less than 85% in Tests 1 and 2. Because the data used

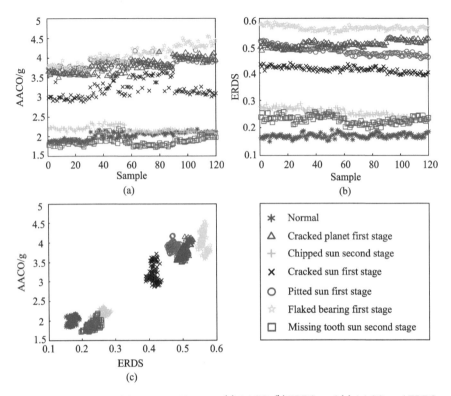

Fig. 2.9 Illustrations of the extracted features: (a) AACO, (b) ERDS, and (c) AACO, and ERDS

are collected by only one accelerometer and data from a single accelerometer may not contain complete feature information for intelligent fault diagnosis.

(4) Test 3: fault diagnosis with the data collected on both stages

Data subset 3 consists of the data acquired by both accelerometer 1 and accelerometer 2. Half of the samples are used for training and the rest are used for testing. Table 2.10 shows that the classification accuracy on data subset 3 is 100%. In other words, the presented method uses a data fusion strategy to correctly distinguish all samples. Compared with Test 1 and Test 2, Test 3 makes full use of the data from the two accelerometers and greatly improves the classification accuracy of health states. In addition, to study the robustness of the presented method under various motor speeds, we set the training samples using data obtained at two lower motor speeds, while the testing samples are composed of data collected at two higher speeds. Then the classification accuracy obtained

Table 2.10 Identification accuracy using the presented method for the three tests

Test number	Test 1	Test 2	Test 3	Test 3 (different speeds)
Accuracy	81.79%	81.43%	100%	95.95%

by the presented method changes from 100% to 95.95%. Although the accuracy drops by 4.05%, the classification accuracy is still acceptable for the application. In other words, the presented method is robust to various motor speeds.

2.3.4 Epilog

This section discusses an SVM-based and RVM-based fault diagnosis method, respectively. In the SVM-based method, multi-scale permutation entropy is used to evaluate various types of failures and severity of the rolling element bearing, firstly. Secondly, for the MPE to work better, the mutual information method and the Cao method are applied to determine the delay time and the incorporation dimension of the MPE. Finally, SVM is used to identify the health conditions of the bearing dataset according to the features extracted by MPE. The fault diagnosis results show that the MPE parameters are crucial for the fault diagnosis accuracy and the presented method is superior in classifying ten bearing health conditions compared to other fault diagnosis methods. As for the RVM-based method, it extracts two features of AACO and ERDS from the collected data, as the input of MRVM. The experiment results indicate the superiority of the RVM-based method. The results also illustrate the effectiveness and robustness of the presented method in distinguishing the health condition of planetary gearboxes.

2.4 Conclusions

In this chapter, the conventional intelligent fault diagnosis methods for mechanical systems are presented. Two representative AI methods are introduced, which are the typical neural network-based and statistical learning-based methods. For neural networks, the RBF network and WNN are described, and the experiments on the bearing fault diagnosis problems are carried out for validations. For statistical learning-based methods, the SVM and RVM methods are provided, and the experiments on the gearbox fault diagnosis problems are implemented for validations. The effectiveness of the conventional intelligent fault diagnosis methods is demonstrated in the basic fault diagnosis tasks. The conventional intelligent fault diagnosis methods generally provide baselines for the fault diagnosis performance. The methods are mostly well-established and the implementations are relatively easy. The presented case studies provide guidelines for the utilizations of the methods. They are well-suited for applications in the real fault diagnosis cases for mechanical systems.

It should be pointed out that despite the effectiveness in the concerned tasks, the conventional intelligent fault diagnosis methods are less capable of modelling complex mechanical system behaviors, especially when the collected condition monitoring data have sophisticated patterns. More advanced intelligent fault diagnosis methods will be introduced in the following chapters to address the challenging problems for mechanical systems.

References

Azamfar M, Singh J, Li X, Lee J (2020) Cross-domain gearbox diagnostics under variable working conditions with deep convolutional transfer learning. J Vib Control 27(7–8):854–864. https://doi.org/10.1177/1077546320933793

Aziz W, Arif M (2005) Multiscale permutation entropy of physiological time series. In: 2005 Pakistan Section Multitopic Conference. IEEE pp 1–6

Bandt C, Pompe B (2002) Permutation entropy: a natural complexity measure for time series. Physical review letters 88 (17):174102

Burges CJ (1998) A tutorial on support vector machines for pattern recognition. Data Min Knowl Disc 2(2):121–167

Cao L (1997) Practical method for determining the minimum embedding dimension of a scalar time series. Physica D 110(1–2):43–50

Chen F, Tang B, Song T, Li L (2014) Multi-fault diagnosis study on roller bearing based on multi-kernel support vector machine with chaotic particle swarm optimization. Measurement 47:576–590

Fraser AM, Swinney HL (1986) Independent coordinates for strange attractors from mutual information. Phys Rev A 33(2):1134

He S, Xiao L, Wang Y, Liu X, Yang C, Lu J, Gui W, Sun Y (2017) A novel fault diagnosis method based on optimal relevance vector machine. Neurocomputing 267:651–663. https://doi.org/10.1016/j.neucom.2017.06.024

Jia F, Lei Y, Lin J, Zhou X, Lu N (2016) Deep neural networks: A promising tool for fault characteristic mining and intelligent diagnosis of rotating machinery with massive data. Mech Syst Signal Process 72:303–315

LeCun Y, Boser B, Denker JS, Henderson D, Howard RE, Hubbard W, Jackel LD (1989) Backpropagation applied to handwritten zip code recognition. Neural Comput 1(4):541–551

Lei Y, He Z, Zi Y (2009) Application of the EEMD method to rotor fault diagnosis of rotating machinery. Mech Syst Signal Process 23(4):1327–1338. https://doi.org/10.1016/j.ymssp.2008.11.005

Lei Y, He Z, Zi Y (2011) EEMD method and WNN for fault diagnosis of locomotive roller bearings. Expert Syst Appl 38(6):7334–7341. https://doi.org/10.1016/j.eswa.2010.12.095

Lei Y, He Z, Zi Y, Chen X (2008) New clustering algorithm-based fault diagnosis using compensation distance evaluation technique. Mech Syst Signal Process 22(2):419–435. https://doi.org/10.1016/j.ymssp.2007.07.013

Lei Y, Jia F, Lin J, Xing S, Ding SX (2016) An intelligent fault diagnosis method using unsupervised feature learning towards mechanical big data. IEEE Trans Industr Electron 63(5):3137–3147

Lei Y, Liu Z, Wu X, Li N, Chen W, Lin J (2015) Health condition identification of multi-stage planetary gearboxes using a mRVM-based method. Mech Syst Signal Process 60–61:289–300. https://doi.org/10.1016/j.ymssp.2015.01.014

Lei Y, Yang B, Jiang X, Jia F, Li N, Nandi AK (2020) Applications of machine learning to machine fault diagnosis: A review and roadmap. Mech Syst Signal Process 138 106587. https://doi.org/10.1016/j.ymssp.2019.106587

Li Y, Yang Y, Wang X, Liu B, Liang X (2018) Early fault diagnosis of rolling bearings based on hierarchical symbol dynamic entropy and binary tree support vector machine. J Sound Vib 428:72–86. https://doi.org/10.1016/j.jsv.2018.04.036

Liu R, Yang B, Zio E, Chen X (2018) Artificial intelligence for fault diagnosis of rotating machinery: A review. Mech Syst Signal Process 108:33–47

Micchelli CA (1984) Interpolation of scattered data: distance matrices and conditionally positive definite functions. In: Approximation theory and spline functions. Springer, pp 143–145

Parker DB (1985) Learning logic technical report tr-47. Center of Computational Research in Economics and Management Science, Massachusetts Institute of Technology, Cambridge, MA

Powell MJ (1987) Radial basis functions for multivariable interpolation: a review. Algorithms for approximation

Rumelhart DE, Hinton GE, Williams RJ (1986) Learning Representations by Back-Propagating Errors. Nature 323(6088):533–536

Rumelhart DE, McClelland JL, Group PR (1988) Parallel distributed processing, vol 1. IEEE Massachusetts

Shi Q, Zhang H (2021) Fault diagnosis of an autonomous vehicle with an improved SVM algorithm subject to unbalanced datasets. IEEE Trans Industr Electron 68(7):6248–6256. https://doi.org/10.1109/TIE.2020.2994868

Smith WA, Randall RB (2015) Rolling element bearing diagnostics using the Case Western Reserve University data: a benchmark study. Mech Syst Signal Process 64–65:100–131. https://doi.org/10.1016/j.ymssp.2015.04.021

Wang G, Yang Y, Xie Q, Zhang Y (2014) Force based tool wear monitoring system for milling process based on relevance vector machine. Adv Eng Softw 71:46–51

Wang Z, Yao L, Cai Y (2020) Rolling bearing fault diagnosis using generalized refined composite multiscale sample entropy and optimized support vector machine. Measurement 156 107574. https://doi.org/10.1016/j.measurement.2020.107574

Werbos P (1974) Beyond regression:" new tools for prediction and analysis in the behavioral sciences. Ph D dissertation, Harvard University

Widodo A, Kim EY, Son J-D, Yang B-S, Tan AC, Gu D-S, Choi B-K, Mathew J (2009) Fault diagnosis of low speed bearing based on relevance vector machine and support vector machine. Expert Syst Appl 36(3):7252–7261

Wu C, Jiang P, Ding C, Feng F, Chen T (2019) Intelligent fault diagnosis of rotating machinery based on one-dimensional convolutional neural network. Comput Ind 108:53–61. https://doi.org/10.1016/j.compind.2018.12.001

Wu Z, Huang NE (2009) Ensemble empirical mode decomposition: a noise-assisted data analysis method. Adv Adapt Data Anal 1(01):1–41

Yan X, Jia M (2018) A novel optimized SVM classification algorithm with multi-domain feature and its application to fault diagnosis of rolling bearing. Neurocomputing 313:47–64. https://doi.org/10.1016/j.neucom.2018.05.002

Yang B, Xu S, Lei Y, Lee C-G, Stewart E, Roberts C (2022) Multi-source transfer learning network to complement knowledge for intelligent diagnosis of machines with unseen faults. Mech Syst Signal Process 162 108095. https://doi.org/10.1016/j.ymssp.2021.108095

Yang L, Chen H (2019) Fault diagnosis of gearbox based on RBF-PF and particle swarm optimization wavelet neural network. Neural Comput Appl 31(9):4463–4478. https://doi.org/10.1007/s00521-018-3525-y

Zhang W, Li X, Ma H, Luo Z, Li X (2021) Universal domain adaptation in fault diagnostics with hybrid weighted deep adversarial learning. IEEE Trans Industr Inf 17(12):7957–7967. https://doi.org/10.1109/TII.2021.3064377

Zhang X, Zhou J (2013) Multi-fault diagnosis for rolling element bearings based on ensemble empirical mode decomposition and optimized support vector machines. Mech Syst Signal Process 41(1–2):127–140

Chapter 3
Hybrid Intelligent Fault Diagnosis

3.1 Introduction

In the related literature (Yu and Liu 2020; Ding et al. 2020; Kaplan et al. 2020; Jia et al. 2018; Siahpour et al. 2020), it can be noticed that the introduction of AI techniques substantially boosts the reliability of fault monitoring and diagnosis systems. Initially, the research works in this field mainly focused on the performance improvement of individual classifiers or the formulization of novel classifiers. Therefore, lots of classification algorithms have been proposed. Those classifiers greatly enriched the application of the AI techniques in fault diagnosis. Nevertheless, the existing methods inevitably encountered their performance bottlenecks due to the intrinsic characteristics. To be specific, one single classifier is unable to reflect the complicated multi-fault problem, which is widely seen in reality. Aiming at further enhancing the fault diagnosis performance, the methodology of combining multiple classifiers have been studied to tackle the limitations of individual classifier and achieve higher classification accuracies (Li et al. 2020). It is revealed that different classifiers using different input feature sets or classification algorithms are usually able to exhibit complementary classification behavior. Therefore, if the classification results of multiple classifiers can be appropriately combined to generate the final classification result, it would be highly promising that the final result is superior to the individually obtained results.

To pursue the superior performance of multiple classifier combination, we need to emphasize two important factors: one is multiple classifiers and the other is input feature sets, as they are the two most critical components influencing the performance of intelligent fault diagnosis. Therefore, a problem arises whether different classifiers or input feature sets will yield complementary classification performance. To explore this problem, this chapter mainly studies three methods of multiple classifier combinations from two different aspects: classifiers and input feature sets. Three types of combination schemes are investigated and analyzed in this chapter, which can be roughly classified into three categories: (1) the combination of different classification engines but with the utilization of the same input feature sets (Sect. 3.2),

(2) the utilization of same classification engines but with different input feature sets (Sect. 3.3) and (3) the utilization of different classification engines as well as different input feature sets (Sect. 3.4). Note that the utilization of different classification engines indicates two-fold meanings: the first indicates that the same classifier but with different parameters, such as multiple weighted KNN with different K. While the other is different classifiers with distinct classification algorithms. With the utilization of three different combination strategies, the multidimensional hybrid intelligent diagnosis method is investigated and complementary behaviors among different classifiers or input feature sets are analyzed in this chapter.

3.2 Multiple WKNN Fault Diagnosis

3.2.1 Motivation

The K nearest neighbor, as one of the commonly used supervised learning algorithms, has attracted wide concerns because it is easy to implement. In many fields regarding pattern recognition, KNN has successfully solved many learning problems. However, in real-world applications, the KNN has two weaknesses. First, it is difficult for KNN to distinguish the samples from different classes when the class number is larger than the number of feature dimensions. Second, the parameter, i.e., the neighborhood parameter K, is usually determined by expert experience, which indicates that the parameter selection and optimization lack specific guidelines. Aiming at the former weakness, researchers come up with some solutions. For example, Zhou and Chen (2006) adopted a weighted distance method to assess the inter-prototype relationship of the input features, by which the invalid features are removed to prevent the curse of feature dimension. In Ref. (Wang et al. 2007), a new KNN construction rule is proposed to significantly improve the classification performance of the standard KNN. According to the existing research, most works focus on developing discrimination measurement metric-induced algorithms to improve the classification accuracy of the KNN model. This is unrealistic to be conducted due to the time complexity of the optimization algorithm. The applications of the KNN model prefer a simpler way with fewer implementation steps. For this purpose, a feature-weighted method is proposed to improve the standard KNN model, which is called the weighted K nearest neighbor (WKNN) (Sharma et al. 2018). WKNN is expected to improve the classification performance more effectively through the weights of the features rather than improving the basic classifier (Wang et al. 2021). The separation operation of the feature optimization from the classifier is easily conducted in real-world applications.

As an usual parameter selection approach, cross-validation helps the users to choose the optimal K, but the optimal one is achieved with the help of both the dataset and the observation of the classification process. If the selected K is not suitable, the classifier may obtain a low classification accuracy on learning tasks.

To overcome the difficulties in parameter selection, this section presents a combination strategy of multiple WKNN. It combines the results of multiple WKNN-based classifiers with different K configurations rather than a single one. After that, the combination of the finite K is used to be approximated to the optimal parameter for the classifier. Finally, the classification performance of the model is improved with the help of multiple parameters. Moreover, such a classification model with multiple classifiers has a more robust performance. The constructed multiple WKNN is further applied for the incipient fault diagnosis of mechanical systems and achieves good performance on the diagnosis cases of rolling element bearings.

3.2.2 Diagnosis Model Based on Combination of Multiple WKNN

3.2.2.1 Multiple WKNN Combination

Before the classification implementation using the KNN model, the training samples are represented into a D-dimensional feature space that is constructed by the extracted D features, and the testing samples are mapped into the same feature space. During the testing procedure, multiple KNN models are determined. Each KNN model will output the classification results of the testing sample based on its parameter setting. The final result of the testing sample is the class that has the most "votes" from the multiple KNN models. The KNN model outputs the classification results by estimating the Euclidean distance between the testing sample and every training sample. Euclidean distance is sensitive to the value scaling of the features, which directly affects the classification accuracy of the KNN model. For this problem, the feature-weighted approach is adopted in the implementation of the KNN model. Compared with the feature selection by the discriminator of a feature to different classes, the feature-weighted method is more general to most classification tasks. Let D and M denote the feature dimension and the training sample amount, respectively. The weighted Euclidean distance can be calculated by

$$d_m^{(wf)} = [\sum_{d=1}^{D} wf_d(Te_d - Tr_{m,d})^2]^{1/2}, d = 1, 2, \cdots, D; m=1, 2, \cdots M \quad (3.1)$$

where wf_d is the weight of the dth feature, $Tr_{m,d}$ is the mth training sample on the dth feature, and Te_d is the testing sample on the dth feature.

The weight $wf_d(d = 1, 2, \cdots, D)$ can be calculated by using the Euclidean compensation distance evaluation technique (CDET). This technique is implemented based on the similar principle of feature selection, which evaluates the discrimination of each feature to different classes. The output discrimination measurements are viewed as the weights of the corresponding features. With the help of weighing features, the performance of the KNN model is improved, and the model induced by the weighted Euclidean distance is called WKNN.

The parameter selection of K is another problem of the KNN model, which also exists in the implementation of the WKNN model. In most works, the WKNN model works with a certain parameter setting but ignores the effects of the data on classification tasks. The parameter setting is commonly achieved using the prior knowledge and experience of experts, and a certain parameter cannot ensure that the WKNN model works well on all the tasks (Li et al. 2019b). To deal with this problem, a combination strategy of multiple WKNN models is adopted. It outputs final results by fusing the sub-results from multiple WKNN models with different settings of the parameter K. However, the combination strategy may not solve the problem, where the feature selection is sensitive to the data and the classification observation of human labor. To improve the effectiveness of the combination of multiple WKNN models, a weighted averaging algorithm is used to facilitate the final results. Suppose that the final results fused by multiple WKNN models are described as follows:

$$\begin{cases} \overline{c_n} = \sum_{g=1}^{G} w_g \overline{c_{n,g}}, \ n = 1, 2, \cdots, N_a; \ g = 1, 2, \cdots, G \\ \sum_{g=1}^{G} w_g = 1, w_g \geq 0 \end{cases} \quad (3.2)$$

where $\overline{c_n}$ and $\overline{c_{n,g}}$ is the classification results of the nth sample through the WKNN combination and the gth WKNN, respectively, w_g is the weight associated with the gth WKNN classifier, N_a represents the sample amount and G is the WKNN classifier amount. According to the commonly used rule in experience, the maximum neighbor parameter $K_{\max}$ is set by following $K_{\max} \leq \sqrt{N_{tr}}$, where N_{tr} is the training sample number. As for the WKNN model sequence, g is selected in the range of $[1, [\sqrt{N_{tr}}]]$ with the interval of one, in which $[\sqrt{N_{tr}}]$ is the round-down integer of $\sqrt{N_{tr}}$. The weight w_g is obtained by optimizing an objective function. The genetic algorithm (GA) can be applied to help optimize the weights for multiple WKNN models. The objective function is given as

$$\begin{cases} f = \dfrac{1}{1 + E_{tr}} \\ E_{tr} = [N_{tr} \sum_{n=1}^{N_{tr}} (c_n - \overline{c_n})^2]^{\frac{1}{2}}, \ n = 1, 2, \cdots, N_{tr} \end{cases} \quad (3.3)$$

where E_{tr} represents the root mean square errors of training samples, and c_n is the actual result of the nth training sample.

The aforementioned combination strategy of multiple WKNN models has two advantages. First, the feature-weighted approach is used to improve the performance of the standard KNN model, in which the feature discrimination to every class is concerned in the construction of the KNN model. Second, the final results are obtained by combining the output of multiple WKNN models so as to reduce the difficulties in the optimal parameter setting when facing different classification tasks. For a diagnosis task, intelligent fault diagnosis is achieved by all the steps including feature extraction and selection as well as the classification. In the next subsection, the fault diagnosis model based on the multiple WKNN models is elaborated.

3.2.2.2 Diagnosis Method Based on Multiple WKNN Combination

In this subsection, we extract ten commonly used time-domain features from the raw vibration signals (Lei et al. 2020). They are the mean value, the standard deviation, the root mean square, the peak value, the crest factor, the clearance factor, the skewness, the kurtosis, the shape factor and the impulse factor. Moreover, four frequency-domain features are extracted, which are the mean frequency, the frequency center, the root mean square frequency, and the standard deviation frequency.

To capture more health information from the condition monitoring data, some signal processing methods are used to divide the vibration data into multiple sub-bands, and then more features can be extracted. The wavelet transform has been studied for many years in the field of condition monitoring and fault diagnosis of mechanical systems (Chen et al. 2016; Yan et al. 2014). In the wavelet transform, the analysis results are related to the selection of the wavelet basis functions. Different basis functions will produce diverse results. For this problem, the lifting scheme (Yang et al. 2019) was proposed to solve the difficulties in the wavelet basis function selection and make them closely associated with the raw vibration data. The wavelet packet transform with the lifting scheme (WPTLS) decomposes the input vibration signal into three levels, and thus produces eight sub-signals in different frequency bands. We further extract ten statistical features of each frequency-band signal, and thus a total of 80 features are obtained. As for the frequency analysis, the demodulation methods can weaken the interaction of frequency components at different frequency bands and help find the low-amplitude features located at the high-frequency bands. Therefore, the sub-signals obtained by WPTLS are demodulated by the Hilbert transform to generate the envelope spectrums. Apart from the commonly used four frequency features, additional three special frequency domain features, i.e., the spectrum peak ratio of bearing outer race, bearing inner race, bearing roller, are extracted from the envelope spectrums of each sub-signals. We finally obtained 150 features from the monitoring data including 90 time-domain features and 60 frequency-domain features.

The above 150 features are able to reveal the health states of mechanical systems in different aspects. However, not all the features are sensitive to fault patterns. Some samples from different fault types may cluster together on certain feature dimensions, which are invalid for the fault classification. Therefore, the features need to be selected before being fed into the classification models. In the feature selection, the irrelevant and redundant features are removed from the original feature set. Such a process can improve the classification performance of the diagnosis model and even prevent the curse of high feature dimensions. Here, the CDET is used to analyze the feature sensitivity to different fault types. CDET works with the criterion that samples on the sensitive features will get larger inter-class distance but small intra-class distance. The features with higher CDET scores will be viewed as the salient ones. These features are further used to train multiple WKNN-based diagnosis models.

Figure 3.1 presents the procedure of the multiple WKNN-induced fault diagnosis model. It mainly consists of five steps. First, the collected monitoring data of mechanical systems are decomposed by WPTLS. After the three-level decomposition, the data are decomposed into eight frequency bands. Through the Hilbert transform, the envelope spectrums of each frequency-band data are obtained. Second, the statistical analysis in the time domain and the frequency domain is conducted on both the original data and the sub-frequency-band data. A total of 150 features are extracted. Third, the CDET is adopted to analyze the feature sensitivity to diverse fault types of mechanical systems so that the salient features are selected to create a new feature set. Forth, the salient features are used to train multiple WKNN-based diagnosis models with different parameter settings of K. Finally, the diagnosis results of multiple WKNN models are fused to generate the final diagnosis results by using the weighted averaging GA. The training of the multiple WKNN is terminated when the training accuracy reaches 100%, or the increased number of salient features cannot improve the training accuracy. Once the training process is done, the obtained model works on the testing data and predicts the corresponding health conditions of mechanical systems. In the next subsection, the performance of the diagnosis model is demonstrated on a diagnosis case of rolling element bearings.

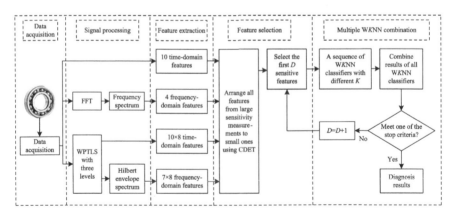

Fig. 3.1 The procedure of the multiple WKNN-based intelligent fault diagnosis

3.2.3 Intelligent Diagnosis Case Study of Rolling Element Bearings

The bearing dataset is provided by the Case Western Reverse University (Smith and Randall 2015), which has been explained in Chap. 2. We make a brief introduction to this dataset here. The data are collected from the motor bearings. During the experiments, the spark-erosion machining is adopted to respectively seed faults on the inner race, the other race, and the roller of the bearings. Each bearing fault has three fault levels. The bearings are tested under four load settings, i.e., 0, 1, 2, and 3 hp. Therefore, the dataset is collected from bearings with different health states and under different working conditions.

We create two types of diagnosis tasks according to the bearing dataset. The tasks respectively demonstrate the combination performance of multiple WKNN on different fault levels of bearings. As shown in Table 3.1, both datasets A and B consist of 240 samples of bearings when the motor works at four working conditions (60 samples for each working condition). Note that the fault samples in dataset A are collected from the fault level of 0.18 mm. The dataset B is from the bearings with a fault level of 0.53 mm. We select 50% samples to train the multiple WKNN-induced diagnosis model, and the rest are used to test the model performance. In addition, there are three other datasets, which will be presented in the following in this section.

According to Subsection 3.2.2.1, the maximum nearest neighbor $K_{\max} \leqslant \sqrt{N_{\text{tr}}}$. We can thus calculate the nearest neighbor amount $K_{\max}$ by $\left[\sqrt{N_{\text{tr}}}\right] = \left[\sqrt{120}\right] = 10$ since the number of training samples is 120. When dataset A is used to train the model, the training process is terminated as the training accuracy reaches 100%. At the moment, the number of the selected features is two, and they are the clearance factor and the shape factor of the fourth frequency-band signal by using the WPTLS. The diagnosis accuracy on the testing samples is 97.67%, as shown in Table 3.2. That means the combination of multiple WKNN could diagnose the incipient bearing faults. To explain the diagnosis method based on the combination of multiple WKNN, the diagnosis accuracy of single KNN-based diagnosis model is further compared.

Table 3.1 Datasets A and B with different fault categories and the same fault levels

Dataset	The number of training samples	The number of testing samples	Fault size of training/testing samples/mm	Fault category	Classification label
A	30	30	0/0	Normal	1
	30	30	0.18/0.18	Outer race	2
	30	30	0.18/0.18	Inner race	3
	30	30	0.18/0.18	Ball	4
B	30	30	0/0	Normal	1
	30	30	0.54/0.54	Outer race	2
	30	30	0.54/0.54	Inner race	3
	30	30	0.54/0.54	Ball	4

The input features of the single KNN are all the extracted features without being selected by CDET. For a fair comparison, the features are randomly selected for ten trials, and the statistical results are recorded in Table 3.2. The testing accuracies of the single KNN classifier are shown in Table 3.3. The results show that the single KNN classifier achieves an average testing accuracy of 91.5%, which is lower than the combination strategy of multiple KNN, verifying the superiority of the feature selection and the hybrid intelligent fault diagnosis method.

As for the demonstration on dataset B, the training process of multiple WKNN combination is done when the training accuracy is 100%. There is only one selected feature, i.e., the crest factor of the raw vibration signal. Thus, the crest factor of the testing sample is fed into the hybrid intelligent diagnosis model. According to the results shown in Table 3.2, a testing accuracy of 95.83% is obtained. The individual KNN-based diagnosis models obtain an accuracy ranging from 84.62% to 87.75% with an average accuracy of 86.20%, which is lower than the combination of multiple WKNN.

Table 3.2 Comparison of the testing accuracies (%) between multiple WKNN combination and individual KNN classifier

Dataset	Single KNN classifier			WKNN combination
	Min	Mean	Max	
A	89.12	91.50	92.42	97.67
B	84.62	86.20	87.75	95.83
C	87.33	89.20	90.33	100
D	87.83	89.18	90.50	100
E	82.75	83.74	84.67	93.33
F	79.76	81.77	84.00	98.67

Table 3.3 Testing accuracies of ten KNN classifiers for different fault categories or levels

Dataset/ Classifier	A /%	B / %	C / %	D / %	E / %
Classifier 1	92.2	84.7	88.2	87.9	82.8
Classifier 2	92.2	84.6	88.2	87.9	82.8
Classifier 3	92.4	85.3	90.0	87.8	83.5
Classifier 4	92.4	86.0	90.3	88.7	84.4
Classifier 5	91.7	85.9	90.0	89.1	83.9
Classifier 6	92.0	86.9	90.2	90.1	84.2
Classifier 7	90.9	86.5	89.4	90.3	84.7
Classifier 8	91.4	86.8	89.8	89.7	84.5
Classifier 9	89.1	87.5	88.6	90.5	83.3
Classifier 10	90.8	87.8	87.3	89.9	83.4

3.2 Multiple WKNN Fault Diagnosis

In addition, we construct other three datasets to verify the combination of multiple WKNN. As shown in Table 3.4, datasets C, D, and E contain different fault types. In each dataset, the samples are from the normal state and the faults with different fault levels, and the diagnosis tasks aim to classify the fault levels. There are 60 samples in the normal state and 60 samples in each fault level. The dataset C includes the normal state and the outer race fault. The datasets D and E respectively contain the inner race fault and the ball fault. Similar to the training/testing sample amount settings in the previous experiments, we select 50% samples to train the model, and the rest are used to test the model performance. For dataset C, the training process selects the crest factor of the third frequency-based signal and the shape factor of the fourth frequency-band signal as the salient features. As for dataset D, there is only one salient feature, i.e., the root mean squared frequency of the raw signals. In terms of dataset E, more features are salient, i.e., the crest factor and the root mean square frequency of the raw signals, and the spectrum peak ratios of bearing outer race, inner race, and roller of the fourth frequency-band signal extracted by WPTLS. As shown in Tables 3.2 and 3.3, the combination of multiple WKNN achieves the average diagnosis accuracy of 100% in dataset C, 100% in dataset D, and 93.3% in dataset E. The individual KNN-based models achieve lower accuracies than the combination of multiple WKNN.

To demonstrate the performance of the combination strategy on the complex diagnosis task, we further construct dataset F, which is shown in Table 3.4. Dataset F contains ten classes including the normal state and three fault types with three fault levels. There are 60 samples in each class, and finally, a total of 600 samples are used to verify the diagnosis performance of multiple WKNN-based model. The training dataset contains 300 samples, while the testing dataset has the remaining 300 samples. Therefore, the dataset F setting is expected to recognize both the fault types and the fault levels by the multiple WKNN combination.

Due to the complexity of the diagnosis task, the training process of the combination of multiple WKNN is achieved by selecting nine salient features, obtaining a training accuracy of 100%. The corresponding testing result of the hybrid intelligent fault diagnosis method is shown in Table 3.2. The specific testing accuracies of the individual KNN models are presented in Table 3.5. The results validate the effectiveness of the multiple WKNN-based diagnosis models.

3.2.4 Epilog

This section presents a hybrid intelligent fault diagnosis method by using the combination of multiple WKNN. The performance of the hybrid intelligent diagnosis method is demonstrated in a diagnosis case of rolling element bearings, which includes the diagnosis tasks for fault types and fault levels. The results of the case study show that the combination of multiple WKNN presents higher diagnosis accuracy than the individual KNN-based diagnosis models. The success of the hybrid intelligent fault diagnosis results from the feature extraction, feature selection, and

Table 3.4 Datasets C, D, E and F with different fault categories and same fault levels

Dataset	The number of training samples	The number of testing samples	Fault size of training/testing samples/mm	Fault category	Classification label
C	30	30	0/0	Normal	1
	30	30	0.18/0.18	Outer race	2
	30	30	0.36/0.36	Outer race	3
	30	30	0.54/0.54	Outer race	4
D	30	30	0/0	Normal	1
	30	30	0.18/0.18	Inner race	2
	30	30	0.36/0.36	Inner race	3
	30	30	0.54/0.54	Inner race	4
E	30	30	0/0	Normal	1
	30	30	0.18/0.18	Ball	2
	30	30	0.36/0.36	Ball	3
	30	30	0.54/0.54	Ball	4
F	30	30	0/0	Normal	1
	30	30	0.18/0.18	Outer race	2
	30	30	0.18/0.18	Inner race	3
	30	30	0.18/0.18	Ball	4
	30	30	0.36/0.36	Outer race	5
	30	30	0.36/0.36	Inner race	6
	30	30	0.36/0.36	Ball	7
	30	30	0.54/0.54	Outer race	8
	30	30	0.54/0.54	Inner race	9
	30	30	0.54/0.54	Ball	10

Table 3.5 Testing accuracies of single KNN classifier for different fault categories and levels (dataset F)

Classifier	K1	K2	K3	K4	K5	K6
Accuracy/%	83.00	84.00	83.53	83.77	82.63	82.40
Classifier	K7	K8	K9	K10	K11	K12
Accuracy/%	81.97	82.23	81.83	81.43	80.97	81.07
Classifier	K13	K14	K15	K16	K17	
Accuracy/%	80.67	80.63	80.00	80.23	79.76	

combination of multiple classifiers. First, the advanced signal processing methods help extract more features to reflect the health information of mechanical systems from the collected data. In this section, the features from the frequency-band signals by WPTLS contribute a lot to the improvement of diagnosis methods. Second, the feature selection techniques can remove the useless features from the large volume of the feature set, which increases the classifier performance. Third, the combination of multiple WKNN classifiers can compensate for the superiority of each one with different parameter settings when the input features are the same. Inspired by the above analysis, we discuss the hybrid intelligent fault diagnosis when the input features are different in the next section.

3.3 Multiple ANFIS Hybrid Intelligent Fault Diagnosis

3.3.1 Motivation

As shown in Sect. 3.2, multiple WKNN combination is utilized for fault class and level diagnosis. As a well-known supervised learning algorithm in pattern recognition, WKNN overcomes the two shortcomings in KNN by introducing a distance-weighted rule that weights close neighbors more heavily by querying their distances. However, the small number of samples available in practical applications limits the model performance. Therefore, the classification accuracy will drop sharply, especially in the case of a small sample size with the curse of high dimensionality. In practice, ANN requires a large number of training samples, which is an extremely time-consuming process. On the contrary, traditional fuzzy systems rely on the knowledge and experience of experts and operators to a large extent. Therefore, it is difficult to obtain satisfactory prediction results, especially when complete information is unavailable. To overcome the above weaknesses, an adaptive network based on a fuzzy inference system, i.e., ANFIS, is developed by Jang in 1993 (Jang 1993). ANFIS is a fuzzy inference system based on the Sugeno model (Zhou et al. 2021; Kuai et al. 2018), which combines the self-learning ability of ANN and the linguistic expression function of fuzzy inference (Rezakazemi et al. 2017).

The combination of multiple classifiers is studied to overcome the limitations of a single classifier and improve accuracy. Based on the advantages of ANFIS, this section presents a hybrid intelligent fault diagnosis method with multiple ANFIS combinations. Different from the diagnosis method in Sect. 3.2, the presented method uses the same classification engine and different input feature sets. The complementary classification performance between different input feature sets is used to improve the classification accuracy as shown in Sect. 3.2. The classification results of each ANFIS are integrated to produce the final classification results, and the final classification performance should be better than the best performance of a single classifier. Multiple ANFIS will be combined by using the weighted average technology, and every single classifier is assigned a non-negative weight. Then, the GA is used to optimize the weight of each classifier (Katoch et al. 2021). Therefore, the hybrid intelligent fault diagnosis method combining multiple ANFIS has better performance than a single ANFIS. At the same time, hybrid intelligent fault diagnosis is explored from another aspect: the same combination of classifiers and different input feature sets.

3.3.2 Multiple ANFIS Combination with GA

The ANFIS maps input membership functions and corresponding related parameters to outputs which are dependent on the output membership functions. As the utilization of multiple classifiers has been verified to achieve promising mechanical fault diagnosis results, it is thus intuitive that many classifiers are aggregated in some way to classify testing samples. The combination of different classifiers will achieve performance improvements. In the literature, the weighted average technique is the simplest technique and is widely used in the application of multi-classifier combination.

This section presents a combination method of multiple ANFIS. In this method, six fault-related feature sets are selected by using EDET, and the number of ANFIS classifiers is determined according to the number of selected salient feature sets. Then the weighted average technique is used to combine the results of the six ANFIS classifiers, and the final classification result is written as follows:

$$\hat{y}_n = \sum_{k=1}^{6} w_k \hat{y}_{n\,k}, \ n = 1, 2, \cdots, N; k = 1, 2, \cdots, 6 \quad (3.4)$$

subject to

$$\sum_{k=1}^{6} w_k = 1 \quad (3.5)$$

where $\hat{y}_n$ and $\hat{y}_{n,k}$ denote the classification results of the nth sample combination and the kth individual classifier respectively, w_k denotes the kth individual classifier weight and N' denotes the number of samples.

In the combination of multiple ANFIS, GA is utilized to optimize the objective function for weight evaluation. The randomly generated genome is used to initialize the population of 10 individuals. In the process of this multiple ANFIS combination, the weights are evaluated by GA to optimize the objective function (Li and Sun 2018). The maximum number of generations is used as the termination condition of the solution process, which is set to 100. Non-uniform-mutation function and arithmetic crossover operator are applied, the mutation probability is 0.01 and the crossover probability is 0.8. The objective function is written as:

3.3 Multiple ANFIS Hybrid Intelligent Fault Diagnosis

$$f = \frac{1}{1+E} \tag{3.6}$$

where E is the root mean squared error:

$$E = [\frac{1}{N''}\sum_{n=1}^{N''}(y_n - \hat{y}_n)^2]^{\frac{1}{2}}, n = 1, 2, \cdots, N'' \tag{3.7}$$

where y_n denotes the result of the nth training sample and N'' denotes the number of the training samples.

This section introduces a multi-ANFIS combination scheme based on GA, but it also includes effective feature extraction and selection in the intelligent fault diagnosis method. The next section will introduce a complete fault diagnosis method based on the combination of multiple ANFIS.

3.3.3 Fault Diagnosis Method Based on Multiple ANFIS Combination

In the beginning, 24 statistical features are extracted from the original vibration signal, including 11 time-domain statistical features and 13 common frequency-domain statistical features (Lei 2017a). The original signal is divided into multiple sub-sequences with 4096 data points. The time domain and frequency domain statistical features are called feature set 1 and feature set 2, respectively. It should be noted that the collected vibration signal contains low-frequency interference. Therefore, high-pass or band-pass filters are used to filter out low-frequency interference components in the signal. In this method, three band-pass (BP1–BP3) and one high-pass (HP) filter are adopted. The band-pass frequencies (in kHz) of the BP1–BP3 filters are selected as BP1 (2.2~3.8), BP2 (3.0~3.8), and BP3 (3.0~4.5). The cut-off frequency of the HP filter is selected as 2.2 kHz. The eleven time-domain statistical features extracted from the four filtered signals form the feature set 3. The Hilbert envelope spectrum is used to extract another feature set composed of common frequency domain statistical features from the four filtered signals and is denoted as feature set 4. In order to extract more feature information, EMD is used to decompose the original signal. The first eight IMFs are selected to further extract features. 11 time-domain statistical features are extracted from each IMF and is denoted as feature set 5. In addition, the Hilbert transform is used to demodulate the IMF and 13 frequency-domain features are extracted from the corresponding envelope signal. This dataset is denoted as feature set 6.

Although the above extracted features can be used to identify mechanical operating conditions, these features have different importance for different types and severity of faults. Some features are closely related to the corresponding fault conditions, but some of them are not. Therefore, it is very necessary to select features that are related to bearing failure from these feature sets, so as to improve the performance of the classifier and avoid the curse of high dimensionality. Among them, EDET can be used for feature selection and evaluation.

Based on the statistics, EMD, EDET, ANFIS, GA, and so on, a new type of multi-ANFIS combination hybrid intelligent fault diagnosis method for mechanical systems is presented as shown in Fig. 3.2. It contains the following main procedures:

(1) The original signal is filtered by the aforementioned four filters. Next, EMD technology is used to decompose the original vibration signal to get the first 8 IMFs. Then, the Hilbert transform is used to calculate the envelope spectrum to further demodulate the filtered signal and IMF. Finally, six feature sets are acquired.
(2) The EDET method is used to evaluate and select the sensitive features in each feature set.
(3) Each selected feature set is fed into the multi-ANFIS ensemble classifier for training and testing.
(4) The output results of the multiple ANFIS are fused using GA weighted average technology to generate the final classification results. The effectiveness of the method is verified through the rolling bearing case in the next section.

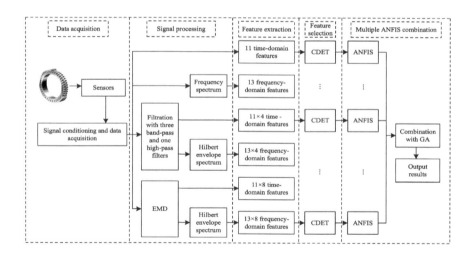

Fig. 3.2 Flowchart of the hybrid intelligent fault diagnosis method based on multiple ANFIS combination

3.3.4 Intelligent Diagnosis Case of Rolling Element Bearings

The rolling element bearing dataset by Case Western Reserve University is utilized to verify the effectiveness of the multiple ANFIS combination method (Smith and Randall 2015). The details of the dataset are presented in Chap. 2. Sensors are used to collect vibration signals under different working loads and bearing conditions. EDM technology is used to set the bearing failure on the drive end of the motor. The four health conditions are: (1) normal condition, (2) with an outer race fault, (3) with an inner race fault and (4) with a ball fault.

In order to evaluate the effectiveness of the hybrid intelligent fault diagnosis method of multiple ANFIS combinations for different fault classes and severity, four different datasets (A-D) from all the datasets of rolling bearings are used. The details of the four datasets are listed in Table 3.6.

Table 3.6 Description of four datasets (A-D)

Dataset	Number of training samples	Number of testing samples	Fault size of training/testing samples/mm	Operating condition	Label of classification
A	30	30	0/0	Normal	1
	30	30	0.18/0.18	Outer race	2
	30	30	0.18/0.18	Inner race	3
	30	30	0.18/0.18	Ball	4
B	The training samples are the same to A above	30	0/0	Normal	1
		30	0.18/0.54	Outer race	2
		30	0.18/0.54	Inner race	3
		30	0.18/0.54	Ball	4
C	The training samples are the testing samples of B	The testing samples are the training samples of B	0/0	Normal	1
			0.54/0.18	Outer race	2
			0.54/0.18	Inner race	3
			0.54/0.18	Ball	4
D	30	30	0/0	Normal	1
	30	30	0.18/0.18	Outer race	2
	30	30	0.18/0.18	Inner race	3
	30	30	0.18/0.18	Ball	4
	30	30	0.36/0.36	Outer race	5
	30	30	0.36/0.36	Inner race	6
	30	30	0.36/0.36	Ball	7
	30	30	0.54/0.54	Outer race	8
	30	30	0.54/0.54	Inner race	9
	30	30	0.54/0.54	Ball	10

Dataset A contains 240 data samples in four types of working conditions: the normal condition, the outer race fault, the inner race fault, and the ball fault. The fault severity of this dataset is with a size of 0.18 mm. Each working condition contains 60 data samples. Dataset A is divided evenly into two subsets: 120 samples for training and 120 samples for testing. Therefore, it is a simple four-class problem, corresponding to four different working conditions. Dataset C is similar to dataset B, and the samples are the same. But the training set of dataset C is set to the testing set of dataset B, and the testing set of dataset C is set to the training set of dataset B. Dataset D contains 600 data samples from four working conditions and loads. Each fault condition contains three fault levels of 0.18, 0.36, and 0.54 mm. In dataset D, 300 data samples are used for training and 300 data samples are used for testing. Generally, the output of a single model is prone to produce false diagnosis results and lower classification performance. In order to apply the multi-ANFIS classifier combination model to improve the effectiveness of generalization ability, the single classifier based on ANFIS was used for comparison in the experiment.

The first four sensitive features selected by the EDET method are used by considering computational time, and then they are fed into the corresponding classifier based on ANFIS. For dataset A, the distance evaluation criteria $\overline{\alpha_j}$ of the six feature sets are plotted in Fig. 3.3. In Fig. 3.3 (a), feature set 1 consists of eleven common time-domain statistical features extracted from raw vibration signals. It can be seen that the selected sensitive features are standard deviation (X_{sd}), root amplitude (X_{root}), root mean square (X_{rms}), and peak (X_{peak}), respectively. In Fig. 3.3 (b), feature set 2 contains thirteen common frequency-domain statistical features from FFT spectra. It can be seen that the selected sensitive features are p_1, p_6, p_9 and p_{13} (Lei 2017a). In addition, feature set 3 consists of four features: three peaks (X_{peak}) by utilizing BP1, BP2, and HP filters and clearance ($X_{clearance}$) by utilizing BP3 filter. The feature set 4 includes two p_{10} and two p_{13} from the filtered signals of BP1 and BP2, respectively. The feature set 5 contains X_{sd}, X_{root}, X_{rms} and X_{peak} from the first IMF1 of empirical mode decomposition. The feature set 6 contains p_1, p_6 and p_{13} from the envelope spectrum of IMF1 and p_6 from the envelope spectrum of IMF2. Obviously, each feature has different sensitivities to different failure types and failure levels. Feature selection can eliminate redundant or useless features and achieve feature dimensionality reduction. In addition, many effective features can be extracted by using signal processing technology from the processed signal. Therefore, the original vibration signal can be preprocessed by signal processing technology to improve classification accuracy. Finally, the classification results of the four datasets are shown in Table 3.7.

From Table 3.7, it can be seen that the six individual classifiers and their combinations have achieved good training and testing accuracies for dataset A as they all reached 100%. However, from Fig. 3.4, the classification error of the combination of multiple classifiers is smaller than the single classifier.

In addition, although the single classifier and the multi-classifier combination have obtained a training accuracy of 100% on the dataset B, the testing accuracy of the single classifier is significantly lower than the multi-classifier combination. The testing accuracy of single classifier ranges from 60.83% to 83.33% and the testing

3.3 Multiple ANFIS Hybrid Intelligent Fault Diagnosis

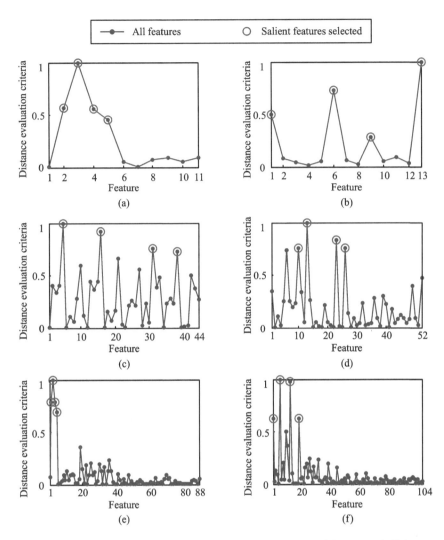

Fig. 3.3 Distance evaluation criteria of six feature sets of dataset A: (a) feature set 1, (b) feature set 2, (c) feature set 3, (d) feature set 4, (e) feature set 5, and (f) feature set 6

accuracy rate of a combination of multiple classifiers is 92.5%. The experimental results show that the combination of multiple classifiers is more effective than the individual classifier.

Next, for dataset C, the training accuracy of all classifiers is 100%. However, compared with the single classifier whose testing accuracy ranges from 55.83% to 81.67%, the combination of multiple classifiers achieves a higher testing accuracy rate of 90.83%. The result shows that a multi-classifier trained by fault samples with larger defect sizes has higher classification accuracy than a single classifier.

Table 3.7 Performance comparison between single classifier and multiple classifier combination for four different datasets (A–D)

Datasets/Classifier		A /%	B /%	C /%	D /%
Classifier 1	Training Accuracy	100	100	100	65.67
	Testing Accuracy	100	66.67	62.5	61
Classifier 2	Training Accuracy	100	100	100	90
	Testing Accuracy	100	75	79.17	87.67
Classifier 3	Training Accuracy	100	100	100	72.67
	Testing Accuracy	100	61.67	74.17	68
Classifier 4	Training Accuracy	100	100	100	80.33
	Testing Accuracy	100	83.33	80	77
Classifier 5	Training Accuracy	100	100	100	67.67
	Testing Accuracy	100	60.83	55.83	67
Classifier 6	Training Accuracy	100	100	100	87.33
	Testing Accuracy	100	72.5	81.67	81
Average of classifiers 1~6	Training Accuracy	100	100	100	77.28
	Testing Accuracy	100	70	72.22	73.61
Multiple classifier combination	Training Accuracy	100	100	100	93.67
	Testing Accuracy	100	92.5	90.83	91.33

Finally, for the dataset D, the training accuracy of all classifiers decreases. However, the combination of multiple classifiers still obtains the highest training accuracy, i.e., 93.67%. For the testing sample, the classification accuracy of the six single classifiers ranges from 61% to 87.67%, and the classification accuracy of the multi-classifier combination has achieved the highest accuracy, i.e., 91.33%. These results show that, compared with a single classifier, a combination of multiple classifiers can identify the type of failure, and the level of failure.

Obviously, the result of a combination of multiple classifiers is better than a single classifier through multiple feature sets. At the same time, the hybrid intelligent fault diagnosis method of multiple ANFIS combination has better generalization ability than a single classifier.

In order to further verify the influence of the feature selection method on the diagnosis accuracy, an experiment is carried out with dataset D as an example. Four features are randomly selected from each of the six feature sets without using the MDET method. Later, these randomly selected features are also fed into the single-classifier and multi-classifier combination models. The experiment was repeated ten times, and the average results are shown in Table 3.8. In order to make a comparison, Table 3.8 also depicts the classification results of the whole classifiers, in which the CDET method is used to select the sensitive features.

From Table 3.8, the accuracies of training and testing processes for the six single classifiers range from 56.5% to 73.73% when the features are randomly selected for

3.3 Multiple ANFIS Hybrid Intelligent Fault Diagnosis

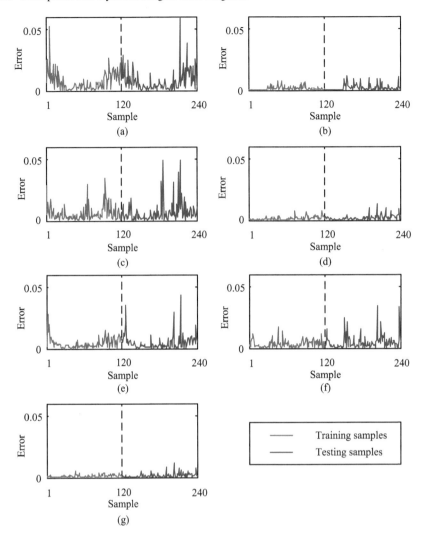

Fig. 3.4 Classification errors of single classifier and multiple classifier combination: (a) classifier 1, (b) classifier 2, (c) classifier 3, (d) classifier 4, (e) classifier 5, (f) classifier 6, and (g) classifier 7

training and testing. Compared with the features selected by EDET, the average training accuracy and testing accuracy reduce by 14.39% and 15.59%, respectively. The training and testing accuracies of the multi-classifier combination drop from 93.67% to 81.19% and 91.33% to 77.15%, respectively. The results indicate that the randomly selected features may contain too much useless or irrelevant information, which limits the model performance. The feature selection method based on EDET can eliminate the useless and insensitive features in the original feature set to achieve better classification accuracy. The multi-classifier combination method is effective in the early fault diagnosis of rolling bearings.

Table 3.8 Classification performance comparison in accuracies (%) for different features

			Salient features	Randomly chosen features
	Classifier 1	Training accuracy	65.67	62
		Testing accuracy	61	58.03
	Classifier 2	Training accuracy	90	69.27
		Testing accuracy	87.67	66.93
	Classifier 3	Training accuracy	72.67	58.5
		Testing accuracy	68	55.1
	Classifier 4	Training accuracy	80.33	73.73
		Testing accuracy	77	66.3
	Classifier 5	Training accuracy	67.67	56.5
		Testing accuracy	67	50.93
	Classifier 6	Training accuracy	87.33	57.37
		Testing accuracy	81	50.8
	Average of classifiers 1~6	Training accuracy	77.28	62.89
		Testing accuracy	73.61	58.02
	Multiple classifier combination	Training accuracy	93.67	81.19
		Testing accuracy	91.33	77.15

3.3.5 Epilog

Inspired by Sect. 3.2, the multi-classifier combination method introduced in Sect. 3.3 emphasizes the influence of different input feature sets on the final classification results. The effectiveness of the presented multi-ANFIS combination hybrid intelligent fault diagnosis method is verified by rolling bearing experiments with different fault types and severities. In addition, the experiment also proves the necessity of feature selection in signal processing, especially when the acquired data are not sufficient enough. In addition, the extracted features may be redundant or irrelevant with the faults, so reasonable techniques or methods are needed for feature selection.

It can be seen from Sects. 3.2 and 3.3 that by combining the complementary effects of different classifiers and input features to construct an intelligent diagnosis method, higher classification performance than the single model will be obtained. Based on this motivation, a multidimensional hybrid intelligent fault diagnosis method concerning complex gear faults instead of bearing faults is further investigated.

3.4 A Multidimensional Hybrid Intelligent Method

3.4.1 Motivation

In Sect. 3.2, a multiple WKNN method is presented for the diagnosis of rolling element bearings, and the classification performance with the same input is investigated. Additionally, Sect. 3.3 examines the diagnosis accuracy for bearings based on multiple ANFIS combination with different inputs. Furthermore, in this section, we will discuss the diagnosis performance for gearboxes based on multiple different classifiers combined with different inputs, which is called the "multidimensional" hybrid intelligent fault diagnosis method (Liu 2019; Li et al. 2019a). Among them, the "multidimensional" means the different input feature sets and the different classifiers (Li et al. 2019a). Therefore, there are two main tasks in this section. On one hand, the effect of different classifiers with different inputs on final classification performance is investigated. On the other hand, the method is verified by the more complicated gearbox fault datasets instead of rolling element bearing fault datasets.

The KNN is the simplest classifier and is widely applied in engineering applications. However, it has two main limitations: one is the difficulty of distinguishing different classes with too much overlapping and the other is the difficulty of determining the crucial parameter K. KNN is considered to be one of the multiple classifiers of the multidimensional hybrid intelligent fault diagnosis method, which is to verify whether multiple classifier combination can address the limitations.

MLP neural networks are the most commonly used neural networks (Zhang et al. 2019). It is characterized by fast operation, easy implementation, and low requirement of training set size. MLP consists of input, hidden, and output layers. The common problem is to determine the appropriate number of neurons in the hidden layer. The irrelevant selection of the number of neurons in the hidden layer will result in poor generalization or overfitting. Moreover, the structural characteristics of MLP are different from the KNN classifier. Thus, it is also selected as one of the multiple classifiers in this section. The purpose of this selection is identical to that of KNN.

The RBF neural network has a similar structure to MLP, but the activation function of RBF is different from that of MLP (Yang and Chen 2019). More importantly, compared with other ANNs, RBF requires less calculation time, and it can effectively solve nonlinear mapping problems from several inputs to one or more outputs. As a result, RBF is chosen as the final member of multiple classifiers.

Although these three classifiers all belong to intelligent algorithms, they have different classification principles or algorithm structures. Therefore, they are supposed to exhibit complementary classification performance and thus overcome their own limitations to achieve better classification performance than each one of them. Finally, the classification results of these three classifiers are integrated by GA to obtain the final diagnosis result.

3.4.2 Multiple Classifier Combination

Generally, the MLP neural network only includes three layers: input layer, hidden layer, and output layer. The number of nodes in the input layer is determined by the number of input features, and the number of the output layer is decided by final output classes. The number of hidden nodes is usually determined by the geometric pyramid rule as the following equation shows (Yang et al. 2002):

$$h = \sqrt{mn} \tag{3.8}$$

where h, m, and n denote the numbers of the nodes of hidden, input, and output layers, respectively. The training process of MLP is to search for the value of the weights. This process is achieved by minimizing the loss function between the network outputs and the actual outputs in the training set. Commonly, the training process is based on back-propagation and the mean square error is selected as the loss function.

The Gaussian function is the commonly used activation function in RBF, which is defined as follows:

$$f(x) = \exp\left(-\frac{(x-c)^2}{2b^2}\right) \tag{3.9}$$

where b and c denote the width and center of the Gaussian function, respectively. The distance criteria between the input features and the data cluster center c are determined by the outputs of the hidden nodes. The physical meaning of b is the radius of the hypersphere. The training process of the RBF is as follows.

(1) The number of nodes of the hidden layer and the parameters of the Gaussian function are decided by the input information.
(2) The connection weights between each layer in RBF are optimized based on the principle of least squares minimization.

The KNN are usually decided by the Euclidean distance between the testing sample and each training sample. Given the testing sample TE_d and the mth training sample $TR_{m,d}$, the Euclidean distance is defined as:

3.4 A Multidimensional Hybrid Intelligent Method

$$D_m = \left[\sum_{d=1}^{D}(TE_d - TR_{m,d})^2\right]^{1/2}, d = 1, 2, \cdots, D; m = 1, 2, \cdots, M \quad (3.10)$$

where D and M denote the numbers of input features and training samples, respectively. As described before, it is difficult to determine the neighborhood parameter K. Usually, the maximum value of K is less than $\sqrt{M}$ according to the rule of thumb (Ghosh et al. 2005). Therefore, the K value is chosen as $[\sqrt{M}]$, where $[\sqrt{M}]$ denotes the round-down integer of $\sqrt{M}$.

Additionally, we also extracted three different feature sets from the time domain, frequency domain, and time–frequency domain. Therefore, this section presents a multidimensional hybrid intelligent diagnosis method by combining MLP, RBF, and KNN with different input feature sets to increase the diagnosis accuracy of gear faults. The output result is calculated by the weighted averaging technique with GA:

$$\begin{cases} y_m = \sum_{c=1}^{9} w_c y_{m,c}, \ m = 1, 2, \cdots, M'; c = 1, 2, \cdots, 9 \\ \sum_{c=1}^{9} w_c = 1, w_c \geq 0 \end{cases} \quad (3.11)$$

where y_m and $y_{m,c}$ denote the classification results of the mth testing sample and the cth individual classifier respectively, w_c denotes the weight related with the cth individual classifier and M' is the total number of the testing samples.

Moreover, the weights w_c are evaluated by applying a GA optimization algorithm. The maximum number of generations is set as 100, and the mutation and crossover probabilities are defined as 0.01 and 0.8, respectively. Finally, the objective function is shown below:

$$\begin{cases} f = \dfrac{1}{1 + E_{tr}} \\ E_{tr} = [\dfrac{1}{M}\sum_{m=1}^{M}(y_m - \overline{y_m})^2]^{1/2}, m = 1, 2, \cdots, M \end{cases} \quad (3.12)$$

where E_{tr} represents the root mean square training errors, $\overline{y_m}$ denotes the actual value of the mth training sample and M denotes the number of the training samples. According to the above algorithm description, the multidimensional hybrid intelligent fault diagnosis method for gearboxes is detailed in the next subsection.

3.4.3 Diagnosis Method Based on Multiple Classifier Combination

In this subsection, some features from time-domain, frequency-domain, and time frequency-domain are firstly extracted to reflect the fault characteristics of gears. Generally, both amplitude and distribution of time-domain signal from gear will change when a fault occurs. Therefore, five informative time-domain features, standard deviation (X_{sd}), root mean square (X_{rms}), kurtosis ($X_{kurtosis}$), crest factor (X_{crest}) and shape factor (X_{shape}), are extracted from raw vibration signals. Moreover, six special time-domain statistical features are also selected, which are specifically designed for gear faults by NASA. These features are FM0, FM4, NA4, NB4, ER, EOP. Thus, a total of 11 time-domain features can be obtained. Meanwhile, four common frequency-domain statistical features are also extracted. They are mean frequency (p_1), frequency center (p_5), root mean square frequency (p_7), and standard deviation frequency (p_{14}), respectively (Lei 2017a).

Furthermore, some advanced signal processing techniques are exploited to obtain more important information. In this subsection, WPT and EMD are utilized to preprocess the vibration signals before feature extraction. Compared with the wavelet transform algorithm that cannot decompose the high-frequency part of the signal, WPT can decompose the high-frequency part and the low-frequency part at the same time. Therefore, WPT is employed to preprocess the vibration signal. Consequently, given a vibration signal x, the decomposed frequency-band signal x_{2^k+m} is produced by employing WPT, where k denotes the number of decomposition levels and m represents the mth frequency-band signal in the kth level, where $m = 0, 1, \cdots, 2^k - 1$.

Although WPT is capable of decomposing a signal into independent frequency bands with equal frequency width and without overlapping or gap, the decomposition results of WPT deeply depend on the selection of wavelet basis function. An inappropriate wavelet basis function will result in inaccurate analysis results. EMD is a self-adaptive decomposition method that does not require any selection of basis function. However, the number of IMF cannot be controlled subjectively. Therefore, WPT and EMD have their own advantages and disadvantages. Consequently, the fault features of gear can be revealed from different aspects by WPT and EMD simultaneously.

The energy of the faulty vibration signal is variable when a fault occurs, and this variation can be quantified by calculating the energy entropy of the frequency-band signals of WPT and IMFs of EMD. To directly demonstrate this variation, Fig. 3.5 shows the waveforms of vibration signals of the normal and faulty gears, the decomposed frequency-band signals of WPT with 3 levels, and the energy distribution, where the wavelet basis function is selected as db10. Figure 3.6 illustrates the decomposed results of EMD. Actually, the energy of the first six IMFs almost covers the energy of the entire signal. Thus, other IMFs and the residual are ignored. It can be seen from the decomposition result of WPT and EMD that there is a clear difference between normal gears and damaged gears. In addition, the energy entropies for

3.4 A Multidimensional Hybrid Intelligent Method

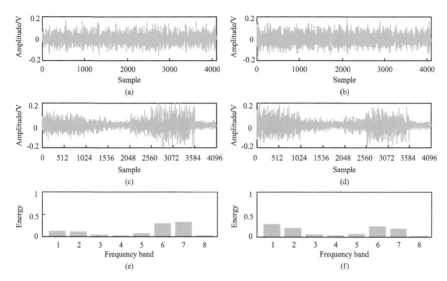

Fig. 3.5 WPT decomposition results and energy distributions: (a) original signal of the normal gear, (b) original signal of the damaged gear, (c) frequency-band signals of the normal gear, (d) frequency-band signals of the damaged gear, (e) energy distribution of different bands for the normal gear, and (f) energy distribution of different bands for the damaged gear

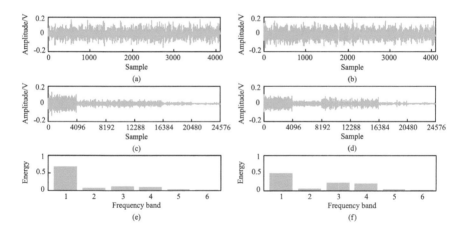

Fig. 3.6 EMD decomposition results and energy distributions: (a) original signal of the normal gear, (b) original signal of the damaged gear, (c) IMFs of the normal gear, (d) IMFs of the damaged gear, (e) energy distribution of IMFs for the normal gear, and (f) energy distribution of IMFs for the damaged gear

WPT on the normal and damaged gears are 1.70 and 1.74, respectively. The energy entropies for EMD on the normal and damaged gears are 1.06 and 1.30, respectively. The calculation formula is Eq. (3.6). Obviously, for WPT and EMD, the energy entropy of a damaged gear is different from that of normal gear. The energy entropy is defined as

$$H_e = -\sum_{j=1}^{J} p_j \log p_j \qquad (3.13)$$

where $p_j = E_j/E$ is the energy percentage of the jth frequency-band signal of WPT or the jth IMF of EMD in the whole signal energy E with $E = \sum_{j=1}^{J} E_j$. Here, the energy E_j can be obtained by using the method of time frequency-domain features in Chap. 2.

According to the above analysis, it is found that the energy distribution of either frequency-band signals or the IMFs will change with the variation of gear health conditions. Therefore, the energy entropy of the frequency-band signals and the IMFs are selected as time-frequency-domain features to reflect gear faults. In this subsection, we choose a db10 wavelet basis with three decomposition levels. In the meantime, the first six IMFs obtained by EMD are selected to represent energy entropy features. Finally, we can obtain 8 energy entropy features from WPT and 6 energy entropy features from EMD. In summary, 29 features have been extracted from the time domain, frequency domain, and time-frequency domain. Due to the small number of features, no further feature selection is required.

Based on the complementary classification performance of different classifiers and different feature sets, a multidimensional hybrid intelligent diagnosis method for gearboxes is presented to improve the accuracy and reliability of the diagnosis results. The method is summarized in Fig. 3.7 and contains the following procedures.

(1) Vibration signals are acquired from gears and processed by exploiting the Hilbert transform.

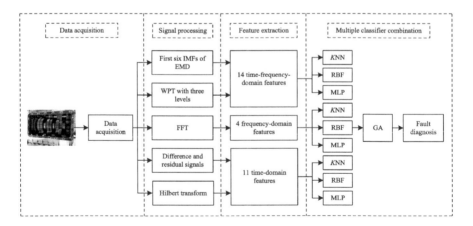

Fig. 3.7 Flowchart of the multidimensional hybrid intelligent method based on multiple classifier combination

3.4 A Multidimensional Hybrid Intelligent Method

(2) 11 time-domain statistical features are extracted from the processed vibration signal, which contains 5 common and 6 special time-domain statistical features. Simultaneously, 4 common frequency-domain statistical features are also extracted from the FFT spectra of vibration signals.
(3) In addition, 14 energy entropy features are extracted, which contain 8 frequency-band signals of WPT and 6 IMFs of EMD.
(4) Finally, the GA weighted averaging technique is used to integrate the outputs of different classifiers and obtain the final diagnosis result of gear health states.

3.4.4 Intelligent Diagnosis Case of Gearboxes

To verify the classification performance of the multidimensional hybrid intelligent diagnosis method, two experiments on a gearbox test rig were performed. One experiment is that gears have a chipped tooth and a missing tooth, and the other experiment is that the cracks have different damage degrees at the root of a gear. Then, vibration signals were collected under the normal condition, the two damage conditions, and the combination of these two damage modes, respectively. The data were utilized to verify the classification performance of this method.

The test rig consists of a gearbox, a motor, and a magnetic brake for loading (Lei 2017b). In addition, the system has three shafts inside the gearbox. Gear 1 on shaft 1 is marked as #1, which has 16 teeth and meshes with gear 2 (#2) of 48 teeth. Gear 3 on shaft 2 is marked as #3, which has 40 teeth and meshes with gear 4 (#4) on the output shaft (shaft 3). Among them, #1 and #4 are the tested gears. #1 with a chipped tooth is labeled as #1'. #4 with a missing tooth is labeled as #4'.

The vibration signals are acquired under the following conditions: (a) all gears are normal, (b) #1 is replaced with #1', (c) #4 is replaced with #4' and (d) both #1 and #4 are replaced by #1' and #4', respectively. In addition, the vibration signals were collected under 3 different loads and 5 different motor speeds from 1200 to 2400 r/min with an increment of 300 r/min. The sampling frequency was set to 5120 Hz, corresponding to 4096 data points for each sample. 4 samples were collected under the same condition. Finally, 60 samples were acquired for every condition, and thereby total of 240 samples were acquired for the four conditions. Vibration signals of four conditions are shown in Fig. 3.8.

Among the 240 samples, 120 data samples are randomly selected as the training set and the remaining 120 samples are for testing. Obviously, this is a classification task corresponding to the four fault types. The multidimensional hybrid intelligent fault diagnosis method is employed to recognize the fault types of the gear. For comparison, the performances of the nine single classifiers are also tested by using the same training set and testing set. The classification accuracies of different methods are displayed in Table 3.9. This experiment is marked as Experiment #1. It is found from this result that the diagnosis accuracy of the 9 single classifiers ranges from 71.67% to 88.33%, with an average of 80.37%. However, for the presented method, the diagnosis accuracy is 98.33%, which is greatly higher than every single classifier.

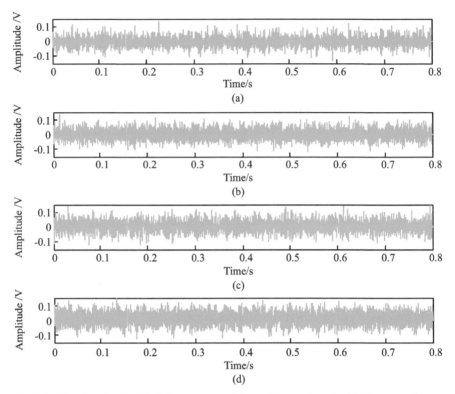

Fig. 3.8 Vibration signals with different gear conditions: (a) normal tooth, (b) chipped tooth, (c) missing tooth, and (d) chipped tooth and missing tooth

The experimental results illustrate that the multidimensional hybrid intelligent fault diagnosis method is superior to an individual classifier for identifying different fault types of gears.

Furthermore, to evaluate the diagnosis performance of the presented method for different fault levels, the test rig was modified and another experiment was carried out (Lei 2017b). The number of teeth of gears 1~4 (#1 ~#4) becomes 48, 16, 24, and 40, and a timing belt was added between the motor and the gearbox. Among them, gear 3 (#3) is the tested gear. In order to simulate the crack failure, the corresponding cracks were machined on the tested gear, and the geometry of the crack failure are illustrated in Table 3.10. The experiment tested three gears under different operation conditions, including one normal gear (F0) and two damaged gears (F1~F2). In F1, there is a very small crack at a tooth root and it can be regarded as an initial crack. In F2, there is a larger but not serious crack at a tooth root and can be regarded as a developing crack.

The vibration signals were acquired for each gear condition. The vibration signals were also captured under three different loads and five different motor speeds from 1200 to 2000 r/min with an increment of 200 r/min. Three different loads, including

3.4 A Multidimensional Hybrid Intelligent Method

Table 3.9 Classification accuracy comparison between the single classifier and the multidimensional hybrid intelligent method

Experiment label	#1	#2		
Dataset	—/%	A/%	B/%	C/%
Classifier K1	73.33	95.56	83.33	61.67
Classifier K2	71.67	100.00	84.72	60.00
Classifier K3	87.50	100.00	90.28	75.00
Classifier K4	72.50	92.22	61.11	85.00
Classifier K5	85.00	100.00	81.94	68.33
Classifier K6	88.33	100.00	87.50	55.00
Classifier K7	80.83	96.67	86.11	61.67
Classifier K8	75.83	100.00	76.39	80.00
Classifier K9	88.33	100.00	83.33	50.00
Average of the 9 classifiers	80.37	98.27	81.63	66.30
The presented method	98.33	100.00	100.00	98.33

Table 3.10 The geometry of the crack faults

Crack fault mode	Geometry of fault			
	Depth/mm	Width/mm	Thickness/mm	Crack angle
F0	0	0	0	—
F1	$(1/4)a$	$(1/4)b$	0.4	45°
F2	$(1/2)a$	$(1/2)b$	0.4	45°

one with no torque and two with different torques, are marked as 0, 1, and 2, respectively. The sampling strategy is the same as before. Four data samples are collected under each operating condition and thereby a total of 180 data samples are obtained in this experiment.

On dataset A described in Table 3.11, the 180 data samples are divided into two equal parts, i.e., 90 samples for the training set and 90 for the testing set. The results of nine single classifiers and the multidimensional hybrid intelligent method are shown in Table 3.9. Since this classification task is relatively simple, except for three single classifiers that fail to achieve the desired results, the remaining classifiers have achieved a classification accuracy of 100%.

In addition, on dataset B, the division of training set and testing set is detailed in Table 3.11. The data samples collected from the lower motor speeds 1200, 1400, and 1600 r/min are used for training, and the data samples of higher motor speeds 1800 and 2000 r/min are used for testing. The examination is able to further verify the robustness of the multidimensional hybrid intelligent method for different motor speeds.

The diagnosis results for both the single classifier and the multidimensional hybrid intelligent method are presented in Table 3.9. It is found that the multidimensional

Table 3.11 Description of the three datasets

Dataset	Number of training/testing samples	Fault types of training/testing	Motor speeds of training/testing samples/(r/min)	Loads of training/testing samples	Label of classification
A	30/30	F0/F0	1200 ~ 2000/1200 ~ 2000	0, 1, 2/0, 1, 2	1
	30/30	F1/F1	1200 ~ 2000/1200 ~ 2000	0, 1, 2/0, 1, 2	2
	30/30	F2/F2	1200 ~ 2000/1200 ~ 2000	0, 1, 2/0, 1, 2	3
B	36/24	F0/F0	1200 ~ 1600/1800, 2000	0, 1, 2/0, 1, 2	1
	36/24	F1/F1	1200 ~ 1600/1800, 2000	0, 1, 2/0, 1, 2	2
	36/24	F2/F2	1200 ~ 1600/1800, 2000	0, 1, 2/0, 1, 2	3
C	30/30	F0/F0	1200 ~ 2000/1200 ~ 2000	0, 1, 2/0, 1, 2	1
	30/30	F2/F1	1200 ~ 2000/1200 ~ 2000	0, 1, 2/0, 1, 2	2

hybrid intelligent method still achieves the highest classification accuracy, i.e., 100%, whereas the accuracies of other single classifiers range from 61.11% to 90.28% with an average of 81.63%. Compared with the highest accuracy of single classifiers, the multidimensional hybrid intelligent method improves the classification accuracy by 9.72%.

Furthermore, on dataset C, the division of the training set and testing set is also detailed in Table 3.11. The training samples are acquired under the gear with a developing crack (F2), whereas the testing samples are acquired under an early crack (F1). The purpose of this examination is to evaluate the generalization capability of the multidimensional hybrid intelligent method for the gear incipient fault.

The examination results are also shown in Table 3.9, which indicate that the multidimensional hybrid intelligent diagnosis method achieves the highest accuracy 98.33% than that of every single classifier. The accuracies of the nine single classifiers range from 50.00% to 85.00% with an average of 66.30%. Obviously, the multidimensional hybrid intelligent method owns a stronger generalization capability for different fault levels.

In experiment #1, the data were collected from different health states of gears. In Experiment #2, three examinations are conducted on three different datasets. The first examination investigates the classification task of normal gears and faulty gears. The second examination studies the robustness of the diagnosis method under different motor speeds. The third examination explores the generalization capability of the diagnosis method for gear incipient fault. By analyzing the above experiment results, it is found that only in the first examination of experiment #2, due to the relatively simple task, there are six single classifiers to obtain the same high classification accuracy as the multidimensional hybrid intelligent method. For all other results, the multidimensional hybrid intelligent method achieves the best result in the gear fault diagnosis, which suggests that it has better robustness under various motor speeds and strong generalization ability for gear incipient faults.

All the above diagnosis results demonstrate that the multidimensional hybrid intelligent method achieves significant improvements in classification accuracy and has better generalization capability in comparison with the single classifier. The success of the multidimensional hybrid intelligent method may be attributed to the following three aspects.

(1) Advanced signal processing techniques, i.e., Hilbert transform, WPT, and EMD are exploited to obtain more important fault-related information.
(2) Different features from time, frequency, and time–frequency domains can fully characterize the health conditions of the gears.
(3) The weighted averaging technique based on GA is exploited to adaptively fuse the classification results.

3.4.5 Epilog

In this section, we investigate the effects of different classifiers and input feature sets on the fault diagnosis results. This multidimensional hybrid intelligent fault diagnosis method consists of MLP, KNN, and RBF, and the input feature sets are extracted from the time domain, frequency domain, and time-frequency domain. Therefore, this method considers the complementary behavior of different classifiers and that of different inputs. Furthermore, the method is verified by gear experiments with different fault classes and levels. The experimental results demonstrate that the classification accuracy of the multidimensional hybrid intelligent fault diagnosis method is superior to those of nine individual classifiers, which further proves the validity and reasonability of the conclusions in Sects. 3.2 and 3.3.

In addition, the feature sets extracted from multidimensional space can reflect the health condition of the gear more effectively. In particular, the features in the time-frequency domain contain more fault-sensitive information than that in time or frequency domains alone. Generally, the multidimensional hybrid intelligent diagnosis method is superior to the diagnosis methods in Sects. 3.2 and 3.3 since the health conditions of gears are more complicated than bearings.

3.5 Conclusions

In this chapter, we extensively introduce and analyze the hybrid intelligent fault diagnosis method. Driven by the three motivations of multiple classifier combination in Sect. 3.1, we firstly introduce the intelligent fault diagnosis method based on multiple WKNN combination with the same input feature sets in Sect. 3.2. Such a method is investigated and analyzed using different bearing experiments. Then in Sect. 3.3, a multiple ANFIS combination method with different input feature sets is introduced for bearing fault diagnosis. In Sect. 3.4, we further examine the multidimensional hybrid intelligent diagnosis method using different classifiers (MLP, RBF, and KNN) combination with different input feature sets for the fault diagnosis of complex gear transmission systems. From the experimental results, it can be concluded that both the combination of different classifiers and the combination of different input feature sets are able to improve the diagnosis accuracy.

In the aforementioned methods, we pay extensive attention to not only combining multiple classifiers but also feature extraction and selection. In practical cases, it is often the case that fault-related information is deeply embedded in the collected vibration signals and therefore many traditional feature extraction methods directly focusing on time or frequency domains are not useful enough. However, lots of effective and suitable signal processing techniques can uncover the fault features embedded in the vibration signals. Driven by this fact, the hybrid intelligent fault diagnosis method should be employed with many advanced signal processing techniques to overcome the feature shortage problem. Additionally, as there would inevitably be some extracted features that are insensitive to faults, and the input with too many features may incur the curse of high dimensionality, feature selection should be performed to only include the salient features. Then, the salient features are fed into multiple classifiers and the final classification results are further obtained. With the help of feature selection, not only the eventual fault diagnosis accuracy will be enhanced, but also the curse of high dimensionality will be overcome. In conclusion, a robust hybrid intelligent fault diagnosis method not only depends on suitable feature extraction and effective feature selection but also is highly related to the compensation between multiple classifiers.

References

Chen J, Li Z, Pan J, Chen G, Zi Y, Yuan J, Chen B, He Z (2016) Wavelet transform based on inner product in fault diagnosis of rotating machinery: a review. Mech Syst Signal Process 70:1–35

Ding J, Xiao D, Li X (2020) Gear fault diagnosis based on genetic mutation particle swarm optimization VMD and probabilistic neural network algorithm. IEEE Access 8:18456–18474

Ghosh AK, Chaudhuri P, Murthy C (2005) On visualization and aggregation of nearest neighbor classifiers. IEEE Trans Pattern Anal Mach Intell 27(10):1592–1602

Jang JR (1993) ANFIS: adaptive-network-based fuzzy inference system. IEEE Trans Syst Man Cybern 23(3):665–685. https://doi.org/10.1109/21.256541

Jia F, Lei Y, Guo L, Lin J, Xing S (2018) A neural network constructed by deep learning technique and its application to intelligent fault diagnosis of machines. Neurocomputing 272:619–628

Kaplan K, Kaya Y, Kuncan M, Minaz MR, Ertunç HM (2020) An improved feature extraction method using texture analysis with LBP for bearing fault diagnosis. Applied Soft Computing 87:106019. https://doi.org/10.1016/j.asoc.2019.106019

Katoch S, Chauhan SS, Kumar V (2021) A review on genetic algorithm: past, present, and future. Multimed Tool Appl 80(5):8091–8126. https://doi.org/10.1007/s11042-020-10139-6

Kuai M, Cheng G, Pang Y, Li Y (2018) Research of planetary gear fault diagnosis based on permutation entropy of CEEMDAN and ANFIS. Sensors 18(3):782

Lei Y, Yang B, Jiang X, Jia F, Li N, Nandi AK (2020) Applications of machine learning to machine fault diagnosis: a review and roadmap. Mech Syst Signal Process 138:106587. https://doi.org/10.1016/j.ymssp.2019.106587

References

Lei Y (2017a) 2 - Signal processing and feature extraction. In: Lei Y (ed) Intelligent fault diagnosis and remaining useful life prediction of rotating machinery. Butterworth-Heinemann, pp 17–66. https://doi.org/10.1016/B978-0-12-811534-3.00002-0

Lei Y (2017b) 4 - Clustering algorithm–based fault diagnosis. In: Lei Y (ed) Intelligent fault diagnosis and remaining useful life prediction of rotating machinery. Butterworth-Heinemann, pp 175–229. https://doi.org/10.1016/B978-0-12-811534-3.00004-4

Li X, Sun J (2018) Signal multiobjective optimization for urban traffic network. IEEE Trans Intell Transp Syst 19(11):3529–3537. https://doi.org/10.1109/TITS.2017.2787103

Li Z, Tan J, Li S, Liu J, Chen H, Shen J, Huang R, Liu J (2019) An efficient online wkNN diagnostic strategy for variable refrigerant flow system based on coupled feature selection method. Energy Build 183:222–237. https://doi.org/10.1016/j.enbuild.2018.11.020

Li X, Zhang W, Ma H, Luo Z, Li X (2020) Deep learning-based adversarial multi-classifier optimization for cross-domain machinery fault diagnostics. J Manuf Syst 55:334–347. https://doi.org/10.1016/j.jmsy.2020.04.017

Li J, Ying Y, Ren Y, Xu S, Bi D, Chen X, Xu Y (2019a) Research on rolling bearing fault diagnosis based on multi-dimensional feature extraction and evidence fusion theory. Royal Soc Open Sci 6(2):181488. https://doi.org/10.1098/rsos.181488

Liu W (2019) Intelligent fault diagnosis of wind turbines using multi-dimensional kernel domain spectrum technique. Measurement 133:303–309. https://doi.org/10.1016/j.measurement.2018.10.027

Rezakazemi M, Dashti A, Asghari M, Shirazian S (2017) H2-selective mixed matrix membranes modeling using ANFIS, PSO-ANFIS GA-ANFIS. Int J Hydrog Energy 42(22):15211–15225. https://doi.org/10.1016/j.ijhydene.2017.04.044

Sharma A, Jigyasu R, Mathew L, Chatterji S (2018) Bearing fault diagnosis using weighted k-nearest neighbor. In: 2018 2nd International conference on trends in electronics and informatics (ICOEI), 11–12 May 2018, pp 1132–1137. https://doi.org/10.1109/ICOEI.2018.8553800

Siahpour S, Li X, Lee J (2020) Deep learning-based cross-sensor domain adaptation for fault diagnosis of electro-mechanical actuators. Int J Dyn Control 8(4):1054–1062. https://doi.org/10.1007/s40435-020-00669-0

Smith WA, Randall RB (2015) Rolling element bearing diagnostics using the Case Western Reserve University data: a benchmark study. Mech Syst Signal Process 64–65:100–131. https://doi.org/10.1016/j.ymssp.2015.04.021

Wang J, Neskovic P, Cooper LN (2007) Improving nearest neighbor rule with a simple adaptive distance measure. Pattern Recogn Lett 28(2):207–213

Wang Q, Wang S, Wei B, Chen W, Zhang Y (2021) Weighted K-NN classification method of bearings fault diagnosis with multi-dimensional sensitive features. IEEE Access 9:45428–45440. https://doi.org/10.1109/ACCESS.2021.3066489

Yan R, Gao RX, Chen X (2014) Wavelets for fault diagnosis of rotary machines: a review with applications. Signal Process 96:1–15

Yang L, Chen H (2019) Fault diagnosis of gearbox based on RBF-PF and particle swarm optimization wavelet neural network. Neural Comput Appl 31(9):4463–4478

Yang D-M, Stronach A, MacConnell P, Penman J (2002) Third-order spectral techniques for the diagnosis of motor bearing condition using artificial neural networks. Mech Syst Signal Process 16(2–3):391–411

Yang W, Su Y, Chen Y (2019) Air compressor fault diagnosis based on lifting wavelet transform and probabilistic neural network. In: IOP conference series: materials science and engineering, vol 1. IOP Publishing, p 012053

Yu J, Liu G (2020) Knowledge extraction and insertion to deep belief network for gearbox fault diagnosis. Knowl Syst 197:105883

Zhang W, Li X, Ding Q (2019) Deep residual learning-based fault diagnosis method for rotating machinery. ISA Trans 95:295–305. https://doi.org/10.1016/j.isatra.2018.12.025

Zhou CY, Chen YQ (2006) Improving nearest neighbor classification with cam weighted distance. Pattern Recogn 39(4):635–645

Zhou J, Li C, Arslan CA, Hasanipanah M, Bakhshandeh Amnieh H (2021) Performance evaluation of hybrid FFA-ANFIS and GA-ANFIS models to predict particle size distribution of a muck-pile after blasting. Eng Comput 37(1):265–274. https://doi.org/10.1007/s00366-019-00822-0

Chapter 4
Deep Transfer Learning-Based Intelligent Fault Diagnosis

4.1 Introduction

In the typical fault diagnosis approaches, the fault features are extracted from the monitoring data through the expert prior knowledge, and then conventional machine learning methods work to construct the nonlinear relationship between the sensitive features and the mechanical system health states. Under the recent development of advanced sensor technologies and the IoT, the monitoring data volume has increased more quickly than ever before. Monitoring big data brings new positive effects and challenges to intelligent fault diagnosis, gradually updating the existing works and applications.

Monitoring big data have become a popular and widely concerned terminology in modern industry. It succeeds the basic characteristics of big data, i.e., volume, velocity, variety, and veracity, in other application areas. In terms of intelligent fault diagnosis, monitoring big data further develop the following domain-specific characteristics (Lei et al. 2020).

(1) Large volume

Many production activities in the modern industry need the cooperation of mechanical system groups to achieve tasks. These mechanical systems usually experience a long-term life cycle, during which the condition monitoring system returns the monitoring data constantly. Furthermore, it is necessary to set high sampling frequency for certain applications so that more health information can be acquired across a large frequency band. Therefore, the volume of monitoring data tends to increase.

(2) Low-value density

In the run-to-failure life cycle of mechanical systems, the healthy state accounts for most of the phase, while the faults experience a short time period compared with the healthy state. Thus, collecting data in the healthy state is easier than that in the fault condition. Besides, it is difficult to make the collected data always hold high quality due to the effects of emergencies such as the sensor anomaly and the transmission interruption. (Xu et al. 2020). For these reasons, the density of the valuable data among the available data is very limited for fault diagnosis.

(3) Multi-source and heterogeneous structures

A mechanical system usually consists of multiple components or sub-systems. To acquire the health information of the whole system, engineers commonly place multiple sensors on key measurement points of mechanical systems. The commonly used sensors include accelerometer, acoustic emission, and current clamp, etc., which could capture the data reflecting complementary information. Moreover, the multi-resource data are stored in diverse structures.

(4) Dynamic data stream

The monitoring data are dynamic time sequences, which contain sufficient real-time information. It is possible to effectively adopt actions once the incipient faults can be recognized. Such an urgent requirement has been widely concerned in the research of big data analytics.

According to the above characteristics, it is beneficial to improve intelligent fault diagnosis through monitoring big data. First, sufficient health information makes intelligent fault diagnosis possible to produce accurate diagnosis results, and promising to make proper decisions for group-level mechanical systems. Second, the big-data era facilitates many progressive computation technologies, such as edge computing and GPU applications. By using the approaches to accelerate computation, intelligent fault diagnosis has the capability of handling the large-volume data and data steam, potentially achieving accurate diagnosis and one-line diagnosis. On the other hand, monitoring big data bring challenges as follows.

In the conventional intelligent fault diagnosis, sensitive features are extracted from the collected data before the implementation of intelligent diagnosis. The diagnosis results greatly rely on the separability of features with respect to the fault patterns. The procedure of feature extraction is conducted manually, in which the engineers or users must design potential algorithms to artificially obtain features. For a large volume of monitoring data, however, extracting specialized features is unrealistic with expert knowledge due to the huge labor cost.

According to the existing research, the intelligent fault diagnosis requires many available data to train the constructed diagnosis models, or the models may perform poorly. The available data refer to the ones that are collected from sufficient typical faults and have been correctly labeled. In engineering scenarios, however, such kind of data are limited. There are two reasons. First, fault data are more difficult to be collected than normal data, resulting in insufficient types of fault patterns. Second, labeling monitoring data requires huge costs. For example, it is unrealistic to frequently inspect the mechanical system health states, which is a way to manually label data. Signal processing provides another way to help label the collected data, but it needs many decision-makers with huge expert knowledge, which requires huge labor cost investment.

4.1 Introduction

Deep transfer learning is potential to overcome the aforementioned challenges by a combination of deep learning and transfer learning. Deep learning could automatically represent features from the input monitoring data, and at the same time identify the mechanical system health states (LeCun et al. 2015; Lei et al. 2016). Such a diagnosis architecture integrates the feature extraction procedure as well as the health state classification procedure. Besides, the features are represented by the correlation between the data and the health states rather than with the guide of expert knowledge. Therefore, deep learning is expected to construct an end-to-end diagnosis model that depicts a direct mapping from the raw monitoring data to the health states.

Transfer learning is a promising way to solve the problem of lacking available data. It refers to two datasets that are respectively collected from the well-studied mechanical systems and other related mechanical systems (Pan and Yang 2010). The former is denoted as the source domain and the latter is the target domain. The goal of transfer learning is to use the diagnosis knowledge of the source domain to achieve the fault diagnosis tasks of other related mechanical systems. Since the data across the source and the target have related knowledge, the diagnosis knowledge contained in the source could help recognize the health states of target data rather than training a new one for the target with massively available data (Li et al. 2020d).

The demand for high model generalization ability in different scenarios promotes a new research topic named deep transfer learning, in which deep learning serves as a feature auto-extractor, and transfer learning improves the performance of diagnosis knowledge when there are limited available data. This chapter systematically studies the deep transfer learning theory that aims at four real-world application issues of intelligent fault diagnosis with monitoring big data:

- How to transfer when very limited data are labeled?
- How to transfer when none of the data is labeled?
- How to transfer when the required diagnosis knowledge is less than the provided?
- How to transfer when the provided diagnosis knowledge cannot meet the requirements?

The solutions to these issues present end-to-end diagnosis models, in which the deep belief network (DBN), deep convolutional network (DCN), and deep residual network (ResNet) are used to cope with raw monitoring data and directly output diagnosis results. Many transfer strategies, such as continual learning, domain adaptation, partial domain adaptation, and open-set domain adaptation, are investigated and analyzed for applications of engineering diagnosis tasks in this chapter.

4.2 Deep Belief Network for Few-Shot Fault Diagnosis

4.2.1 Motivation

The successes of intelligent fault diagnosis require many labeled samples to train the diagnosis model. So does the deep learning-based intelligent diagnosis. In engineering scenarios, however, mechanical systems usually work at the healthy state, while the faults account for a small phase of the long-term life cycle of mechanical systems. Besides, the regular maintenance for a mechanical system possibly makes a small number of samples labeled. As a result, very limited samples are available to train the diagnosis model although massive monitoring data are accumulated. The model cannot learn sufficient diagnosis knowledge from the limited samples, resulting in poor diagnosis performance on the given diagnosis tasks.

The above problem refers to the few-shot learning issues, and is called few-shot fault diagnosis (Xing et al. 2021b). Different from the original few-shot learning, few-shot fault diagnosis assumes massive samples in the healthy state and very limited samples in faults rather than all the classes have a small number of samples. For the tasks of few-shot fault diagnosis, existing works commonly target the imbalance classification methods, such as the data augmentation and the sample resampling. This kind of method has some weaknesses. For example, data augmentation cannot ensure the authenticity of the generated samples. The sample resampling methods may compromise the diversity of the samples and even reduce the effective information for diagnosis.

It is the insufficient knowledge contained in the limited training data that essentially remains a significant challenge to achieve a few-shot fault diagnosis. This is the reason why imbalanced classification methods cannot adequately achieve such tasks with the very long tail effect. Transfer learning provides a promising way for the few-shot fault diagnosis, by which the diagnosis knowledge learned from existing tasks (the source) can complement the knowledge required by the few-shot diagnosis tasks (the target). To achieve the purpose of knowledge transfer, the deep belief network with continual learning (DBNCL) is presented (Xing et al. 2021b). DBNCL stores the knowledge from the source into the connection weights of DBN. The weights are shared among the multiple diagnosis tasks in the source, and continually updated task by task. Such a process makes the deep learning-based model continually store the diagnosis knowledge from existing source tasks. Finally, the knowledge is transferred to the few-shot diagnosis task in the target.

4.2.2 Deep Belief Network-Based Diagnosis Model with Continual Learning

4.2.2.1 Problem Formularization

Given M source diagnosis tasks. For the kth task S^k, and $k = 1, 2, \cdots, M$, there are n_{S_k} labeled samples $X^{S_k} = \{(x_i^{S_k}, y_i^{S_k}) \mid i = 1, 2, \cdots, n_{S_k}\}$, where $x_i^{S_k} \in \mathbb{R}^N$ is the ith vibration sample with N dimension, and $y_i^{S_k}$ is the health state of the ith source sample. The target diagnosis task is denoted as t. It includes a set of labeled samples $X^t = \{(x_i^t, y_i^t) \mid i = 1, 2, \cdots, n_t\}$. Note that $n_t = n_t^N + n_t^F$, wherein n_t^N and n_t^F are the number of samples from the normal state and the faults respectively, and $n_t^F \ll n_t^N$ (at least one sample for one fault type). When the dataset X^t is used to train a diagnosis model, a large generalization error will be produced on other unlabeled samples because the model cannot learn sufficient manifold information of fault types from very limited fault samples. A few-shot fault diagnosis is expected to improve the diagnosis performance of the model against the severe long tail effect in the training dataset.

Multiple source diagnosis tasks contain massive monitoring data, which possibly provide sufficient knowledge to represent features from diverse fault types. A combination of knowledge from the source diagnosis tasks is promising to complement the lack of manifold information in few-shot diagnosis tasks. Such a knowledge transfer is addressed by continual learning (Parisi et al. 2019), which trains a model task by task in the source to learn the data manifold, i.e., the knowledge to represent features. The knowledge is stored in the model parameters, and further used to represent features of samples in the target. The data manifold of the source tasks complements the lack of that in the target. Consequently, the diagnosis model that is trained with very limited fault samples can also work on other unseen samples.

4.2.2.2 Diagnosis Model

The DBNCL is constructed for the few-shot fault diagnosis tasks, and the model flowchart is shown in Fig. 4.1. DBNCL is achieved by three steps (Xing et al. 2021b). In the first step, the samples from multiple source tasks are used to pre-train local-connected restricted Boltzmann machines (RBMs) so that the parameters can store the knowledge to represent features of source samples. After that, the learned knowledge is transferred to the few-shot diagnosis tasks in the target, by which the manifold information of source data complements the lack of fault samples in the target. Finally, an ensemble of multiple classifiers is trained by using the resampled target data. The above three steps are details in the following subsections.

(1) Pre-training of local-connected RBMs with multiple source tasks

The vibration samples are viewed as one-dimensional vectors with respect to the shift-invariant features (Jia et al. 2018). In order to extract the shift-invariant features,

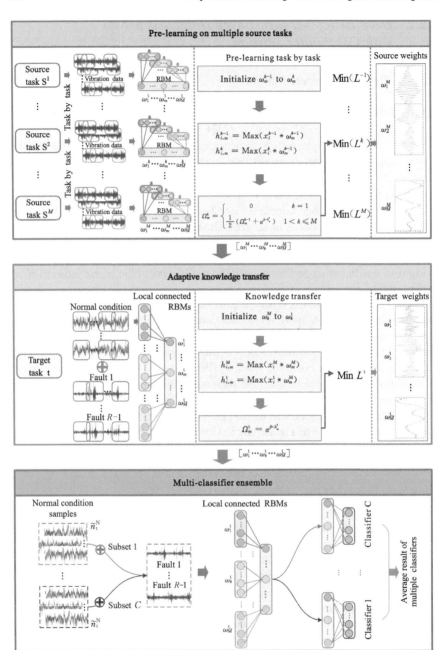

Fig. 4.1 Flowchart of DBNCL

4.2 Deep Belief Network for Few-Shot Fault Diagnosis

the local-connected RBMs are used to handle the source samples task by task. The architecture of the local-connected RBMs is shown in Fig. 4.2. For the kth source diagnosis task S^k, the local-connected RBMs represent features by the following steps (Xing et al. 2021a).

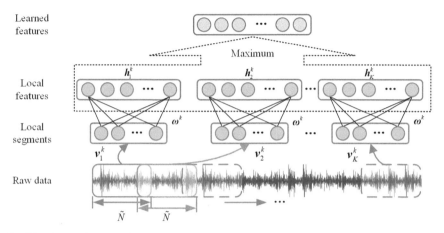

Fig. 4.2 The architecture of the local-connected RBMs

In the first step, the ith sample $x_i^{S_k} \in \mathbb{R}^N$ is split into K segments through a sliding window with length $\tilde{N}$. After that, a segment set $\{v_{i,1}^{S_k}, v_{i,2}^{S_k}, \cdots, v_{i,K}^{S_k}\}$ is obtained. Second, multiple RBMs are constructed for the segments, and one RBM is for one segment. Each RBM consists of a visible layer and a hidden layer. Given the visible layer v with $\tilde{N}$ neurons and the hidden layer h with $\tilde{M}$ neurons, the loss function of the jth RBM working on the jth segment is shown as follows (Hinton 2012):

$$E_j\left(v^{S_k}, h^{S_k} | \theta^{S_k}\right) = \sum_{n=1}^{\tilde{N}} \frac{\left(v_{i,j,n}^{S_k} - a_n^{S_k}\right)^2}{2\left(\sigma_n^{S_k}\right)^2} - \sum_{m=1}^{\tilde{M}} b_m^{S_k} h_{i,j,m}^{S_k} - \sum_{n=1}^{\tilde{N}} \sum_{m=1}^{\tilde{M}} \omega_{n,m}^{S_k} \frac{v_{i,j,n}^{S_k}}{\sigma_n^{S_k}} h_{i,j,m}^{S_k} \tag{4.1}$$

where a_n is the bias of the nth visible neuron, b_m is the bias of the mth hidden neuron, $\omega_{n,m}$ means the connection weights between the nth visible neuron and the mth hidden neuron, and σ_n is the standard deviation of visible neurons. The training parameters for the source task S^k are collected as $\theta^{S_k} = \{a^{S_k}, b^{S_k}, \omega^{S_k}\}$. For the local connected RMBs, the loss function can be calculated by

$$E\left(v^{S_k}, h^{S_k} | \theta^{S_k}\right) = \sum_{j=1}^{K} E_j\left(v^{S_k}, h^{S_k} | \theta^{S_k}\right) \tag{4.2}$$

The learned features of the input samples are the vectors that include the maximum values of hidden neurons among the segment samples, and that is:

$$h_{i,m}^{S_k} = \max_j(h_{i,j,m}^{S_k}) \qquad (4.3)$$

where $h_{i,m}^{S_k}$ is the output of the mth hidden neuron for the input sample $x_i^{S_k}$.

When the local-connected RMBs are re-trained task by task through the continual learning, the above loss function is regularized by

$$L^{S_k} = E(v^{S_k}, h^{S_k}|\theta^{S_k}) + \gamma \sum_{m=1}^{\tilde{M}} \Omega_m^{S_k} \|\omega_m^{S_k} - \omega_m^{S_{k-1}}\|_2^2 \qquad (4.4)$$

where $\omega_m^{S_k} \in \mathbb{R}^{\tilde{N}}$ is the connection weights between the mth hidden neuron and all the visible neurons by pre-training with the source task S^k, $\omega_m^{S_{k-1}} \in \mathbb{R}^{\tilde{N}}$ is the weights pre-trained by the source task S^{k-1}, and γ presents the penalty parameter. The second term in Eq. (4.4) copes up with continual learning. When the local connected RBMs are applied to the source task S^k, the training parameters $\omega_m^{S_k}$ is initialized by the parameters $\omega_m^{S_{k-1}}$ from the previous task S^{k-1}. Continual learning aims to train the parameters that are similar to those from the previous task. The weights $\Omega_m^{S_k}$ are defined by

$$\Omega_m^{S_k} = \begin{cases} 0, & k = 1 \\ \frac{1}{2}\left(\Omega_m^{S_{k-1}} + \alpha^{\beta \cdot h_m^{S_k}}\right), & 1 < k \leq M \end{cases} \qquad (4.5)$$

where

$$h_m^{S_k} = \left(\frac{1}{n_{S_k}^N}\sum_{i=1}^{n_{S_k}^N} h_{i,m}^{S_k} - \frac{1}{n_{S_{k-1}}^N}\sum_{j=1}^{n_{S_{k-1}}^N} h_{j,m}^{S_{k-1}}\right)^2 \qquad (4.6)$$

The first term in Eq. (4.4) is minimized by using the contrastive divergence (CD) algorithm (Fischer and Igel 2014; Hinton 2012), and the second term is minimized through the gradient decent algorithm. Note that the settings of $\alpha > 0$ and $\beta < 1$ make $\Omega_m^{S_k}$ large when there is a small $h_m^{S_k}$.

(2) Adaptive knowledge transfer to the target

The pre-training of local-connected RBMs can learn the prior knowledge to represent the source features task by task. The knowledge is further transferred to the target by using the loss function as follows:

$$L^t = E(v^t, h^t|\theta^t) + \gamma \sum_{m=1}^{\tilde{M}} \Omega_m^t \|\omega_m^t - \omega_m^{S_K}\|_2^2 \qquad (4.7)$$

where ω_m^t are the weights of the mth hidden neuron connected with all the visible neurons, and it is initialized by the weights $\omega_m^{S_K}$ of the last source task. The weights

4.2 Deep Belief Network for Few-Shot Fault Diagnosis

for the hidden neurons are calculated by

$$\Omega_m^t = \alpha^{\beta \cdot h_m^t} \text{ and } h_m^t = \left(\frac{1}{n_t^N} \sum_{i=1}^{n_t^N} h_{i,m}^t - \frac{1}{n_{s_K}^N} \sum_{j=1}^{n_{s_K}^N} h_{j,m}^{S_K} \right)^2 \quad (4.8)$$

After the training by minimizing the loss shown in Eq. (4.8), the knowledge to represent the source features is transferred to the target for feature extraction.

(3) Ensemble of Multi-classifier

The learned features of the target samples are mapped into the health state set by adding a classification layer after the hidden layer. The target samples are used to train the classification layer through the supervised learning strategy, by which the health states of unlabeled target samples are predicted. For the target task, there are massive samples in the normal state and a small number of fault samples. The severe imbalance of sample amount makes the classifier be overfitting on the fault samples easily. For this problem, the ensemble learning of multiple classifiers is constructed. We create C subsets of target samples. Each subset contains $\tilde{n}_t^N$ ($\tilde{n}_t^N < n_t^N$) normal samples randomly from the whole and all the fault samples, and the number of samples in the cth subset is $\tilde{n}_t^c = \tilde{n}_t^N + n_t^F$. Give C classifiers, and one subset will train one classifier. The cross-entropy of the cth classifier is calculated by

$$L_{\text{clf}}^c = -\frac{1}{\tilde{n}_t^c} \sum_{i=1}^{\tilde{n}_t^c} \sum_{j=1}^{R} I(y_i^{t,c} = j) \log \frac{\exp(\theta_j^c h_i^{t,c})}{\sum_{j=1}^{R} \exp(\theta_j^c h_i^{t,c})} \quad (4.9)$$

where $\theta^c = \{\theta_j^c | j = 1, 2, \cdots, R\}$ is the training parameters of the cth classifier, $I(\cdot)$ is the indication function, and it returns one if the condition is satisfied.

The final diagnosis decision is the average value of the total C classifiers, which is expressed as

$$\hat{y}_i^t = \frac{1}{C} \sum_{c=1}^{C} \frac{1}{\sum_{j=1}^{R} \exp(\theta_j^c h_i^{t,c})} \begin{bmatrix} \exp(\theta_1^c h_i^{t,c}) \\ \exp(\theta_2^c h_i^{t,c}) \\ \vdots \\ \exp(\theta_R^c h_i^{t,c}) \end{bmatrix} \quad (4.10)$$

where $\hat{y}_i^t$ represents the one-hot label of the ith target sample. The predicted label picks out the index of the maximum element in $\hat{y}_i^t$.

(4) Training process

The training process of the DBNCL includes the following three steps. First, the local-connected RBMs are trained task by task through

$$\min_{\theta^{S_k}} L^{S_k}, k = 1, 2, \cdots, M \qquad (4.11)$$

The continual learning is adopted in Eq. (4.11) to obtain the parameters after the pre-training of the last source task. In the second step, the local-connected RBMs are trained by using the labeled samples in the target, which is to minimize

$$\min_{\theta^t} L^t \qquad (4.12)$$

Finally, the learned features from the target samples are pushed into multiple classifiers after random sampling from the whole. The subnets train the classifiers by

$$\min_{\theta^c | c=1,2,\ldots C} \sum_{c=1}^{C} L_{\text{clf}}^c \qquad (4.13)$$

A combination of the CD algorithm and the gradient descent algorithm is used to update the training parameters by

$$\begin{aligned} \theta^{S_k} &\leftarrow \theta^{S_k} - \eta_s \cdot \nabla_{\theta^{S_k}} L^{S_k} \\ \theta^t &\leftarrow \theta^t - \eta_t \cdot \nabla_{\theta^t} L^t \\ \{\theta^c\} &\leftarrow \{\theta^c - \eta_c \cdot \nabla_{\theta^c} L_{\text{clf}}^c\} \end{aligned} \qquad (4.14)$$

where η_s, η_t, η_c are the learning rate settings. To be specific, the training process is detailed in Algorithm 4.1.

Algorithm 4.1. Mini-batch training of DBNCL by using CD and gradient descent.

Input: Source data X^{s_k}, and $k=1,2,\cdots,M$, target data X^t.

Output: Diagnosis results of unlabeled target data.

Randomly initialize a set of training parameters θ^{s_0}.

For source task s_k, and $k=1,2,\cdots,M$ **do**

4.2 Deep Belief Network for Few-Shot Fault Diagnosis

1. Initialize the training parameter θ^{s_k} by $\theta^{s_{k-1}}$

For epoch $t = 1, 2, \cdots, T$ **do**

/ Feed-forward propagation is omitted /

2. Draw m samples from the dataset of the kth source task.

3. Calculate loss by Eq. (4.4).

4. Calculate the gradient $g_{\theta^{s_k}} \leftarrow \nabla_{\theta^{s_k}} L^{s_k}$

5. Update parameters $\theta^{s_k} \leftarrow \theta^{s_k} - \eta_s \cdot \text{CD\&SGD}(\theta^{s_k}, g_{\theta^{s_k}})$

End

End

For epoch $t = 1, 2, \cdots, T$ **do**

6. Initialize the training parameter θ^t by θ^{s_K}

/ Feed-forward propagation is omitted /

7. Randomly draw m normal samples and all the fault samples from the target.

8. Calculate loss by Eq. (4.7).

9. Calculate the gradient $g_{\theta^t} \leftarrow \nabla_{\theta^t} L^t$

10. Update parameters $\theta^t \leftarrow \theta^t - \eta_t \cdot \text{CD\&SGD}(\theta^t, g_{\theta^t})$

End

For epoch $t = 1, 2, \cdots, T_{\text{clf}}$ **do**

11. Randomly initialize the training parameter $\{\theta^c \mid c = 1, 2, \cdots, C\}$.

/ Feed-forward propagation is omitted /

12. Draw m normal samples and all the fault samples from the target.

13. Calculate loss by Eq. (4.9).

14. Calculate the gradient $\{g_{\theta^c}\} \leftarrow \{\nabla_{\theta^c} L^c_{\text{clf}}\}$

15. Update parameters $\{\theta^c\} \leftarrow \{\theta^c - \eta_c \cdot \text{SGD}(\theta^c, g_{\theta^c})\}$

4.2.3 Few-Shot Fault Diagnosis Case of Industrial Robots

(1) Dataset Descriptions

The RV reducer is one of the most critical components in the industrial robots. The DBNCL is demonstrated on a dataset that is collected from the single-joint RV reducer fault test rig, as shown in Fig. 4.3(a). The test rig is composed of a servo motor, support, an RV reducer of BX-40E-121 and a swing arm. To simulate the actual running state of the industrial robot reducer, the swing arm is driven by the servo motor to swing back and forth between 0 and 180°. The mass blocks are added to the swing arm to simulate the load on the robot joint. During the test, the loads were set to 0 kg and 24 kg, respectively, and the rotating speed of the swing arm was set to 45°/s and 90°/s. The dataset contains four kinds of health states, i.e., the normal (N) state, wear in the sun gear (SW), crack in the planetary gear (CPG) and wear in the planetary gear (WPW). The pictures of the fault gears are shown in Fig. 4.3b. To obtain the vibration data of the normal state and three fault states, an accelerometer is installed on the top of the RV reducer. During the test, the sampling frequency is set as 6250 Hz and there are 100 samples for each health state under every load and speed setting. Each sample contains 62500 sampling points, by which the arm swing angle of a sample is greater than one reciprocating cycle of 360°. The details of the dataset and diagnosis tasks are shown in Table 4.1.

(2) Parameter analysis

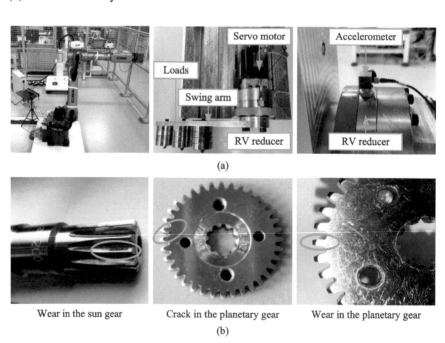

Fig. 4.3 (a) The fault test rig of the single-joint RV reducer, and (b) the fault gears of the RV reducer

4.2 Deep Belief Network for Few-Shot Fault Diagnosis

The key parameters of the pre-learning step are shown in Table 4.2. During the process of continual learning, the number of neurons in the hidden layer $\tilde{M}$ is recommended to be large so that local-connected RBMs can accumulate much prior knowledge on source tasks. Besides, Eq. (4.5) shows that $0 < \alpha < 1$ and $0 < \beta < 1$. If α and β are close to 0, the adaptive weighting factor $\Omega_m^{S_k}$ is calculated to be very small, and thus it is difficult to update the training parameters. The penalty factor γ is determined by the cross-validation between multiple auxiliary tasks.

Table 4.1 Dataset and diagnosis tasks

Diagnosis tasks	Rotating speed/ (°/s)	Load/kg	Health states	Number of samples
A	45	0	N/SW/CPG/WPW	100 × 4
B	45	24		100 × 4
C	90	0		100 × 4
D	90	24		100 × 4

Table 4.2 Learning parameters of DBNCL

Parameters	Value	Parameters	Value
Visible neuron amount $\tilde{N}$	100	Iteration number of pre-learning	500
Hidden neuron amount $\tilde{M}$	1000	α	0.9
Sliding window stride	60	β	0.9
Number of sub-classifiers	10	Penalty factor γ	10^{-4}
Learning rate η_s	0.001	Learning rate η_t	0.0005
Learning rate η_c	10^{-6}	Iteration number of knowledge transfer	100

In the steps of the pre-training of local-connected RBMs and the ensemble of multiple classifiers, $\tilde{n}_t^N$ samples in the normal state are randomly selected from the target task. For the fault states, only one sample from every health state is included in the target dataset. If $\tilde{n}_t^N$ is very small, the model cannot learn sufficient knowledge from the normal state. The opposite setting makes a serious imbalance degree of sample amount across the normal state and the faults. Therefore, it is necessary to discuss different settings of $\tilde{n}_t^N$. Given a diagnosis task, whereethe sample amount is denoted by S in every health state, and the feature dimension is F. According to the reported research (Fink 2005), the diagnosis task is viewed as the few-shot diagnosis task when $S \leq \log(F)$. Based on this, the number of normal samples is set by $\tilde{n}_t^N = \log(F)$. To verify the setting, tasks A, B, C are created as the source and task D is viewed as the target. For the dataset in task D, each fault state only contains one sample. When the number of normal samples is searched from 1 to 12, the statistical diagnosis results are recorded in Fig. 4.4. Each setting conducts ten repeated trials. With the setting $\tilde{n}_t^N = 3$, the model gets the highest diagnosis accuracy with the lowest standard deviation. The experiment result is consistent with the theoretical calculation.

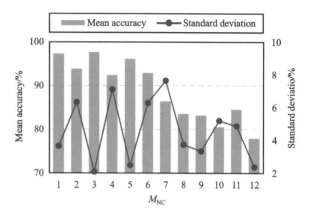

Fig. 4.4 DBNCL diagnosis mean accuracies and standard deviation with different M_{NC}

(3) Diagnosis results

To demonstrate the performance of every module in DBNCL, the performance of DBNCL is compared with its ablation versions. As shown in Table 4.3, Version 1 is a diagnosis model under the paradigm of isolated learning. It only uses the samples of the target task for the training of local-connected RMBs, and uses the features extracted by local-connected RMBs as the inputs of the Softmax classifier for fault recognition. Version 2 firstly uses local-connected RMBs for continual pre-learning on multiple source tasks. After that, it uses the target data samples to fine-tune the classifier. Version 3 fine-tunes multiple classifiers after the continual pre-learning with multiple source tasks. The outputs of multi-classifiers are integrated as the final result. Compared with the presented DBNCL, Version 4 recognizes faults by using a single classifier.

The methods are demonstrated on a transfer task with the source combination of A, B, C and the target setting of D. Note that the fault sample amount in the dataset D decreases. The details of the dataset setting are shown in Table 4.4. The ablation versions 1 to 4 cannot handle the tasks with an imbalanced target dataset. For fair comparisons, give $\tilde{n}_t^N = 3$ if $n_t^F \leqslant 3$, and $\tilde{n}_t^N = n_t^F$ if $n_t^F > 3$. Other parameter settings of the ablation versions are consistent with the DBNCL. Ten repeated trials are conducted in each setting, and Fig. 4.5 shows the statistical results.

DBNCL achieves the highest accuracy and the lowest standard deviation as the number of training samples decreases, especially in a one-shot diagnosis. The average

Table 4.3 Description of ablation settings of DBNCL

Versions	Available modules		
	Pre-learning with source tasks	Knowledge transfer	Classifier ensemble
1			
2	✓		
3	✓		✓
4	✓	✓	
DBNCL	✓	✓	✓

4.2 Deep Belief Network for Few-Shot Fault Diagnosis

Table 4.4 Details of training and testing samples in DBNCL and its ablation settings

Target task	Training samples		Testing samples	
	$\tilde{n}_t^N / n_t^N$	n_t^F	n_t^N	SW/CPG/WPW
D	3/50	1~3	50	All the rest fault samples
	4/50~10/50	4~10	50	
	15/50	15	50	
	20/50	20	50	

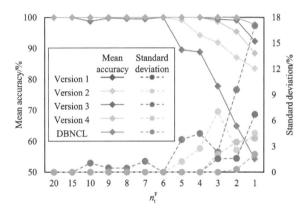

Fig. 4.5 Mean accuracies and standard deviations with decreasing n_t^F

accuracy of Version 1 drops rapidly as the target fault sample amount decreases. For the one-shot diagnosis task, Version 1 especially achieves the diagnosis accuracy of 42.1%, which is much lower than the DBNCL. Due to the positive effects of the continual pre-learning with source tasks, the average accuracies of Version 2 and Version 3 are improved by 79.12% and 73.07%, respectively. They transfer the knowledge to the target without continual learning, getting a limited increase of accuracies for the one-shot diagnosis task. Version 4 improves the average accuracy by 93.18% for the one-shot diagnosis, but the standard deviation of 7.23% is larger than DBNCL.

DBNCL is further compared with other few-shot diagnosis methods. Method 1 is a traditional data augmentation method. According to (Li et al. 2020b), the samples generated by phase-shifting can obtain the highest diagnosis accuracy. Therefore, it generates fault training samples by phase-shifting to supplement the target task. Method 2 is a data augmentation method based on machine learning, which generates fault samples by VAE (Wang et al. 2020). For a fair comparison, after generating fault samples by Method 1 and Method 2, the generated samples and the real collected samples are mixed to train the local-connected RBMs. The fault recognition is achieved by the Softmax classifiers. The training parameters of local-connected RBMs and the Softmax classifiers in the above two methods are consistent with the presented method. Method 3 is a structure simplified method. It introduces the autoencoder into the capsule network and constructs the capsule auto-encoder (CaAE).

The parameters in Method 3 follow those reported in (Ren et al. 2020). Method 4 is a representative transfer learning model based on domain adaptation called deep adaptation network (DAN) (Long et al. 2015). The methods are implemented on four few-shot diagnosis tasks, as shown in Table 4.5. For the implementation of DBNCL, the datasets listed in Table 4.1 are set as the target, respectively, and the rest are viewed as the source tasks.

Table 4.5 Details of few-shot diagnosis tasks for comparisons with other methods

Target tasks	Training datasets		Testing datasets	
	N	SW/CPG/WPW	N	SW/CPG/WPW
A	50	1 × 3	50	297 (99 × 3)
B	50	1 × 3	50	297 (99 × 3)
C	50	1 × 3	50	297 (99 × 3)
D	50	1 × 3	50	297 (99 × 3)

The comparison results are presented in Fig. 4.6. DBNCL achieves the highest accuracy among the given methods. Method 1 fills the data gap caused by phase shifting with zeros after shifting the vibration data along the time axis. In Method 2, VAE is trained with the target data. Both of the methods are difficult to get various samples with subject to only one sample for each fault. Data augmentation algorithm can increase the training sample amount. Different from Methods 1 and 2, Method 3 reduces the training parameters by simplifying the structure of the capsule network, lowering the number of training parameters required by the model. However, the simplified structure easily leads to a decrease in the ability of feature extraction, and hinders the improvement of the diagnosis accuracy. Therefore, the diagnosis accuracies of Method 3 are mostly less than 90% in presence of one-shot tasks.

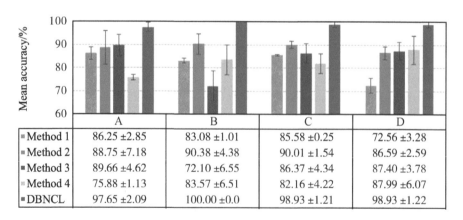

Fig. 4.6 One-shot diagnosis accuracies of DBNCL and existing methods

Method 4 achieves transfer learning by adapting the distribution between source and target datasets. The data distribution of the target task is hard to be evaluated accurately when there is only one sample for each fault. As a result, Method 4 achieves the diagnosis accuracy about 80%, which is lower than DBNCL. DBNCL copes up with the transfer learning by continually pre-training the local-connected RBMs task by task. It obtains higher diagnosis accuracies without the help of distribution adaptation.

4.2.4 Epilog

This section presents the DBNCL for the few-shot fault diagnosis task, in which very limited fault samples are available during the model training. DBNCL is implemented in three steps. The main performance is achieved by using continual learning, which can learn knowledge of feature representation from multiple source tasks and transfer knowledge to the target. After that, the target dataset is resampled into multiple subsets, which train an ensemble of multiple classifiers. According to the results on datasets of industrial robots with different working conditions, continual learning greatly improves the diagnosis of the model on the target when only very limited fault samples are available for the model training.

4.3 Multi-Layer Adaptation Network for Fault Diagnosis with Unlabeled Data

4.3.1 Motivation

The labeled data are precious for intelligent fault diagnosis due to the huge labor cost in labeling data. For a new mechanical system in the factory, particularly, there is no labeled data in history. This leads to the fact that very limited data or even none of the data are labeled although the monitoring data are continuously collected to be massive. The data are not enough and hence they fail to fine-tune a reliable diagnosis model.

In a laboratory, it is convenient to simulate diverse fault types on experimental mechanical systems, and then to collect sufficient labeled data. Due to the similar working principle of the laboratory mechanical systems with those working in the practical engineering scenarios, the collected data may contain related diagnosis knowledge to the real-case mechanical systems. Therefore, it is possible to reuse the diagnosis model that is trained with easily obtained data in the laboratory to recognize the health states of mechanical systems used in the engineering scenarios (Yang et al. 2019). However, the data from the laboratory mechanical systems and real-case mechanical systems follow different distributions, which are drawn from the differences in the physical structures, working conditions, measurement environments, etc. The distribution discrepancy breaks the basic condition (independent identical distribution, i.e., i.i.d), which makes the intelligent fault diagnosis fail.

To be specific, the training data and the testing data for a diagnosis model should be subject to i.i.d, or the model will produce poor generalization performance on the testing data (Guo et al. 2019; Yang et al. 2021b). Affected by the difference in distributions of the laboratory data and the real-world scenario data, the diagnosis model only trained with the laboratory data cannot directly work with the practical engineering data, and results in a large misdiagnosis rate on real-world mechanical systems.

For a successful transfer, a possible solution is to correct the distribution discrepancy. To do this, many works are developed as the categories of instant-based transfer approaches, parameter-based transfer approaches, and feature-based transfer approaches. Among them, feature-based transfer approaches are widely concerned because they can correct severe distribution discrepancy and meanwhile achieve transfer tasks where none labeled data are available in the target domain (Long et al. 2016). As shown in Fig. 4.7, the intelligent fault diagnosis with the feature-based transfer learning is achieved by four steps (Yang et al. 2019). First, the collected data are divided into the source domain and the target domain. The source domain is the one that provides the diagnosis knowledge, like the data from laboratory mechanical systems. The target domain refers to the one that the diagnosis knowledge is applied to, which contains the data from real-world mechanical systems. In the second step, a nonlinear mapping is constructed to extract features both from the source and the target data. It should be noted that the cross-domain data are processed simultaneously through the same feature mapping. In the third step, the distribution difference of the extracted domain features is assessed by distance metrics such as Euclidean distance, maximum mean discrepancy (MMD) (Gretton et al. 2012), CORAL (Sun and Saenko 2016), etc. The measure results are further used to update the parameters of the nonlinear mapping, and the optimization objective is to minimize the distribution discrepancy. Finally, the domain-shared classifier that is fully trained by using the source features can also work with the target features like the performance to the source.

Fueled by deep learning, the hieratical networks can serve as the nonlinear mapping to automatically represent features both from the source and target data.

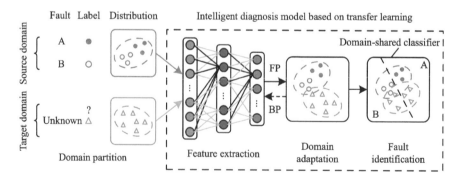

Fig. 4.7 Steps of applications of feature-based transfer learning to intelligent fault diagnosis

Moreover, deep learning has the capability of learning domain-invariant features. The adaptation to such feature distribution is beneficial to slightly correct the discrepancy, and improve the transfer performance of the models (Long et al. 2016). Inspired by the deep learning framework and the basic principle of distribution adaption, a multi-layer adaptation network is presented to achieve diagnosis tasks of real-world application mechanical systems by transferring the diagnosis knowledge from the laboratory.

4.3.2 Multi-Layer Adaptation Network-Based Diagnosis Model

4.3.2.1 Problem Formularization

Let D^s and D^t be the source domain and the target domain respectively. Denote by $X^s \in D^s$ and $X^t \in D^t$ the sample spaces that are from the source and the target domains. The label space $Y = \{1, 2, \cdots, R\}$ contains R kinds of health states. We concern a transfer from laboratory mechanical systems to real-world mechanical systems. It is assumed that the samples from the source and the target are subject to the marginal probability distributions P_s and P_t. The source domain and the target domain are defined as follows (Pan and Yang 2010; Yang et al. 2019):

- The source domain $D^s = \{X^s, P_s\}$ contains n_s labeled samples $X^s = \{(x_i^s, y_i^s) | i = 1, 2, \cdots, n_s\}$ from the laboratory mechanical system, which provide diagnosis knowledge.
- The target domain $D^t = \{X^t, P_t\}$ has n_t unlabeled samples $X^s = \{x_j^t | j = 1, 2, \cdots, n_t\}$. It is aimed that the samples can be correctly classified through the source diagnosis knowledge.
- For a successful transfer, the source domain should bring sufficient diagnosis knowledge to the target, which requires the label space of the source should cover that of the target, i.e., $Y^t \subseteq Y^s \subseteq Y$.

The data in the source and the target are obtained from different mechanical systems, and thus there is serious distribution discrepancy, i.e., $P_s \neq P_t$. Figure 4.8(a) presents the effects of the discrepancy. The domain-shared diagnosis boundary $h(\cdot)$ is fine-trained by using the source data with the optimization objective to minimize the source risk $R_s = E_{x \sim P_s}[h(x) \neq y^s]$. When the classifier is reused on the target, the generalization error of $h(\cdot)$ is unsatisfactory due to the distribution discrepancy, misclassifying the target data. Therefore, there aims at an intelligent diagnosis model that could extract domain features with similar distribution across domains, and finally minimize the risk of $h(\cdot)$ on the target, i.e., $R_t = E_{x \sim P_t}[h(x) \neq y^t]$. As shown in Fig. 4.8(b), the distribution discrepancy of the learned features is reduced so that the domain-shared classifier can correctly recognize the health states of the target.

Fig. 4.8 Illustrations of feature-based transfer learning in intelligent fault diagnosis: (a) without distribution adaptation, and (b) with distribution adaptation

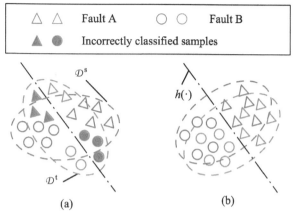

4.3.2.2 Model Architecture

A multi-layer adaptation network (MLAN) is presented to achieve the diagnosis task of transferring diagnosis knowledge from the laboratory to real cases. The architecture of MLAN is shown in Fig. 4.9 (Yang et al. 2019). MLAN contains three parts. First, it uses a domain-shared convolutional neural network (CNN) to extract domain features from the source and target samples. Second, the multi-layer domain adaptation is constructed to correct the distribution shift of the layer-wise features. Third, the pseudo labels are generated for the unlabeled target data so that the data could help improve the model performance. The three parts are detailed as follows.

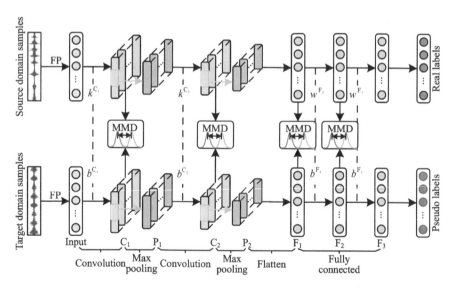

Fig. 4.9 The architecture of MLAN for transfer diagnosis

4.3 Multi-Layer Adaptation Network for Fault Diagnosis with Unlabeled Data

(1) Domain-shared CNN

Two CNNs are individually set for the source and the target respectively, but they share the architecture parameters. The parameters of the domain-shared CNN are shown in Table 4.6. The domain-shared CNN includes three parts, i.e., the convolutional layers, the pooling layers, and the fully connected layers.

Table 4.6 Architecture parameters of the domain-shared CNN

Layers	Shared parameters	Activation functions	Output size
Input	—	—	1200 × 1
C_1	5 × 1 × 20	ReLU	1196 × 20
P_1	2	—	598 × 20
C_2	5 × 20 × 20	ReLU	594 × 20
P_2	2	—	297 × 20
F_1 (Flatten)	—	—	5940 × 1
F_2	5940 × 256	ReLU	256 × 1
F_3 (Classification)	256 × R	Softmax	R × 1

The convolutional layers handle the raw vibrations data in the source and the target by using the shared kernel $k^l \in \mathbb{R}^{H \times L \times P}$, where H, L, and P are respectively the height, the length, and the depth of the kernels. The vibration data in the time domain are one-dimensional vectors, and thus L is set as 1. Denote by $x_i^{l-1,D} \in \mathbb{R}^{N \times M}$ the domain features from the $(l-1)$th layer. The output features of the lth layer can be expressed as follows:

$$x_i^{l,D} = \sigma_r \left(x_i^{l-1,D} * k^l + b^l \right)$$

$$\left(x_i^{l-1,D} * k^l \right)_{j,d} = \sum_{m=1}^{M} \sum_{h=1}^{H} x_{(i),j+h-1,m}^{l-1,D} \cdot k_{h,d}^l \quad (4.15)$$

where $D = \{s, t\}$ is the superscript of domains, $x_i^{l,D} \in \mathbb{R}^{(N-H+1) \times P}$ is the extracted features from the ones of the $(l-1)$th layer, b^l is the bias, and $\sigma_r(\cdot)$ is the activation function of the rectified linear unit (ReLU) (Nair and Hinton 2010). Figure 4.10(a) presents the basic principle of the convolutional process.

The pooling layer is stacked after the convolutional layer, which can reduce the dimension of the learned features. By this process, the number of the training parameters in the net is reduced to possibly overcome the over-fitting problem. The commonly used pooling process includes the max-pooling and the mean-pooling types. We use the max-pooling process in the net. It first divides the features into many non-overlapping segments. After that, the maximum value of each segment is returned. The basic principle of the max-pooling process is briefly explained in Fig. 4.10(b). We further formul中ize the max-pooling process as:

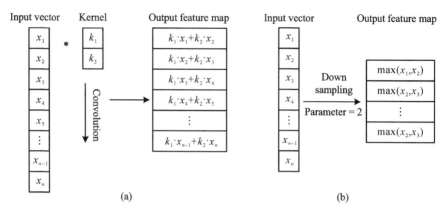

Fig. 4.10 Basic principles of (a) convolutional process, and (b) max-pooling process

$$\left(v_i^{l,D}\right)_{k,m} = \max\left\{x_{(i),j,m}^{l-1,D} \mid j \in N^+, s(k-1)+1 \leqslant j \leqslant sk, \forall x_{(i),j,m}^{l-1,D} \in x_i^{l-1,D}\right\}$$
(4.16)

where s is the height of each non-overlapping segment.

By stacking the convolutional layer and the pooling layer in turn, the constructed network can represent the deeper-layer features. The features are further mapped into the label space by using a fully connected subnet with three layers. The input layer of this subnet is the deeper-layer features that are flattened as 1D vectors. Let $x_i^{F_1,D}$ be the input features of layer F_1, which are obtained by flattening features $v_i^{P_2,D}$. Through the basic connection of multi-layer network, the output of the layer F_2 is obtained as

$$x_i^{F_2,D} = \sigma_r\left(w^{F_2} x_i^{F_1,D} + b^{F_2}\right) \quad (4.17)$$

where w^{F_2} and b^{F_2} are respectively the tied weights and bias of layer F_2. Layer F_3 achieves the fault classification, in which the Softmax function is used to predict the probability of classifying a sample into each health state. The number of neurons in layer F_3 equals the size of the source label space, and the output probability distribution is

$$h_i^{F_3,D} = \left[P\left(y_i^D = 1 \mid x_i^{F_2,D}\right) \cdots P\left(y_i^D = j \mid x_i^{F_2,D}\right) \cdots P\left(y_i^D = R \mid x_i^{F_2,D}\right)\right]$$

$$P\left(y_i^D = j \mid x_i^{F_2,D}\right) = \frac{\exp\left[\left(w_j^{F_3}\right)^T x_i^{F_1,D} + b^{F_2}\right]}{\sum_j \exp\left[\left(w_j^{F_3}\right)^T x_i^{F_1,D} + b^{F_2}\right]}$$
(4.18)

where $\boldsymbol{w}^{F_3}$ and $\boldsymbol{b}^{F_3}$ are the shared weights and bias in layer F_2.

All the source data are labeled so as to train the domain-shared CNN through supervised learning. After that, the net could adequately learn the diagnosis knowledge, i.e., the relationship between the input data and the output health state. Given source samples $\boldsymbol{x}_i^s \in X^s$, and $i = 1, 2, \cdots, n_s$, the difference between the predicted source labels and actual labels is estimated by the cross-entropy loss function:

$$L_{\text{clf}} = -\sum_{i=1}^{n_s}\sum_{j=1}^{R} I\left(y_i^s = j\right) \log h_{i,j}^{F_3,s} \quad (4.19)$$

where $I(\cdot)$ is the indication function, and $h_{i,j}^{F_3,s}$ is the predicted probability of classifying the ith source sample into the jth health state.

(2) Multi-layer domain adaptation

The source and target data are subject to distribution discrepancy. Such discrepancy also exists in the learned domain features when the cross-domain data are handled by the domain-shared CNN. A nonparametric distance metric named MMD is adopted to measure the discrepancy, which is defined as (Gretton et al. 2012)

$$D_H\left(X^s, X^t\right) := \sup_{\Phi \in H} \left(E_{\boldsymbol{x}^s \in X^s}[\Phi(\boldsymbol{x}^s)] - E_{\boldsymbol{x}^t \in X^t}[\Phi(\boldsymbol{x}^t)]\right) \quad (4.20)$$

where $\sup(\cdot)$ means the supremum of the input aggregate, and $\Phi(\cdot)$ is the nonlinear mapping in the reproducing kernel Hilbert space (RKHS) H. MMD aims to find a mapping, by which the mean distance of the domain features is measured in RKHS. Through the kernel mean embedding, RKHS is created by kernel functions, such as Gaussian kernels, polynomial kernels, and Laplace kernels. The MMD can be empirically estimated by (Gretton et al. 2012)

$$\begin{aligned}\hat{D}_H^2\left(X^s, X^t\right) &= \left\| \frac{1}{n_s}\sum_{i=1}^{n_s}\Phi(\boldsymbol{x}_i^s) - \frac{1}{n_t}\sum_{j=1}^{n_t}\Phi(\boldsymbol{x}_j^t) \right\|_H^2 \\ &= \frac{1}{n_s^2}\sum_{i=1}^{n_s}\sum_{j=1}^{n_s}k(\boldsymbol{x}_i^s, \boldsymbol{x}_j^s) - \frac{2}{n_s n_t}\sum_{i=1}^{n_s}\sum_{j=1}^{n_t}k(\boldsymbol{x}_i^s, \boldsymbol{x}_j^t) + \frac{1}{n_t^2}\sum_{i=1}^{n_t}\sum_{j=1}^{n_t}k(\boldsymbol{x}_i^t, \boldsymbol{x}_j^t)\end{aligned} \quad (4.21)$$

where $k(\cdot, \cdot)$ represents the characteristic kernels.

The distribution discrepancy of domain features is derived from two parts when handled by the domain-shared CNN. First, the discrepancy is enlarged on the deeper layers because the deep neural network tends to extract domain-specific features. Thus, it is necessary to adapt the distributions of domain features in fully connected layers (Long et al. 2015). Second, the distributions of the learned features are dynamically changed when updating the training parameters of domain-shared CNN. The discrepancy is derived from the shift of internal covariate (Yosinski et al. 2014;

Ioffe and Szegedy 2015). To correct the shift, the distributions of domain features in layers C_1 and C_2 need to be adapted. From the above two parts, we can calculate the multi-layer MMD of domain features by

$$\hat{D}_H^2(Z^{L,s}, Z^{L,t}) = \sum_{i=1}^{n_s}\sum_{j=1}^{n_s}\sum_{l \in L} \kappa_l \cdot k\left(x_i^{l,s}, x_j^{l,s}\right) - 2\sum_{i=1}^{n_s}\sum_{j=1}^{n_t}\sum_{l \in L} \kappa_l \cdot k\left(x_i^{l,s}, x_j^{l,t}\right) +$$
$$\sum_{i=1}^{n_t}\sum_{j=1}^{n_t}\sum_{l \in L} \kappa_l \cdot k\left(x_i^{l,t}, x_j^{l,t}\right), \kappa_l = 1 - \frac{\hat{D}_H^2(Z^{l,s}, Z^{l,t})}{\hat{D}_H^2(Z^{L,s}, Z^{L,t})}$$

(4.22)

where $l \in L = \{C_1, C_2, F_1, F_2\}$ indicates the layers of domain-shared CNN. It is noted that CNN tends to extract domain-specific features. This makes the distribution discrepancy of the shallower-layer features larger than that of the deeper-layer features. There is an even larger difference in the value magnitude. Thus, the tradeoff parameter κ_l is added to weigh the MMD value of domain features in every layer. The parameter follows a strategy that the larger MMD will get smaller weights. The opposite is for the layers with smaller MMD.

The calculation to MMD shown in Eq. (4.22) suffers two problems. First, it is sensitive to the kernel parameter (Yang et al. 2020). Take the commonly used Gaussian kernel function $k(x^s, x^t) = \exp(-\|x^s - x^t\|^2/2\gamma^2)$ for example. The kernel bandwidth γ controls the results of MMD. If $\gamma \to 0$ or $\gamma \to \infty$, the MMD of the source and target domain features will approach zero, resulting in an invalid measurement for distribution discrepancy. Besides, the small change of γ may produce the MMD values with different magnitudes so that the measurement is unconvinced. To solve this problem, the median heuristic strategy is recommended to choose γ (Garreau et al. 2017). This strategy adequately calculates the kernel bandwidth by $\gamma = E_{x^s \in X^s, x^t \in X^t} \|x^s - x^t\|^2$, which is the median distance among all pairs of domain features. Second, the calculation in Eq. (4.22) requires the time complexity of $\mathcal{O}\left[|L| \times (n_s + n_t)^2\right]$, which is quite high that much computation resource is to be consumed. Unbiased estimation of MMD is used to reduce the time complexity, which is calculated as (Gretton et al. 2012)

$$\hat{D}_H^2(Z^{l,s}, Z^{l,t}) = \frac{1}{n}\sum_{i=1}^{n}\left[\sum_{l \in L} \kappa_l \cdot k\left(x_{2i-1}^{l,s}, x_{2i}^{l,s}\right) + \sum_{l \in L} \kappa_l \cdot k\left(x_{2i-1}^{l,t}, x_{2i}^{l,t}\right)\right] -$$
$$\frac{1}{n}\sum_{i=1}^{n}\left[\sum_{l \in L} \kappa_l \cdot k\left(x_{2i-1}^{l,s}, x_{2i}^{l,t}\right) + \sum_{l \in L} \kappa_l \cdot k\left(x_{2i}^{l,s}, x_{2i-1}^{l,t}\right)\right]$$

(4.23)

which assumes $2n = n_s = n_t$. The calculation of MMD using Eq. (4.23) reduces the time complexity of multi-layer MMD by $O(|L| \times n)$, accelerating the training process effectively.

4.3 Multi-Layer Adaptation Network for Fault Diagnosis with Unlabeled Data

(3) Pseudo label learning

All the target samples are unlabeled, which cannot directly train the net through supervised learning. According to the idea of semi-supervised learning, however, the unlabeled samples could improve the classification performance of models. To improve the domain-shared CNN with the help of unlabeled target samples, pseudo label learning is introduced into the training process.

The pseudo label of an unlabeled sample is to pick up the label that has the maximum prediction probability by using a classifier (Lee 2013). The classifier could be fine-tuned by using the labeled source samples. For example, layer F_3 could output the probability of target samples belonging to every health state (label). After that, the pseudo labels are generated based on the results shown in Eq. (4.18), and the procedure is expressed as follows:

$$\tilde{y}_i^t = \begin{bmatrix} \tilde{y}_{i,1}^t & \tilde{y}_{i,2}^t & \cdots & \tilde{y}_{i,j}^t & \cdots & \tilde{y}_{i,R}^t \end{bmatrix}$$

$$\tilde{y}_{i,j}^t = \begin{cases} 1 & \text{if } j = \arg\max_j h_{i,j}^{F_3,t} \\ 0 & \text{otherwise} \end{cases} \quad (4.24)$$

where $\tilde{y}_i^t$ is the pseudo label (one-hot format) of the ith target sample.

From Eq. (4.24), the difference between predicted labels of target samples and pseudo labels is measured by cross-entropy as

$$L_{pl} = -\sum_{i=1}^{n_t}\sum_{j=1}^{R} I(\tilde{y}_i^t = j) \log h_{i,j}^{F_3,t} \quad (4.25)$$

where $h_{i,j}^{F_3,t}$ means the probability of classifying the ith target sample as the jth health state.

(4) Training process

With a combination of losses shown in Eqs. (4.19), (4.23), and (4.25), the optimization objective of MLAN is shown as

$$\min_{\theta} L = L_{clf} + \alpha L_{pl} + \beta \hat{D}_H^2(Z^{l,s}, Z^{l,t}) \quad (4.26)$$

where θ collects the training parameters of the layers C_1, C_2, F_1, F_2, and F_3, and α, β are respectively the tradeoff parameter for the pseudo label learning and multi-layer MMD. From the optimization objective shown in Eq. (4.26), MLAN targets to minimize three terms. The first term reduces the error between the predicted labels of source samples and their actual labels so that the model could adequately learn the source diagnosis knowledge. The second term focuses on reducing the cross-entropy between the predicted labels and pseudo labels of the target samples. As reported by research (Lee 2013; Yang et al. 2019), this regularization term is promising to reduce the intra-class distance and enlarge the inter-class distance so that the decision boundary could pass through the low-density region of the target easily. The third term achieves the core function to correct the distribution discrepancy of the domain features, by which the domain-shared CNN could represent domain features subject to similar distributions. With the cooperation of the above three regularization terms, the model that performs well on the source could also work at the target.

In the training of MLAN, the Adam optimization algorithm (Kingma and Ba 2014) is used to update the training parameters by

$$\theta \leftarrow \theta - \eta \cdot \nabla_\theta \left(L_{\text{clf}} + \alpha L_{\text{pl}} + \beta \hat{D}_H^2 \right) \tag{4.27}$$

where η is the learning rate. Algorithm 4.2 details the training process.

Algorithm 4.2. Mini-batch training of MLAN by using Adam.

Input: Labeled source data X^s, unlabeled target data X^t.

Output: Diagnosis results of unlabeled target data.

Initialize parameters θ and set the tradeoff parameters α and β.

For epoch $t=1,2,\cdots,T$ **do**

> / Feed-forward propagation is omitted /
>
> 1. Draw m samples simultaneously from the source and the target.
>
> 2. Calculate classification loss of the source by Eq. (4.19).
>
> 3. Estimate multi-layer MMD of domain features in layer F2 by Eq. (4.23).
>
> 4. Generate pseudo labels for the target samples through Eq. (4.24).
>
> 5. Obtain the pseudo label learning loss in Eq. (4.25).
>
> 6. Calculate gradient $g_\theta \leftarrow \nabla_\theta (L_{\text{clf}} + \alpha L_{\text{pl}} + \beta \hat{D}_H^2)$.
>
> 7. Update training parameters $\theta \leftarrow \theta - \eta \cdot \text{Adam}(\theta, g_\theta)$

End

Return the predicted labels of target data as the diagnosis results.

4.3.3 Fault Diagnosis Case of Locomotive Bearings with Unlabeled Data

4.3.3.1 Dataset Description

The performance of MLAN is demonstrated using two bearing datasets for knowledge transfer. More detail of the datasets can be found in Table 4.7.

The dataset GBearing is collected from a multi-stage transmission rig, as shown in Fig. 4.11. The rig consists of four parts, i.e., a motor, a fixed-shaft gearbox, a planetary gearbox, and a magnetic powder brake. The rig is to transmit the power of the motor through the two-stage gearbox transmission system, and finally drive the magnetic powder brake. The motor speed is controlled by changing the electric power frequency, and the input current of the magnetic powder brake changes the loaded torque of the system. The tested bearings (LDK UER204) are installed into the right end of the intermediate shaft of the fixed-shaft gearbox. We simulate four kinds of health states of bearings in the laboratory, which are N, IF, RF, and OF respectively. When the bearings are fine-installed into the rig respectively, an accelerometer mounted on the end house of the gearbox collects the vibration data. During the test, the motor speed is set at 1200 r/min, and the sampling frequency is set as 12.8 kHz. The dataset GBearing contains 404 samples that are balanced across every state, and each sample has 1200 sampling points (Yang et al. 2020).

The dataset LBearing is collected from locomotive bearings that are tested in a hydraulic motor-driven system. There are four kinds of health states in the tested bearings, which includes the normal state and the abraded faults respectively on the inner race, the roller, and the outer race. During the test, the output speed of the hydraulic motor is controlled as about 500 r/min. Besides, the tested bearings are loaded by a hydraulic cylinder that could provide the maximum radial load of 20 kN. The vibration data are collected at the sampling frequency of 12.8 kHz. There are 101 samples in each health state. In terms of each sample, there are 1200 sampling points (Lei 2017).

Table 4.7 Details of the datasets of the gearbox bearings and the locomotive bearings

Datasets	Specification	Health states	Number of samples	Working conditions
GBearing	LDK UER204	N	4 × 101	1200 r/min
		IF		
		RF		
		OF		
LBearing	552732QT	N	4 × 101	490 ~ 530 r/min
		IF		
		RF		
		OF		

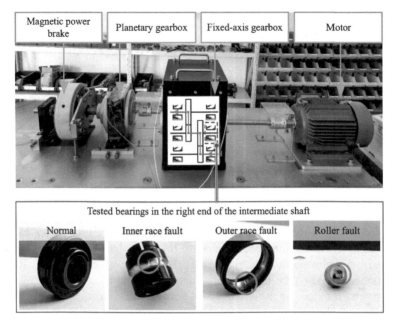

Fig. 4.11 Picture of the multi-stage transmission rig

Based on the two datasets, the transfer diagnosis task GBearing → LBearing is created. This task concerns the fact that the data from bearings in real-world applications are seldom labeled, while the data from laboratory bearings are controlled. Therefore, the diagnosis knowledge contained in the laboratory bearings is expected to recognize the health states of locomotive bearings. It is assumed that all the samples in dataset GBearing are labeled, but the LBearing dataset is unlabeled.

4.3.3.2 Diagnosis Results

In the training of MLAN, the tradeoff parameter α is searched from the unit of $\{0.01, 0.05, 0.1, 0.5, 1\}$, and β is chosen from $\{0.01, 0.05, 0.1, 0.5, 1, 5, 10, 50\}$. With the optimal combination of the parameters, MLAN is implemented on the transfer diagnosis task GBearing → LBearing. Figure 4.12 records the statistical results of 10 trials of MLAN. The average accuracy of MLAN is 74.81%. The results are further compared with other methods including CNN, transfer component analysis (TCA) (Pan et al. 2011), domain adaptation for fault diagnosis (DAFD) (Lu et al. 2017), deep domain confusion (DDC) (Tzeng et al. 2014), and MLAN ($\alpha = 0$). CNN is the version that MLAN removes the multi-layer adaptation submodule and the pseudo label learning submodule. The source data are used to train the domain-shared CNN, and then it is directly tested on the target data. TCA is a typical transfer learning method, which obtains the features with similar distributions after the discrepancy is

4.3 Multi-Layer Adaptation Network for Fault Diagnosis with Unlabeled Data

corrected in a low-dimensional subspace. After that, the SVM-based classifier trained with the source features will work at the target features. In the method, search for the tradeoff parameter for the regularization term from the range {0.01, 0.1, 1, 10, 100}, and the subspace dimension is selected from {2, 4, 8, $\cdots$, 128}. DAFD is an intelligent diagnosis model that represents domain features subject to similar distributions by a combination of auto-encoders and MMD. This method has reported pretty results on transfer diagnosis tasks across different working conditions of an individual bearing. Two tradeoff parameters need to be concerned in DAFD, and they are both searched from {$10^{-5}, 10^{-4}, \cdots, 10^{5}$}. DDC is a deep transfer learning method. Different from MLAN, it only adapts distributions of domain features of the deepest layer. MLAN ($\alpha = 0$) is the version that the MLAN removes the regularization term of pseudo label learning. It is noted that TCA and DAFD have no capability of handling raw vibration data, and thus their inputs are frequency spectrum data of the raw vibration signals. For fair comparisons, CNN, DDC, and MLAN ($\alpha = 0$) are respectively constructed with the same architecture back bone as MLAN. Every method is implemented on the transfer diagnosis task with the optimal parameter combination.

CNN only achieves an average accuracy of 26.73% due to a lack of distribution adaptation of domain features. This accuracy is approximated to 25%, which means a random gauss occurred in the model. TCA and DAFD adopt a shallow network rather than the deep one to extract features. The nonlinear transformation of the net is weak to just extract general features from the input data. This is the reason why TCA and DAFD get the lower accuracy (25% and 45.09% respectively) than MLAN although there is a distribution adaptation procedure in the two methods. DDC obtains an average accuracy of 58.41%, which is lower than MLAN ($\alpha = 0$) because DDC only adapts the distribution of domain features of one layer. The result verifies the effectiveness of multi-layer distribution adaptation. The accuracy that MLAN ($\alpha = 0$) obtains is slightly smaller than MLAN, which explains the positive effects of pseudo label learning in improving the model performance.

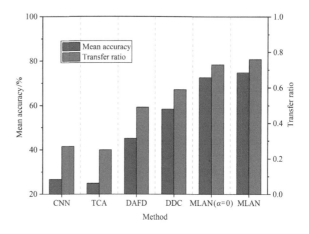

Fig. 4.12 Comparisons of the diagnosis results of MLAN with other methods

The transfer performance of models is compared through the metric of transfer ratio. Before getting the metric, denote by err (S, T) and err_b(T, T) the transfer error and the in-domain error, respectively. The transfer error depicts the performance of a model first trained with the data of source S and then tested on the data of target T. The in-domain error is defined as the generalization error of a model that performs both the training process and the testing process on the target. With the above definitions, TR = $E\left[(1 - \text{err}(S_i, T_i))/(1 - \text{err}_b(T_i, T_i))\right]$ calculates the transfer ratio. This metric shows the general performance of a transfer learning-based model on transfer diagnosis tasks. The larger the transfer ratio is, the better transfer performance is obtained by the model. We use a CNN-based diagnosis model for the calculation of in-domain error. In the dataset LBearing, 50% of samples are randomly selected to train the diagnosis model, and then the model is tested on the rest samples. By the repeated trials for 10 times, the average accuracy of 99.02% is obtained, and thus err_b(T, T) is 0.98%. Based on this, the transfer ratio is calculated for every comparison method. The results are shown in Fig. 4.12. The MLAN respectively obtains the transfer ratio of 0.76 on the given transfer diagnosis task, which is the best results among the methods. This demonstrates that the MLAN has achieved the highest transfer performance than the other methods.

4.3.3.3 Feature Visualization

The t-distributed stochastic neighbor embedding (t-SNE) algorithm (Van der Maaten and Hinton 2008) is used to visually explain the performance of MLAN. This algorithm is a way to reduce the feature dimensions, by which the learned features from the source and the target data can be mapped into a plane. The feature distributions are visually presented through the scattered figures. The learned features respectively by CNN, DDC, and MLAN are plotted in Fig. 4.13. Furthermore, the diagnosis results of the three methods are presented as confusion matrix that are shown in Fig. 4.14.

The conventional CNN-based diagnosis model lacks the distribution adaptation module. Thus, the learned features from the source and the target are subject to serious distribution overlapping, as shown in Fig.4.13(a). This leads to the poor diagnosis accuracy of the model performed on the target, as shown in Fig. 4.14(a), in which only partial IF target samples are correctly recognized. In the method of DDC, the distribution discrepancy of learned features in the deepest layer is reduced to improve the performance of the source model on the target. Figure 4.13(b) presents that the learned features of the target are more similar to those of the source than using CNN. Thus, the diagnosis accuracy is improved, but it is found in Fig. 4.14(b) that most of the target samples are still misclassified due to the under-adaptation of DDC. MLAN integrates the multi-layer distribution adaptation module and the pseudo label learning module. The powerful capability of distribution adaptation makes the learned features more similar across domains than DDC achieves, as shown in Fig. 4.13(c). The diagnosis accuracy of MLAN is thus higher than other methods.

4.3 Multi-Layer Adaptation Network for Fault Diagnosis with Unlabeled Data

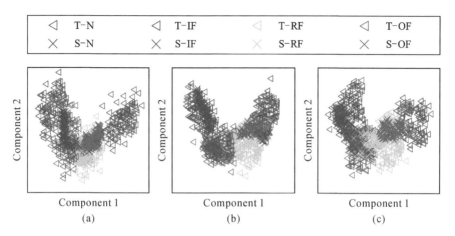

Fig. 4.13 The visualization of learned features of the source and the target data respectively by: (a) CNN, (b) DDC, and (c) MLAN

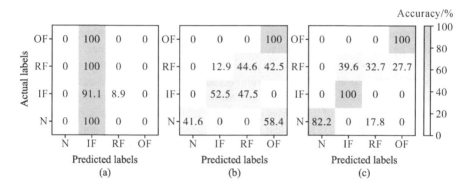

Fig. 4.14 Confusion matrix of diagnosis results by using: (a) CNN, (b) DDC, and (c) MLAN

4.3.4 Epilog

Since the labeled data are insufficient and even none of the data are labeled in real-world application mechanical systems, it is difficult to train a diagnosis model to produce a result with high diagnosis accuracy. A transfer diagnosis method named MLAN is presented to achieve the diagnosis tasks of real cases by using the diagnosis knowledge contained in source mechanical systems. MLAN is constructed with the deep learning architecture of CNN that could automatically represent domain features both from the source and the target. After that, MLAN comes up with the core function by multi-layer domain adaptation module and the pseudo label learning module. The former has the capability of reducing the distribution discrepancy of the learned domain features, and the latter makes the target features easily discriminated. From the results of a transfer from the gearbox bearings to the locomotive bearings, MLAN presents higher transfer performance than other methods, such as conventional deep learning and typical transfer learning. The pretty diagnosis results are from the strong correction of MLAN to distribution discrepancy. The transferability between the source and the target is significant for the success of the transfer learning-based method, especially for the tasks with none of the labeled target data. However, the transferability assessment is still an open problem in the field, which remains an important issue in further works.

4.4 Deep Partial Adaptation Network for Domain-Asymmetric Fault Diagnosis

4.4.1 Motivation

The applications of deep transfer learning to mechanical system fault diagnosis are usually subject to a common assumption of domain symmetry, by which there are two constraints on the source and target domains. First, the number of target samples in every health state of mechanical systems is balanced to each other. Second, the target domain always requires the diagnosis knowledge that is consistent with the one from the source domain. Such requirements are unrealistic for real-case scenarios (Yang et al. 2021a). For the first one, mechanical systems in real-world applications work normally at most of the phases in the long-term useful life cycle. Consequently, the number of fault data is much less than the data collected from the normal state. In terms of the second constraint, there is incomplete health information in the target-domain dataset because some typical faults may not happen in the service life of real-case mechanical systems. As a result, partial diagnosis knowledge in the source domain is available for the target domain. With respect to the domain asymmetry both in the sample amount and the health state, the standard deep transfer learning may produce poor diagnosis results.

To explain the above negative effects, it is assumed that the source and the target are subject to the domain asymmetry, which actually exists in the consistent number of health states across domains and the imbalanced degree of sample amount in the target. To be specific for the source domain, sufficient data are balanced across each health state. As for the target domain, the unlabeled data are possibly imbalanced, and their health states are a subset of those of the source. This assumption is derived from the fact that the source could provide desired data bringing sufficient diagnosis knowledge for the target. For example, in a laboratory, we can conveniently simulate various typical faults of laboratory-used mechanical systems, and the collected dataset is easy to be created with balanced samples on health states. According to the illustration in Fig. 4.15, there is domain asymmetry between the source and the target. The common health states are called the shared states, while the states that are only experienced in the source are the outlier health states. In addition, the target sample amount drawn from the normal state is larger than that from the fault. Through the

4.4 Deep Partial Adaptation Network for Domain-Asymmetric Fault Diagnosis

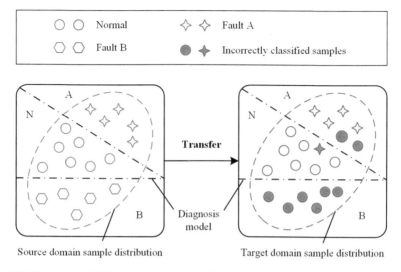

Fig. 4.15 Illustration of failed transfer diagnosis with the effects of domain asymmetry

conventional transfer learning methods, some target samples in the normal state will be aligned with the source samples in the states of Faults A and C. This leads to severe misdiagnosis under the knowledge transfer from the source to the target.

To fertilize the applications of deep transfer learning on fault diagnosis with respect to domain asymmetry, this section presents an importance-weighted distance metric to evaluate the difference of cross-domain sample distributions in presence of domain asymmetry. Based on the distance metric, a deep partial transfer learning network (DPTLN) is further constructed to adapt partial distributions of the source and the target, selectively transferring diagnosis knowledge across domains (Yang et al. 2021a).

4.4.2 Deep Partial Transfer Learning Net-Based Diagnosis Model

4.4.2.1 Problem Formularization

Denote by D^s the source domain. The source dataset X^s has n_s labeled samples by following the distribution P_s. The label space is $Y^s = \{1, 2, \cdots, R\}$, which includes R health states. A target domain D^t has n_t unlabeled samples with the distribution P_t, and $P_t \neq P_s$. Note that the target label space is assumed to be a subset of the source, i.e., $Y^t \subset Y^s$. Such the assumption makes the source cover the diagnosis knowledge required by the target, and thus facilitates a successful transfer across domains. The intersection of the cross-domain label spaces $Y^t \cap Y^s$ contains the samples from the shared health states, which provide invalid diagnosis knowledge. The difference Y^s/Y^t consists of samples from the outlier health states. Moreover, the source sample amount is balanced on every health state due to the controlled settings in the source domain. On the other hand, the target sample amount is imbalanced across all the health states. In presence of the above asymmetric label spaces and sample amount across the source and the target, the partial transfer learning is to reduce the distribution discrepancy $P_t \neq P_s$ so that the diagnosis model for the source is expected to work at the target (Li et al. 2020c).

To achieve the aforementioned purpose, a promising solution of partial transfer learning is to reweigh the source samples in the procedure of distribution adaptation. This is explained by Fig. 4.16. As shown in Fig. 4.16(a) the conventional transfer learning methods give equal weights to samples drawn from the source domain and the target domain. With the effects of such weights, the source samples from the outlier health states, such as the Faults A and C, will provide invalid information for the distribution adaptation. This is the reason why the target samples are misaligned towards the source outlier states, producing a low diagnosis accuracy with the model transferred across domains. As for the partial transfer learning, the source samples are reweighed, and one possible condition is illustrated in Fig. 4.16(b). It is found that the source samples from the outlier states get very small weights so as to block the negative effects on distribution adaptation. Moreover, the source from the state with massive target samples, i.e., the normal states, achieve larger weights than those from the state with limited samples, i.e., the Fault B. As a result, the target distribution and the partial source distribution can be adequately adapted to make a successful diagnosis knowledge transfer.

4.4.2.2 Domain Asymmetry Factors Weighted MMD

The generative adversarial network (GAN) can be used to assess the asymmetry degree between the source and the target domains. The optimization objective of Wasserstein GAN is shown as follows (Arjovsky et al. 2017):

$$\min_{G} \max_{f_D} E_{x \sim P_s}[f_D(G(x))] - E_{x \sim P_t}[f_D(G(x))] \tag{4.28}$$

where $G(\cdot)$ means the generator function, and $f_D(\cdot)$ is a domain discriminator that removes the activation function in the output layer. Wasserstein GAN games the generator and the discriminator through Eq. (4.28) to make the cross-domain samples undistinguished. When the training parameters of $G(\cdot)$ are frozen, the optimal discriminator can output (Yang et al. 2021a):

$$f_D^*(z) = \arg\max_{f_D} \int_z [P_s(z) - P_t(z)] f_D(z) dz \tag{4.29}$$
$$\text{s.t. } \|f_D\|_L \leq K$$

4.4 Deep Partial Adaptation Network for Domain-Asymmetric Fault Diagnosis

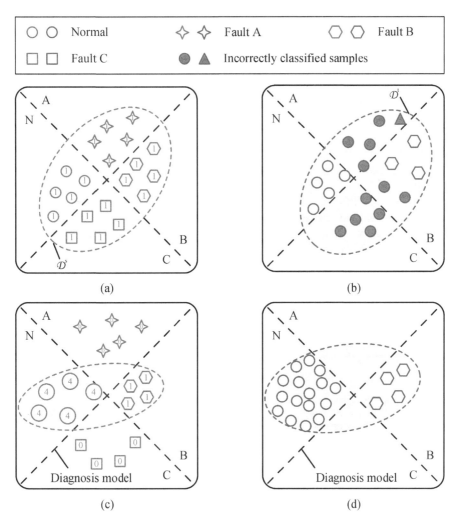

Fig. 4.16 The basic principles of: (a) source sample weights by conventional transfer learning, (b) conventional transfer learning on the target samples, (c) source sample weights by partial transfer learning, and (d) partial transfer learning on the target samples

where $z = G(x) = \{z^s, z^t\}$ are the learned features from the source and the target, f_D is bounded with the respect to the Lipschitz continuity, and $\|f_D\|_L \leq K$ means that the output of the discriminator holds the K-Lipschitz continuity and the Lipschitz constant is K (Arjovsky et al. 2017). According to Eq. (4.29), we can infer the relationship between the cross-domain probability density and the optimal output of the discriminator. If the source samples are from the health states with a large target sample amount, there is $P_s(z^s) > P_t(z^s)$, and the optimal output $f_D^*(z^s)$ tends to the upper bound of f_D. On the contrary, there is $P_s(z^s) < P_t(z^s)$ when the source samples from the states that have a limited amount of target samples, and thus $f_D^*(z^s)$

approaches to the lower bound of f_D. The sigmoid function is further used to clip the $f_D^*(z^s)$ into the range of (0, 1). The factor $\sigma_s[f_D^*(z^s)]$ depicts the asymmetry degree across the source and the target domain samples. To be specific, the large value means that the source samples are drawn from the states that have limited target samples, and the small value indicates the opposite.

Based on the above two conditions, we can construct the domain-asymmetry factors for the source samples by

$$W^s = \frac{1 - \sigma_s[f_D^*(z^s)]}{E_{P_s}\left[1 - \sigma_s[f_D^*(z^s)]\right]} \quad (4.30)$$

The very small factors are given to the source samples drawn from the outlier health states, and the source samples in the shared states will get large values. The more target sample amount is in a certain shared state, the larger values are given to the source samples from this state.

The domain asymmetry factors can be further used to weight the calculation of MMD shown in Eq. (4.7). The importance weighted MMD is inferred as

$$\hat{D}_H^2\left(W^s z^s, z^t\right)$$
$$= \frac{1}{n_s^2} \sum_{i=1}^{n_s} \sum_{j=1}^{n_s} W_i^s W_j^s k\left(z_i^s, z_j^s\right) - \frac{2}{n_s n_t} \sum_{i=1}^{n_s} \sum_{j=1}^{n_t} W_i^s k\left(z_i^s, z_j^t\right) + \frac{1}{n_t^2} \sum_{i=1}^{n_t} \sum_{j=1}^{n_t} k\left(x_i^t, x_j^t\right)$$
(4.31)

where $W^s = \{W_i^s | i = 1, 2, \cdots, n_s\}$ represents the domain-asymmetry factors that are assigned to the source samples. The commonly used kernel functions in Eq. (4.21) includes the Gaussian kernels and polynomial kernels. According to the reported research (Yang et al. 2020; Zellinger et al. 2017), the Gaussian kernel-induced MMD has three shortcomings. First, it requires many computation resources due to the high time complexity, especially for massive data. Second, it calculates the mean distance of the given two distributions in the RKHS, but the calculation does not add the results on the high-order moments. Third, the change of the kernel width greatly affects the measurement results. Therefore, the polynomial kernel-induced MMD is preferred to assess the distribution difference of the given samples from the source and the target. Given the polynomial kernel functions as

$$k(x, y) = \left(ax^T y + b\right)^c, c = 1, 2, \cdots \quad (4.32)$$

where a, b, and c are respectively the slope, the intercept, and the order of the polynomial kernels. When the RKHS is built by using the polynomial kernels, Eq. (4.31) can be inferred as follows (Yang et al. 2021a):

4.4 Deep Partial Adaptation Network for Domain-Asymmetric Fault Diagnosis

$$\hat{D}_{\text{P-H}}^2\left(\boldsymbol{W}^s \boldsymbol{z}^s, \boldsymbol{z}^t\right)$$
$$= \sum_{q=0}^{c} \binom{c}{q} a^q b^{c-q} \left[\frac{1}{n_s^2} \sum_{i=1}^{n_s} \sum_{j=1}^{n_s} W_i^s W_j^s \left\langle \boldsymbol{z}_i^s, \boldsymbol{z}_j^s \right\rangle^q - 2 \sum_{i=1}^{n_s} \sum_{j=1}^{n_t} W_i^s \left\langle \boldsymbol{z}_i^s, \boldsymbol{z}_j^t \right\rangle^q + \sum_{i=1}^{n_t} \sum_{j=1}^{n_t} \left\langle \boldsymbol{x}_i^t, \boldsymbol{x}_j^t \right\rangle^q \right] \tag{4.33}$$

The domain-asymmetry factor performs to assess the partial distribution discrepancy of the shared health states when weighted the source samples. The contributions of source samples in outlier states are removed by using very small weights. The factor weighted MMD further assesses the distribution discrepancy between the balanced source and the imbalanced target as accurate by the adaptively changed weights for the source.

4.4.2.3 Diagnosis Model Architecture

The DPTLN is presented to adapt partial distributions of the source and the target data, and then uses the diagnosis knowledge in the source to recognize the health states of the target. The architecture of DPTLN is shown in Fig. 4.17 (Yang et al. 2021a). It consists of three parts: a domain-shared deep ResNet, a domain discriminator, and an importance weighted distribution adaptation module. The domain-shared ResNet is to represent the deeper-layer features from both the source and target domain samples. The learned features are then pushed into the domain discriminator to assess the domain asymmetry degree. The discriminator further produces the domain-asymmetry factors, which weights the MMD-based distribution adaptation of the learned features. The above three parts are detailed in the following subsections.

(1) Domain-shared ResNet

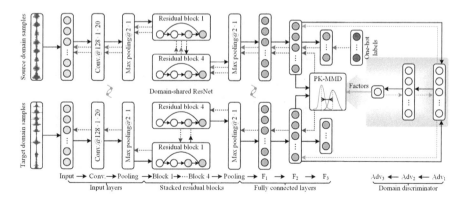

Fig. 4.17 The architecture of DPTLN

The domain-shared ResNet is used to simultaneously handle the samples from the source and the target domains. Noted that ResNet shares the training parameters for the domains. Table 4.8 lists the architecture parameters of the domain-shared ResNet. It includes the input layer, the stacked residual blocks, and the fully connected layers.

Given the source and target domain samples $x_i^D = \{x_i^s, x_i^t\} \in \mathbb{R}^N$ with N dimension. The cross-domain samples are pushed into the input layer, in which a convolutional layer is first used to handle them through

$$x_i^{D,\text{Inp}} = f_{\text{Inp}}(x_i^D; \theta^{\text{Inp}}) = \sigma_r(x_i^D * k^{\text{Inp}} + b^{\text{Inp}}) \quad (4.34)$$

where $\theta^{\text{Inp}} = \{k^{\text{Inp}}, b^{\text{Inp}}\}$ represents the training parameters in the input layer. To reduce the number of training parameters, the dimension of the output features is then reduced by a max-pooling layer, and then obtain

$$v_i^{D,\text{Inp}} = \text{down}(x_i^{D,\text{Inp}}, s) \quad (4.35)$$

where $\text{down}(\cdot)$ is the down-sampling function shown in Eq. (4.16), and s is the parameter controlling the output feature dimension.

To represent the deeper layer features, multiple residual blocks are stacked to process the output of the features from the input layer. Figure 4.18 presents the diagram of the residual block (He et al. 2016).

Table 4.8 Architecture parameters of the domain-shared ResNet

Layers	Parameter size	Output Size	Activation
Input	—	N	—
Convolutional	$128 \times 1 \times 20$ (Zero padding)	$N \times 20$	ReLU
Max-pooling	2	$N/2 \times 20$	—
Residual blocks	$2 \times$ Conv.: $3 \times 20 \times 20$ (Zero padding)	$N/2 \times 20$	ReLU
Max-pooling	2	$N/4 \times 20$	—
F_1	—	$5N$	—
F_2	$5N \times 256$	256	Sigmoid
F_3	$256 \times R$	R	Softmax
Adv_1	256×128	128	ReLU
Adv_2	128×128	128	ReLU
Adv_3	128×1	1	—

4.4 Deep Partial Adaptation Network for Domain-Asymmetric Fault Diagnosis

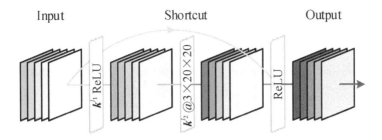

Fig. 4.18 Diagram of the residual block

In a residual block, two convolutional layers are used to process the learned features from the previous block by

$$f_{RB}\left(x_i^{D,l-1}; \theta^l\right) = \left[\sigma_r\left(x_i^{D,l-1} * k^{l_1} + b^{l_1}\right)\right] * k^{l_2} + b^{l_2} \quad (4.36)$$

where $\theta^l = \{k^{l_1}, k^{l_2}, b^{l_1}, b^{l_2}\}$ means the training parameters of the lth residual block. The output of the lth residual block can be expressed as follows:

$$x_i^{D,l} = \sigma_r\left[f_{RB}\left(x_i^{D,l-1}; \theta^l\right) + x_i^{D,l-1}\right], l = 1, 2, \cdots, L \quad (4.37)$$

where $x_i^{D,0} = v_i^{D,\text{Inp}}$. From the above equation, the residual block adopts a smart shortcut connection to calculate the sum of the features respectively before and after the process of the stacked convolutional layers. Such a framework releases the performance degradation of networks in presence of stacking many layers (He et al. 2016).

The fully connected layers map the output of the features from the stacked residual blocks into the label space. The input of layer F_1 is the features $x_i^{D,L}$, which are handled by another max-pooling process. The produced features are then reshaped as a one-dimensional vector. The hidden layer F_2 outputs the features of the deepest layer before classification. Layer F_3 uses the Softmax function to predict the probability of the learned features when they are classified into certain health states in the source. Denote by $f_{FC}(\cdot)$ the nonlinear mapping of the fully connected layers. The output of the fully connected layers can be expressed as

$$P\left(y_i^D = j\right) = f_{FC}\left(v_i^{D,L}; \theta^{FC}\right) \quad (4.38)$$

where $P(y_i^D = j)$ is the probability of the ith cross-domain samples classified into the jth health state, and $v_i^{D,L}$ is the input one-dimensional features of layer F_1, and θ^{FC} is the training parameters in the fully connected layers. The labeled source samples train the domain-shared ResNet by minimizing the cross-entropy loss L_{clf} shown in Eq. (4.19) so that the model can learn the source diagnosis knowledge.

(2) Domain discriminator

The perceptron with three layers serves as the domain discriminator, and the architecture parameters are shown in Table 4.8. The input of the domain discriminator is the domain features of layer F_2. Let $f_{\text{Adv}}(\cdot)$ be the nonlinear mapping that is served by the domain discriminator, and the output of the layer Adv_3 is

$$x_i^{D,\text{Adv}_3} = f_{\text{Adv}}\left(x_i^{D,F_2}; \theta^{\text{Adv}}\right) \tag{4.39}$$

where θ^{Adv} represents the training parameters of the domain discriminator. Note that the domain discriminator removes the activation of layer Adv_3. According to Eq. (4.28), the Wasserstein loss below is maximized to train the discriminator.

$$L_{\text{Adv}} = \frac{1}{n_s}\sum_{i=1}^{n_s} x_i^{s,\text{Adv}_3} - \frac{1}{n_t}\sum_{j=1}^{n_t} x_j^{t,\text{Adv}_3} \tag{4.40}$$

After a certain training iteration of the domain discriminator, the domain-asymmetry factors can be calculated through Eq. (4.30) and obtained as

$$W_i^s = \frac{1 - \sigma_s\left(x_i^{s,\text{Adv}_3}\right)}{\frac{1}{n_s}\sum_{i=1}^{n_s}\left[1 - \sigma_s\left(x_i^{s,\text{Adv}_3}\right)\right]} \tag{4.41}$$

This factor is further used to weight source samples, which is achieved in the following module to adapt the distribution.

(3) Importance weighted distribution adaptation module

In presence of the domain asymmetry, the distribution discrepancy of the learned domain features in layer F_2 is estimated by using the factors weighted MMD. The MMD is calculated by using the polynomial kernels. According to Eq. (4.33), the distribution discrepancy with respect to the domain asymmetry can be calculated by:

$$L_{\text{DA}} = \hat{D}_{\text{P-H}}^2\left(W^s X^{s,F_2}, X^{t,F_2}\right) \tag{4.42}$$

where X^{s,F_2} and X^{t,F_2} collect the learned features of layer F_2 respectively in the source domain and the target domain.

(4) Training process

The training process of DPTLN includes two steps. First, the Wasserstein loss shown in Eq. (4.40) is maximized to train the discriminator:

$$\max_{\theta^{\text{Adv}}} L_{\text{Adv}} \tag{4.43}$$

4.4 Deep Partial Adaptation Network for Domain-Asymmetric Fault Diagnosis

The training parameters θ^{Adv} must follow the K-Lipschitz continuity during the training process, which is from the basic principle of Wasserstein GAN. When this condition does not hold, the exploding gradient will be produced and an unstable discriminator will be trained. Therefore, the training parameters are clipped into $[-\xi, \xi]$ after each training iteration of the discriminator parameters (Arjovsky et al. 2017). The gradients generated by Eq. (4.43) individually train the discriminator parameters so as to generate the domain-asymmetry factors, and are not propagated towards the domain-shared ResNet. The opposite corrects the Wasserstein distance of the learned features without weighting the source (Arjovsky et al. 2017). As a result, the model performance is still subject to domain asymmetry. Besides, the reversed gradients for the ResNet have a contradictory performance with the constructed importance weighted adaptation module.

Second, the domain-shared ResNet is optimized by the objective below to adapt partial distributions of the source and target features:

$$\max_{\theta} L_{\text{clf}} + \lambda \cdot L_{\text{DA}} \tag{4.44}$$

where $\theta = \{\theta^{\text{Inp}}, \theta^{l=1:L}, \theta^{\text{FC}}\}$ represents the training parameter set of the domain-shared ResNet, and λ means the tradeoff parameter to balance the source classification loss and the distribution adaptation. Equations (4.43) and (4.44) are implemented by the RMSProp optimization algorithm in turn, and the parameters are updated by

$$\begin{cases} \theta^{\text{Adv}} \leftarrow \theta^{\text{Adv}} + \eta_{\text{Adv}} \cdot \nabla_{\theta^{\text{Adv}}} (L_{\text{Adv}}) \\ \theta \leftarrow \theta - \eta_{\text{da}} \cdot \nabla_{\theta} (L_{\text{clf}} + \lambda \cdot L_{\text{DA}}) \end{cases} \tag{4.45}$$

To be specific, the training of DPTLN is detailed in Algorithm 4.3.

Algorithm 4.3. Mini-batch training of DPTLN by using RMSProp.

Input: Labeled source data X^s, unlabeled target data X^t.

Output: Diagnosis results of unlabeled target data.

Initialize parameters θ and θ^{Adv}. Set the tradeoff parameter λ.

While the parameters have not converged do

> **For** iteration $n = 1, 2, \cdots, n_{Adv}$ **do**
>
>> 1. Draw a batch with m samples from the source and the target
>> / Feed-forward propagation is omitted /
>> 2. Calculate the gradient $g_{\theta^{Adv}} \leftarrow \nabla_{\theta^{Adv}}(L_{Adv})$
>> 3. Update the parameters $\theta^{Adv} \leftarrow \theta^{Adv} + \eta_{Adv} \cdot \text{RMSProp}(\theta^{Adv}, g_{\theta^{Adv}})$
>> 4. $\theta^{Adv} \leftarrow \text{clip}(\theta^{Adv}, -\xi, \xi)$
>
> **End**
>
> 5. Draw another batch-size sample from the target.
> 6. Calculate the domain-asymmetry factors for the source by Eq. (4.27).
> 7. Obtain the gradient $g_\theta \leftarrow \nabla_\theta(L_{clf} + \lambda \cdot L_{DA})$
> 8. Update training parameters $\theta \leftarrow \theta - \eta_{da} \cdot \text{RMSProp}(\theta, g_\theta)$

End

Return the predicted labels of target data as the diagnosis results.

4.4.3 Partial Transfer Diagnosis of Gearboxes with Domain Asymmetry

4.4.3.1 Dataset Description

A partial transfer diagnosis of the gearbox is created to verify the DPTLN. The task is expected to improve the model performance against the varying working conditions

4.4 Deep Partial Adaptation Network for Domain-Asymmetric Fault Diagnosis

with the settings of domain asymmetry. More details of the datasets are detailed in Table 4.9.

Datasets PGearbox-A and PGearbox-B are obtained from a powertrain dynamic simulator device (Jia et al. 2016). This device consists of a motor, a planet gearbox, a fixed-shaft gearbox, and a magnetic powder brake. The motor drives the magnetic powder brake by the power transmission through the fixed-shaft gearbox and the planet gearbox. In the first stage of the planet gearbox, five types of the gear health states are simulated, which are the normal state, the tooth crack in the sun gear (CS), the wear in the sun gear (WS), the tooth CPG, and the bearing fault in the planet gear (BFP). During the test, datasets PGearbox-A and PGearbox-B are collected under the motor speed settings of 2100 r/min and 3000 r/min respectively, and at the sampling frequency of 5120 Hz. A total of 1000 samples are included in the datasets, and each sample contains 2560 sample points. For PGearbox-A, the samples are balanced across all the health states, but for PGearbox-B, the number of fault samples is controlled to be imbalanced by $m\% \in (0, 1)$ and part of fault types are missing.

The transfer diagnosis tasks created by datasets in Table 4.9 are expected to transfer diagnosis knowledge of PGearbox-A to PGearbox against the discrepancy from the change of motor speed. Table 4.10 presents the created three partial transfer diagnosis tasks. The target label space is the subset of the source. It is noted that 30% of target fault samples are randomly selected to make the target dataset imbalanced. The DPTLN is demonstrated on the tasks T_1, T_2, and T_3 at the increasing domain asymmetry degree since the settings increase the number of the removed target health states gradually.

Table 4.9 Details of planet gearbox datasets

Datasets	Health states	Sample amount	Working conditions
PGearbox-A	N	1000 (200 × 5)	2100 r/min
	CS		
	WS		
	CPG		
	BFP		
PGearbox-B	N	200	3000 r/min
	CS	800 (Note that partial fault states are missing and $m\%$ is used to control the number of available fault samples)	
	WS		
	CPG		
	BFP		

Table 4.10 Partial transfer learning tasks

Tasks	T_1	T_2	T_3
Target domain (PGearbox-B)	N, WS, CPG, BFP	N, WS, BFP	N, BFP
Source domain (PGearbox-A)	N, CS, WS, CPG, BFP		

4.4.3.2 Parameter Analysis

The performance of DPTLN is affected by two parameters, i.e., the clipped range $[-\xi, \xi]$ for domain discriminator parameters and the tradeoff parameter λ. The parameters are analyzed by implementing DPTLN on task T_1. Search for the truncated range parameter ξ from $\{0.0001, 0.001, 0.01, 0.1, 1\}$ with the setting of $\lambda = 1$. The learning rate of the RMSProp algorithm is set as 0.0005, and the value is changed by following the exponential decay. Under different truncated ranges, the log-normalized gradients $\log \left\| \nabla_{\theta^{\text{Adv}}} (L_{\text{Adv}}) \right\|$ are recorded in Fig. 4.19(a) With the larger truncated ranges, such as $\xi = 1$ and $\xi = 0.1$, severely exploding gradients can be found, which may result in difficult convergence of the domain discriminator. The shrinking ranges, such as $\xi = 0.001$ and $\xi = 0.0001$, will produce vanishing gradients in the training. Therefore, the truncated range is set as $[-0.01, 0.01]$. In addition, search for the tradeoff parameter λ from the range $\{0.01, 0.05, 0.1, 0.5, 1, 5, 10, 50\}$, and Fig. 4.19(b) presents the total loss $L_{\text{clf}} + L_{\text{DA}}$ of different λ settings. The small λ results in the high loss because the model cannot adapt to the distribution of learned features adequately. With the increase of λ, the total loss decreases, and thus improve the transfer diagnosis accuracy. However, the very large settings, such as $\lambda = 10$ and $\lambda = 50$, remove the benefits because the large loss of distribution adaptation hinders the training process of the domain-shared classifier. As a result, the classifier will misclassify some target samples.

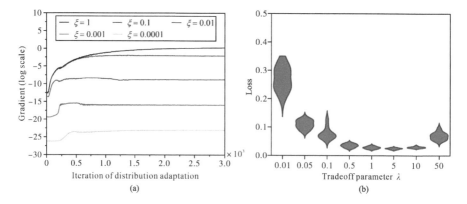

Fig. 4.19 Gradient and loss under different parameter settings: (a) normalized gradient of the domain discriminator under different truncated range, and (b) box-plot of the total loss $L_{\text{clf}} + L_{\text{DA}}$ under different tradeoff parameters λ

4.4.3.3 Analysis of Domain-Asymmetry Factors

The domain-asymmetry factors are automatically learned by DPTLN, which further affect the performance of the model on partial transfer diagnosis tasks. For the given tasks T_1, T_2, and T_3, Fig. 4.20 plots the learned domain-asymmetry factors and the confusion matrix of diagnosis results. As the results of Fig. 4.20(a) the source sample in the CS state are judged as the outlier health states because there is no sample in the state. The DPTLN assesses the domain-asymmetry degree. Very small factors are given to the source samples in CS. Furthermore, the target sample amount in the normal state is larger than that of the fault. DPTLN further assigns larger weights to the source samples in the normal states than those from the faults. Through a distribution adaptation weighted by the domain-asymmetry factors, DPTLN achieves high diagnosis accuracy on the target. For tasks T_2 and T_3, part of fault types is gradually removed from the target. DPTLN can adequately change the weights for the source to block the negative effects of outlier health states (CS and CPG states in Fig. 4.20(b) and CS, WS, CPG states in Fig. 4.20(c) in distribution adaptation. The weights for the source samples in the normal state are increased when the fault types in the target are gradually removed. The high diagnosis accuracies are presented on the two tasks.

4.4.3.4 Comparison with Other Methods

The performance of DPTLN is compared with other methods. They are the standard deep transfer learning methods including ResNet, P-ResNet (Yang et al. 2020), and WDANN, and the partial transfer learning methods including SAN (Cao et al. 2018) and IWAN (Zhang et al. 2018). For fair comparisons, all the methods are conducted with the same domain-shared ResNet architecture, which is shown in Table 4.8. The training parameters including the tradeoff parameters and the learning rate are chosen to make the methods have the optimal performance. The ResNet is the method without a distribution adaptation module. The source samples are used to train a domain-shared ResNet, and the performance of the model is tested on the target samples directly. P-ResNet adapts the distributions of the learned domain features of layer F_2 by using the polynomial kernel-induced MMD. WDANN constructs the Wasserstein GAN architecture to cope up with the transfer diagnosis tasks T_1, T_2, and T_3. According to the reported research, this method reduces the Wasserstein distance of the learned cross-domain features. SAN has been used in image recognition tasks. In this method, the class-wise domain discriminators (one discriminator for one health state in the source) are used rather than a single discriminator for the standard GAN. IWAN learns the domain-asymmetry factors to weigh the adversarial loss of the standard GAN. Due to the imbalanced sample amount in the target, the imbalanced classification metrics, i.e., the F-score, the mean average precision (mAP), and the area under the receiver operation characteristic curve (AUC) (Fawcett 2006), are used to present the diagnosis results of the target data. Each method is repeated in

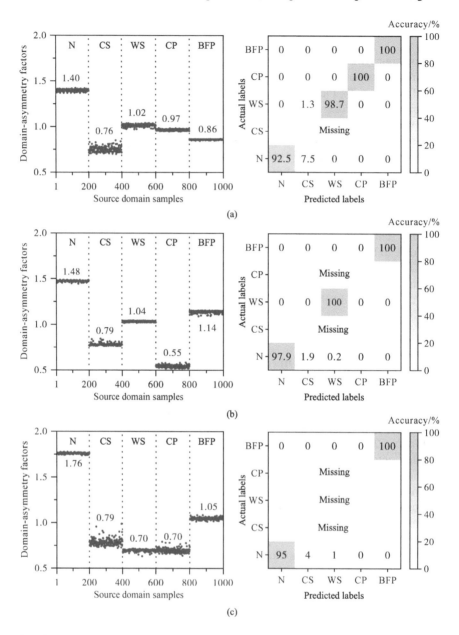

Fig. 4.20 Domain-asymmetry factors and confusion matrix on partial transfer learning tasks: (a) T_1, (b) T_2, and (c) T_3

4.4 Deep Partial Adaptation Network for Domain-Asymmetric Fault Diagnosis

ten trials, and the statistical results of the diagnosis accuracies are recorded in Table 4.11.

From the results shown in Table 4.11, DPTLN obtains the highest imbalance classification metrics among the listed methods. Besides, the standard deviation is small, which means the stability of DPTLN on tasks with imbalanced target samples. There is no distribution adaptation module in the method-ResNet. The distribution discrepancy of the learned cross-domain features reduces the diagnosis performance of the model on given tasks. P-ResNet gives the same weights to all the source samples in the distribution adaptation procedure, which ignores the negative effects of domain asymmetry. As a result, the diagnosis performance is improved but lower than partial transfer learning methods. WDANN can reduce the Wasserstein distance of the learned features. In this method, the problem of domain asymmetry is not addressed, and thus similar results to P-ResNet are obtained. SAN constructs class-wise discriminators to achieve partial transfer learning tasks under the guide of pseudo labels assigned to the target samples. If the pseudo labels are not actual truth, SAN will perform poorly on the tasks. This is the possible reason why SAN gets lower results than DPTLN. IWAN overcomes the weaknesses of SAN, which learns the domain-asymmetry factors to weigh the training of the standard GAN. Thus, the diagnosis performance is improved by overcoming the negative effects of domain asymmetry. However, the standard GAN suffer from unstable gradients, which may lower the diagnosis results of IWAN. DPTLN trains the domain discriminator by adopting the Wasserstein loss. The generated domain-asymmetry factors further weigh source samples for distribution adaptation. Such a diagnosis framework presents higher results than IWAN.

Table 4.11 Statistical diagnosis results of imbalanced target data on tasks T_1, T_2, and T_3 [mean (standard deviation)]

Methods	Task T_1			Task T_2			Task T_3		
	F-score	mAP	AUC	F-score	mAP	AUC	F-score	mAP	AUC
ResNet	0.265 (0.112)	0.486 (0.033)	0.545 (0.042)	0.121 (0.026)	0.418 (0.005)	0.499 (0.007)	0.204 (0.013)	0.388 (0.015)	0.518 (0.016)
P-ResNet	0.864 (0.029)	0.829 (0.016)	0.823 (0.014)	0.845 (0.019)	0.779 (0.014)	0.778 (0.011)	0.745 (0.017)	0.666 (0.022)	0.683 (0.014)
WDANN	0.872 (0.027)	0.824 (0.018)	0.813 (0.025)	0.853 (0.062)	0.815 (0.068)	0.781 (0.078)	0.709 (0.078)	0.657 (0.063)	0.661 (0.057)
SAN	0.715 (0.071)	0.871 (0.106)	0.862 (0.098)	0.619 (0.108)	0.826 (0.129)	0.817 (0.109)	0.601 (0.224)	0.526 (0.265)	0.624 (0.211)
IWAN	0.954 (0.033)	0.951 (0.041)	0.942 (0.048)	0.912 (0.087)	0.941 (0.033)	0.934 (0.038)	0.872 (0.109)	0.834 (0.145)	0.829 (0.143)
DPTLN	0.975 (0.014)	0.963 (0.026)	0.954 (0.033)	0.982 (0.017)	0.972 (0.023)	0.966 (0.034)	0.971 (0.025)	0.959 (0.033)	0.951 (0.041)

To visually explain the performance of DPTLN, the similarity matrix of the learned features of layer F_2 on task T_1 is plotted in Fig. 4.21. The similarity matrix is calculated by using $G(x_i^s, x_j^t) = \exp(-\|x_i^s - x_j^t\|^2 / 100)$. As shown in Fig. 4.21(a), the learned features of the target are dissimilar to those of the source. Thus, the source classifier will misclassify most of the target samples. P-ResNet can reduce the distribution discrepancy of learned features. Figure 4.21(b) shows that the cross-domain features in the states WS, CPG, BFP are well adapted. However, due to the effects of the outlier states in the source, i.e., CS state, the method makes target samples from the normal state similar to the source samples in states CS and WS. Similar results can be found in Fig. 4.21(c) for WDANN. SAN is failed to learn domain features with similar distributions in the state WS, as shown in Fig. 4.21(d), and thus the lower results are obtained than DPTLN. As for IWAN, the similarity of feature distributions in most of the health states is improved as shown in Fig. 4.21(e), Only the features of the state WS are misaligned. From Fig. 4.21(f), DPTLN obtains the domain features with high distribution similarity so that the highest results are obtained among the six methods.

4.4.4 Epilog

The deep transfer learning applications in engineering scenarios are commonly affected by the domain asymmetry, where the target health state set is a subset of the source and the target sample amount is imbalanced across every state. For addressing these issues, the DPTLN-based transfer diagnosis model is presented. The core of DPTLN is to evaluate the domain asymmetry degree by training the domain discriminator separately. As a result, the obtained domain-asymmetry factors can distinguish whether the input domain features are from the outlier health states or the shared ones. In addition, according to the imbalanced degree of the target samples, the factors can be automatically changed. In the distribution adaptation procedure, the source samples are weighed by the learned domain-asymmetry factors so that the negative effects of the source samples in the outlier states can be blocked, improving the partial distribution of features adequately. Compared with conventional deep transfer learning and existing partial transfer learning methods, DPTLN presents the better performance of tasks against the domain asymmetry problem. The diagnosis knowledge from different source domains may produce different transfer performances on the target. To obtain the optimal diagnosis results on the target, it is necessary to select or create a source domain bringing the most useful knowledge for the target. This will be an important work in the future.

4.4 Deep Partial Adaptation Network for Domain-Asymmetric Fault Diagnosis 143

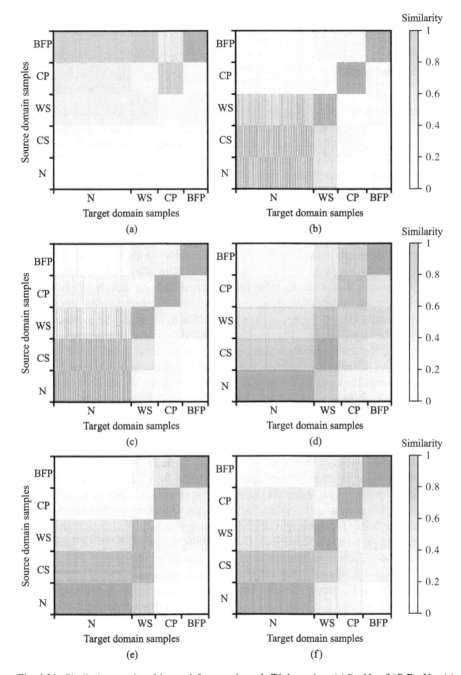

Fig. 4.21 Similarity matrix of learned features in task T1 by using: (a) ResNet, (b) P-ResNet, (c) WDANN, (d) SAN, (e) IWAN, and (f) DPTLN

4.5 Instance-Level Weighted Adversarial Learning for Open-Set Fault Diagnosis

4.5.1 Motivation

In the current literature, the transfer learning method has been widely used to solve the problem of fault diagnosis under different working conditions. However, one of the main concerns of most transfer learning methods is the same label space in different domains. This means that training and testing data should include the same set of mechanical system failure modes. In this case, fault diagnosis knowledge can usually be easily transferred from the source domain to the target domain.

However, in actual industries, it is usually difficult to meet this condition. Collecting labeled data can be very expensive and difficult, and for model training purposes, acquiring data under all mechanical system conditions is usually not a good idea. This results in the source domain often including limited health states. On the other hand, for the target domain, there are usually invisible faults in the mechanical system testing scenario, and these faults have never been included in the source domain. Therefore, in practice, there is usually a mismatched domain label space. More specifically, the label space of the source domain can usually be included in the label set of the target domain. This problem is called the open set migration learning problem, and the intelligent fault diagnosis for this problem is called the open set fault diagnosis (Zhang et al. 2021a). Take bearing fault diagnosis for example. The normal state and the outer race fault can be included in the source domain data for training with a rotating speed of 1000 r/min. In the model testing case, other than the two health states, the inner race fault can occur in the target domain, which is new to the source, and is supposed to be identified as the unknown state (Zhang and Li 2021).

In the actual industry, however, it is usually difficult to meet this condition. Collecting labeled data can be very expensive and difficult, and for model training purposes, acquiring data under all mechanical system conditions is usually not a good idea. This results in the source domain often including limited health states. On the other hand, for the target domain, there are usually invisible faults in the mechanical system testing scenario, and these faults have never been included in the source domain. Therefore, in practice, there is usually a mismatched domain label space. More specifically, the label space of the source domain can usually be included in the label set of the target domain. This problem is called the open set migration learning problem, and the intelligent fault diagnosis for this problem is called the open set fault diagnosis. Take bearing fault diagnosis as an example. The health and outer loop fault status can be included in the source domain data for training at a speed of 1000 r/min. In the model testing case, in addition to these two health states, the target domain may experience an inner circle failure, which is new to the source and should be identified as an unknown condition (Zhang et al. 2021b).

As illustrated in Fig. 4.22, for the new class of the target domain, the popular transfer learning method cannot handle this problem well. This is because existing

4.5 Instance-Level Weighted Adversarial Learning for Open-Set Fault Diagnosis

methods usually focus on edge data distribution. Using deep neural networks, these algorithms aim to learn a representation subspace in which all high-level features of the source and target domains can be merged together. However, due to the existence of additional classes in the target domain, simply merging all the features together will inevitably lead to cross-domain mismatches (Panareda Busto and Gall 2017). Therefore, model performance will degrade in most cases. In response to this problem, this section introduces instance-level weighted adversarial learning (ILWAL) (Zhang et al. 2021a). This method is expected to accurately identify invisible faults in the target, thereby preventing their influence on the adaptation of the sample distribution from the shared health state. As a result, diagnosis knowledge can be transferred from the source to the target.

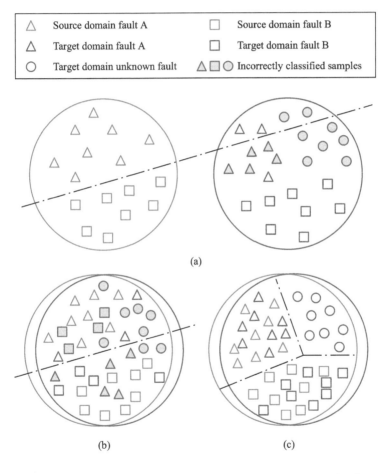

Fig. 4.22 Illustrations of: (a) the fault diagnosis task with open set domain discrepancy, (b) conventional domain adaptation methods, and (c) open set domain adaptation methods

4.5.2 Instance-Level Weighted Adversarial Learning-Based Diagnosis Model

4.5.2.1 Problem Formulation

In this part, the open set problem in fault diagnosis is investigated. Let $X^s = \{(x_i^s, y_i^s) | i = 1, 2, \cdots, n_s\}$ denote data sample and label of the ith source sample, and n_s is the number of the source-domain samples. $X^t = \{x_i^t | i = 1, 2, \cdots, n_t\}$ represent the target domain samples and n_t means the number of the target samples. It can be assumed that the source domain label set, that is denoted as Y^s, is included in the target-domain label set, that is denoted as Y^t. That suggests the source mechanical system health states can be sampled from the known classes. Nonetheless, the target domain may contain additional new classes that are never shown in the source domain. In addition, the domain discrepancy still exists that the source and target data are subject to different distributions.

It is expected that a data-driven fault diagnosis model will be established, which can learn generalized features from different fields for fault diagnosis. It can automatically learn the shared features from the source domain and the target domain, and can classify the target shared class well. The model can also identify unknown faults in the target domain. One of the main settings of this research is that the data should be available for model training. By learning the data of the source and target domains, we can learn generalized high-level features and use them for fault diagnosis. We can also identify the health status of the target abnormal mechanical system.

4.5.2.2 Adversarial Learning in Transfer Diagnosis

In adversarial learning, a feature extractor $G(\cdot|\theta_g)$ can be generally adopted to learn the features of the input data. Next, a domain discriminator $f_D(\cdot|\theta_d)$ can be used to separate the origins of the data from source and target domains. The major process of the adversarial learning lies in that, the feature extractor tries to learn the generalized features from source and target domains which cannot be easily separated. On the other hand, the domain discriminator is expected to accurately identify the origins of the high-level representations of the source and target data. Through the application of the adversarial learning between the feature extractor and domain discriminator, the learned features are supposed to be more and more generalized regarding different domains. Therefore, the domain shift issue can be well solved. In most cases, a classifier module $f_C(\cdot|\theta_c)$ can be attached to the high-level features for the health state prediction. The general loss function of adversarial learning in transfer diagnosis can be expressed as (Ganin et al. 2016):

$$L = L_{\text{clf}} + \lambda L_{\text{adv}} = E_{(x_i, y_i) \in X^s} L_{\text{ce}}[f_C(G(x_i)), y_i] + \lambda \left[E_{x_i \in X^s} \log f_D(G(x_i)) + E_{x_i \in X^t} \log(1 - f_D(G(x_i))) \right] \quad (4.46)$$

where L_{clf} denotes the classification loss of the source samples, L_{adv} represents adversarial loss in order to make the source domain and the target domain undistinguished, and λ denotes the penalty parameter. The training parameters of the net can be optimized as follows:

$$\begin{aligned}(\theta_c^*, \theta_g^*) &= \arg\min_{\theta_c, \theta_g} L(\theta_c, \theta_g, \theta_d^*) \\ \theta_d^* &= \arg\max_{\theta_d} L(\theta_c^*, \theta_g^*, \theta_d)\end{aligned} \quad (4.47)$$

where θ_c^*, θ_g^*, θ_d^* represent the optimal values of θ_c, θ_g, θ_d respectively. θ_c, θ_g, θ_d denote the training parameters of the feature extractor, the classifier and the domain discriminator respectively.

4.5.2.3 Mode Architecture

Figures 4.23 and 4.24 show a general overview of ILWAL in the training and testing processes, as well as the network structures. It includes 4 modules, including feature extractor, state classifier, domain discriminator and anomaly classifier. In the feature extractor block, there are three convolutional layers, and the number of filters is 128, 64, and 32, respectively. Afterwards, a flat layer is used, and a fully connected layer with 128 neurons represents the high-level features of the learned data. The domain discriminator block has two convolutional layers, with 64 and 32 filters in each layer. Next, two fully connected layers with 128 and 2 neurons are used respectively. At the end of the network, the Softmax function is used to classify failure modes.

For the detailed neural network architecture, the state classifier and the anomaly classifier have similar configurations. Two neurons are used at the end of the block to indicate visible and invisible failure modes. For the state classifier, finally use N_c neurons to reflect the number of source domain classes. For all convolution operations

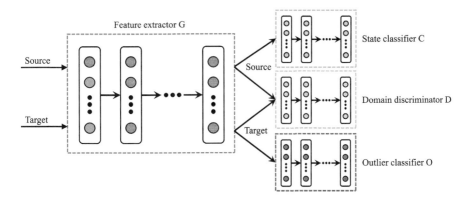

Fig. 4.23 Overview of the ILWAL in the training process

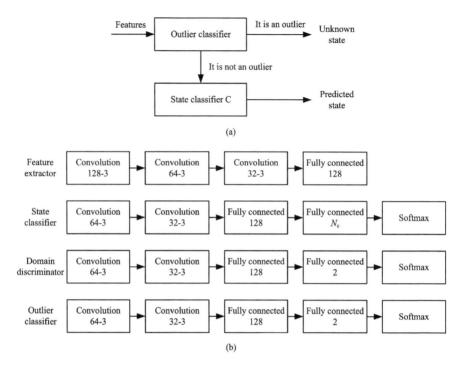

Fig. 4.24 The details of ILWAL: (a) testing process, and (b) network details

in the model, a filter with a size of 3 and a leaky rectified linear unit function is usually used.

Aiming at the domain shift problem with the open set problem, an instance-level weighting method is used for the target domain samples. Specifically, the target domain samples are attached with corresponding weights, reflecting the similarity with the source domain failure mode. Recall that in the adversarial learning framework, the feature extractor aims to learn shared features, while the domain discriminator tries to identify the source of the data. Therefore, for the target domain anomaly classes that are more different from the source class, the domain discriminator can identify them more easily. Correspondingly, the domain prediction error of the domain discriminator can be used to reflect the weight of the target sample.

In this part, the well-established cross-entropy loss is adopted, which measures the errors between the domain label actual truths and the predictions. With respect to the detailed descriptions, the loss function $L_{d,i}^s$ for the ith source instance and $L_{d,i}^t$ for the ith target instance can be expressed as:

4.5 Instance-Level Weighted Adversarial Learning for Open-Set Fault Diagnosis

$$L_{d,i}^{s} = -\sum_{k=1}^{2} I(d_i^s = k) \log \frac{\exp x_{d,i,k}^{s}}{\sum_{m=1}^{2} \exp x_{d,i,m}^{s}}$$
$$L_{d,i}^{t} = -\sum_{k=1}^{2} I(d_i^t = k) \log \frac{\exp x_{d,i,k}^{t}}{\sum_{m=1}^{2} \exp x_{d,i,m}^{t}}$$
(4.48)

where d_i^s and d_i^t denote the source and target domain labels respectively, $x_{d,i,k}^{s}$ and $x_{d,i,k}^{t}$ represent the kth point of the output from the end layer of the domain prediction module for the source and target instances respectively. Using this method, the initial weights can be learned. Let $w_{r,i}$ denote the initial weight of the ith target domain instance, which can be given as $L_{d,i}^{t}$. That indicates the fact that the target-domain instances are supposed to be better separated with low classification errors, and the low weights can be thus used.

In addition, normalization is supposed to be used in order to scale the initial weights. The min–max method is used for normalization:

$$w_i = \gamma \frac{w_{r,i} - \min_r w_{r,i}}{\max_r w_{r,i} - \min_r w_{r,i} + \varepsilon}$$
(4.49)

where w_i represents the normalized value of the initial weight $w_{r,i}$, γ denotes the parameter for scaling, and ε means a small positive value. For the experiments in this study, γ is used as one for convenience. It can be also changed based on specific demands in the application cases.

After the weight item is normalized, the target sample can be added with the weight in the adversarial learning accordingly. A larger weight value indicates that the samples are more likely to come from the health state of the shared mechanical system, while a smaller weight value means that they are more likely to come from the target domain outlier class. Therefore, outliers can be filtered out in adversarial learning to generate cross-domain shared features.

In the model training process, three classification modules are used in the presented instance weighting algorithm. Train a state classifier to classify mechanical system health using source domain label data. The popular cross-entropy loss function is used, and the loss for the ith source sample is denoted as $L_{c,i}$.

The domain discriminator should classify the data according to the original domain. Therefore, it is also a binary classification problem, and the popular cross-entropy loss function is also used to express the error. The loss function for the ith source data is denoted as $L_{d,i}^{s}$, and that for the ith target data is denoted as $L_{d,i}^{t}$.

In addition, the anomaly classifier should accurately classify the anomaly classes in the target domain. However, since this is an unsupervised learning problem, no labeled target domain data is available, and it is impossible to train an outlier classifier in a traditional supervised way. To solve this problem, we suggest using pseudo-outlier labels for samples in the target domain.

As described previously, the domain classification error $L_{d,i}^t$ can be used to reflect the outlier possibilities. The target domain data from the shared class is usually difficult to separate, and large domain prediction errors can be obtained. On the other hand, outliers in the target domain are usually easy to identify, and classification errors are small.

Concretely, in the model training process, for the ranking of the normalized target weights w_i, the samples whose weights are above the $(1-\rho)$ percentile can be regarded as from the shared fault modes across domains. The target samples below the ρ percentile can be considered the outliers. Using this method, we can attach pseudo-outlier labels to unlabeled data from the target domain in order to train the outlier classifier module in a supervised manner. In this method, only a limited part of the target domain data participates in the training process, so the prediction confidence is very strong. In this way, the pseudo labeled data can be used for training, and the cross-entropy loss function $L_{o,i}^t$ is also used in this case.

Although the pseudo-label learning method can be used for outlier recognition, due to noise and uncertainty, the convergence of the model cannot be guaranteed during optimization. In this section, we present an additional technique for unsupervised learning, the principle of entropy minimization (Grandvalet and Bengio 2005). This technique promotes the decision-making boundary to be located in the low-density area of the relevant data by minimizing the entropy of the class condition distribution. To be detailed, the loss function $L_{e,i}^t$ for the ith target sample can be defined as

$$L_{e,i}^t = -\sum_{k=1}^{2} x_{o,i,k} \log x_{o,i,k} \qquad (4.50)$$

By reducing the entropy loss function of all target domain samples, it is possible to train outliers to identify decision-making boundaries in low-density areas of unlabeled data. In this way, the stability of the classifier in outlier classification can be greatly enhanced.

In general, the model optimization objectives in this section can be expressed as

$$\begin{aligned} L_c &= \frac{1}{n_s} \sum_{i=1}^{n_s} L_{c,i}^s \\ L_d &= \frac{1}{n_s} \sum_{i=1}^{n_s} L_{d,i}^s + \frac{1}{n_t} \sum_{i=1}^{n_t} w_i L_{d,i}^t \\ L_e &= \frac{1}{n_t} \sum_{i=1}^{n_t} L_{e,i}^t \\ L_o &= \frac{1}{|S_{\text{pseudo}}|} \sum_{i \in S_{\text{pseudo}}} L_{o,i}^t \end{aligned} \qquad (4.51)$$

where S_{pseudo} represents the set of the target domain instances that are considered in the pseudo label learning process.

The neural network model is updated in order to obtain

$$\begin{aligned} \hat{\theta}_G &= \arg\min_{\theta_G} L_c + \alpha_e L_e + \alpha_o L_o - \alpha_d L_d \\ \hat{\theta}_D &= \arg\min_{\theta_D} \alpha_d L_d \\ \hat{\theta}_C &= \arg\min_{\theta_C} L_c \\ \hat{\theta}_O &= \arg\min_{\theta_O} \alpha_o L_o + \alpha_e L_e \end{aligned} \quad (4.52)$$

where α_d, α_e, and α_o denote the weight parameters for L_d, L_e and L_o respectively. θ_O represents the parameters of the outlier classifier module and $\hat{\theta}_O$ is the optimal value of θ_O. The model training process is carried out using the stochastic gradient descent-based algorithms, and the parameters can be optimized at each epoch as

$$\begin{aligned} \theta_G &\leftarrow \theta_G - \eta \cdot \left(\frac{\partial L_c}{\partial \theta_G} + \alpha_e \frac{\partial L_e}{\partial \theta_G} + \alpha_o \frac{\partial L_o}{\partial \theta_G} - \alpha_d \frac{\partial L_d}{\partial \theta_G} \right) \\ \theta_D &\leftarrow \theta_D - \eta \cdot \alpha_d \frac{\partial L_d}{\partial \theta_D} \\ \theta_C &\leftarrow \theta_C - \eta \cdot \frac{\partial L_c}{\partial \theta_C} \\ \theta_O &\leftarrow \theta_O - \eta \cdot \left(\alpha_o \frac{\partial L_o}{\partial \theta_O} + \alpha_e \frac{\partial L_e}{\partial \theta_O} \right) \end{aligned} \quad (4.53)$$

By adopting the gradient reversal layer, the sign of the gradient can be changed during the optimization process (Li et al. 2020a), and all the parameters of the model can be updated at the same time. Therefore, the open set domain adaptation problem is expected to be well solved. Not only can accurately identify target outliers, but also can classify cross-domain shared failure modes well.

4.5.3 Fault Diagnosis Case of Rolling Bearing Datasets

4.5.3.1 Transfer Diagnosis Tasks of Motor Bearings

(1) Dataset description

The ILWAL is demonstrated on the motor bearing dataset. The details of the datasets are shown in Table 4.12. The motor bearing dataset is supported by CWRU. More details are presented in Chap. 2. In general, 10 bearing health states are included, containing healthy, inner race fault, ball fault and outer race fault. For each

Table 4.12 Details of the motor bearing datasets

Health states	Fault severity/mm	Label
N	—	1
IF	0.18	2
	0.36	3
	0.54	4
RF	0.18	5
	0.36	6
	0.54	7
OF	0.18	8
	0.36	9
	0.54	10

fault location, 3 different levels of severities are considered. The motor speeds of 1730 and 1797 r/min are used in this section for simulating the transfer learning scenario.

According to the datasets shown in Table 4.12, we can create open-set transfer diagnosis tasks that are listed in Table 4.13. A total of 100 labeled instances are used for the source domain fault modes, and 100 unlabeled instances are used for the target-domain fault modes. In the tasks, all the ten health states are considered in the target domain, and the source domain only includes partial classes. The tasks are expected to transfer diagnosis knowledge from the source to the target against different working conditions.

(2) Comparison settings

The ILWAL is verified on the tasks created by the motor bearing dataset. Table 4.14 shows the parameters of the model, which are decided based on the validation task. In the task, the motor speed of 1730 r/min is used as the source domain and the speed

Table 4.13 Open-set transfer diagnosis tasks of the motor bearings

Task	Source to Target	Source Classes
A_1	1730 r/min of the source to 1797 r/min of the target	All classes
A_2		1, 2, 3, 4, 5, 6, 7, 8, 9
A_3		1, 2, 3, 4, 8, 9, 10
A_4		1, 4, 5, 6, 8, 9
A_5		1, 2, 3, 8, 9, 10
A_6		1, 2, 7, 8
A_7		1, 5, 6
A_8	1797 r/min of the source to 1730 r/min of the target	1, 2, 5, 6, 7, 8, 9, 10
A_9		1, 2, 3, 4, 7, 8
A_{10}		1, 8, 9, 10

4.5 Instance-Level Weighted Adversarial Learning for Open-Set Fault Diagnosis

Table 4.14 Parameter settings of ILWAL for Open-set transfer diagnosis

Parameter	Value	Parameter	Value
Batch size	256	Epochs	5000
α_d	1	ρ	0.1
α_e	1	γ	1
α_o	1	N_{input}	256

of 1750 r/min is the target. We generate a random open-set problem with source domains containing classes 1, 2, 4, 6, 7 and 10. The testing accuracy is adopted as the metric for deciding the parameters.

The diagnosis results are further compared with other methods.

Basic is a traditional method based on deep learning. The traditional supervised learning paradigm is implemented without transfer learning technology. Only the cross-entropy loss function of the state classifier is considered.

The widely used domain adaptation algorithm is also evaluated in the open set task. The MMD-DA method refers to a popular domain adaptive method, which aims to minimize the maximum average difference between the source data and the target data in the representation subspace (Lu et al. 2017). In this way, the characteristics of the two domains are fused together. The deep CORAL method is also used, which stands for correlation alignment method (Sun and Saenko 2016). The domain transfer problem is solved by the alignment of the second-order statistics of the two domain features, which is good for unsupervised transfer learning problems. The DANN method is a promising transfer learning method, which is short for Domain Adversarial Neural Network (Ganin et al. 2016). Through the adversarial learning between the feature extractor and the domain discriminator, cross-domain shared features can be learned. DANN also considers the alignment of marginal data distribution across domains.

The open set support vector machine (OSVM) method is designed to detect unseen classes (Jain et al. 2014), and it uses a threshold to identify whether the data are an outlier. In this section, the neural network model is first trained using labeled data. Next, consider the learned advanced features and use the OSVM method. In order to combine the transfer learning technology with OSVM, the OSVM-MMD and OSVM-DANN methods are implemented. For details, the MMD metric is minimized in the first stage to narrow the source and target features, which is called OSVM-MMD. By combining DANN and OSVM methods, the OSVM-DANN method is realized, which can realize knowledge transfer and detect unseen classes.

This research introduces entropy minimization technology to improve the performance of unsupervised learning. To check its benefits, the NoEntropy method was implemented, in which the entropy minimization algorithm was discarded from the presented framework.

(3) Comparison results with other methods

In this study, only limited classes are included in the source domain for simulating the open set problem. To examine the performance, all the target outlier fault modes are treated as the unknown fault in the tasks. The experimental results are presented in Fig. 4.25 using different approaches with respect to the CWRU data. The general testing accuracies are provided to indicate the effectiveness for all the target domain samples. To be specific, the testing accuracies for the known and unknown fault modes are also presented in Fig. 4.26. That shows the model performance for the shared fault modes across domains, as well as the target-domain outlier fault modes.

For the specific tasks A_3, the confusion matrix is provided to show the detailed cross-domain fault diagnosis results, and that is shown in Fig. 4.27. For comparisons, the popular domain adaptation method DANN is also evaluated to show the effectiveness of the presented method.

From the results, it can be seen that generally, ILWAL achieves the best performance against other approaches. The testing accuracies can be above 80% in different

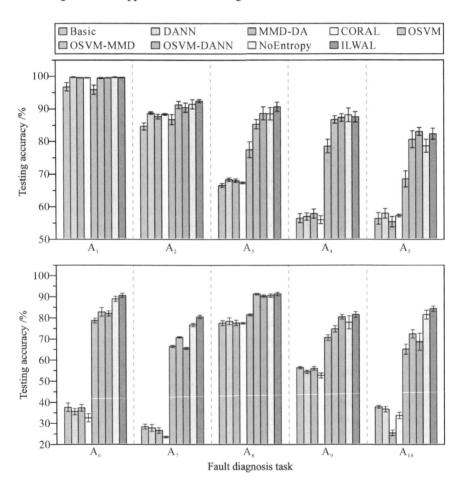

Fig. 4.25 Experimental results in the open set fault diagnosis problems using different methods on the CWRU data

4.5 Instance-Level Weighted Adversarial Learning for Open-Set Fault Diagnosis

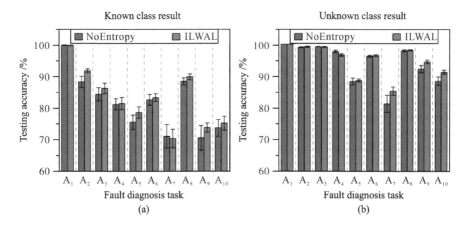

Fig. 4.26 Experimental results in the open set cases on the CWRU data: (a) the results of the known classes, and (b) the results of the unknown classes

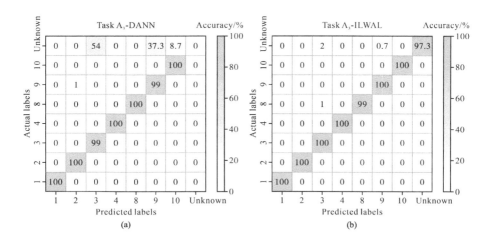

Fig. 4.27 Confusion matrix in task A_3 by using: (a) DANN method, and (b) ILWAL method

tasks using the presented instance-level weighted adversarial algorithm. Since the open set problem is pretty difficult for domain adaptation, the Basic method generally loses effectiveness with only the traditional supervised learning paradigm. The MMD-DA, DANN and CORAL approaches are effective for knowledge transfer, whose effectiveness has also been validated in different application scenarios. Nearly 100% testing accuracies are achieved in the closed set problem where no target outlier is considered. However, taking into consideration of the target outlier fault modes, the performances of all the concerns methods drop to some extent. That is because they mostly pay attention to the marginal distribution merging, and thus neglecting the outlier disturbance. As a result, their performances are not satisfactory in the open set problems.

The OSVM algorithm is designed to detect the unseen classes, and higher accuracies are obtained in comparisons with the basic methods in different tasks. Nonetheless, the cross-domain fault diagnosis performance is not promising. By combing the transfer learning skills with the outlier detection method OSVM, the developed OSVM-MMD and OSVM-DANN show many improvements. On the one hand, they can address domain shift issues by domain adaptation strategies. On the other hand, the outliers can be well detected. However, since the marginal distribution alignment is still focused on, their performance is not as good as the presented method. The detailed results as shown by the confusion matrix also indicate the superiority of the presented method in different scenarios.

This difficult open set problem can be generally well addressed by the ILWAL. For the shared classes between source and target domains, the testing accuracies remain at a high level in most cases, and the unseen fault mode identification accuracies are also promising. To evaluate the improvements by the entropy minimization algorithm in ILWAL, the NoEntropy method is evaluated. Based on the results in different tasks, it clearly shows that the ILWAL achieves higher testing accuracies than the NoEntropy method in different open set cases.

(4) Feature visualization

We visualize the learned features by different methods in the open set cases for validation. The automatically learned features by the feature extractor are focused on and the popular t-SNE technique is used for dimension reduction and visualization. The results in cases A_3 and A_6 are presented in Fig. 4.28.

It is noted that in case of A_3, the learned features by the ILWAL are promising for domain alignment with respect to different classes. The features from the shared classes between source and target domains are projected into the same regions in the high-level sub-space, while the target-domain unseen faults are clearly isolated in different regions. Therefore, the shared knowledge can be effectively transferred across domains, and the outliers can be also identified. On the other hand, with respect to the Basic and DANN methods, many data overlapping across different fault modes are noted, which leads to less promising knowledge transfer effects. For task A_6, the patterns are similar. In summary, the effectiveness of the ILWAL in the open set problems has been well validated through feature visualizations.

Afterwards, we aim at analyzing the learned instance-level weight values for the target samples in different tasks, and Fig. 4.29 shows the results. It is found that the target samples with the same fault mode labels with the source domain data are generally attached with the larger weights, and smaller weights are usually learned for the target domain outliers. For example, in task A_3, the target domain unseen samples with classes 5, 6 and 7 have small weight values. In this way, they are neglected to some extent in the adversarial learning process for generating shared features. The rest target domain samples with the shared fault modes mostly have larger weight values, and they are more focused on the adversarial learning process.

4.5 Instance-Level Weighted Adversarial Learning for Open-Set Fault Diagnosis

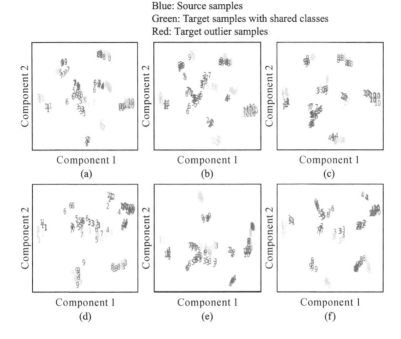

Fig. 4.28 The feature visualization results in tasks A_3 and A_6 using different approaches. The number denotes the fault mode class label:(a)basic method for the task A_3,(b) DANN method for the task A_3,(c)ILWAL method for the task A_3,(d) basic method for the task A_6,(e)DANN method for the task A_6, and (f) ILWAL method for the task A_6

This fact has also been validated using the statistics of the learned weights. The experimental results in task A_6 also show similar patterns.

Next, the effects of the model parameters on the performance are studied, and the experimental results in the validation case are shown in Fig. 4.30. The effects of many network parameters are evaluated. It is shown that the specific neural network structure poses a limited effect on the general model behavior. If the network capacity is large enough for learning, the performance can be stable in different scenarios. On the other hand, if the network capacity is not large enough, poor model performance is resulted in. In addition, the influence of the threshold parameter ρ in the pseudo label learning part and the multiple weight parameters α_d, α_e and α_o in the optimization part are also evaluated. The results show that better performance can be generally achieved with smaller ρ. The weight parameters have limited influence on the model performance if they are in a common and reasonable value range.

4.5.3.2 Transfer Diagnosis Tasks of Train Bogie Bearings

(1) Dataset description

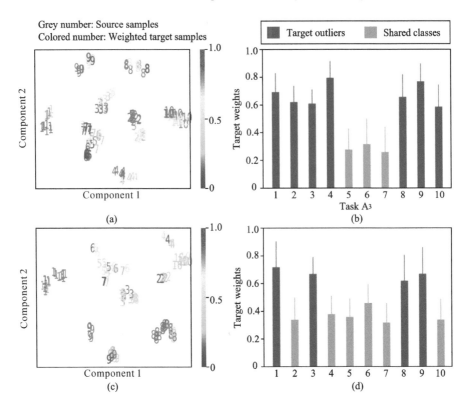

Fig. 4.29 (a) and (b): the visualization results of the learned features by ILWAL in the cases A_3 and A_6, respectively. The numbers mean the class labels of the samples. (c) and (d): the means and standard deviations of the learned weights for the samples in different fault modes in the cases A_3 and A_6, respectively

The performance of ILWAL is further verified by the train bogie dataset, which is collected from a high-speed train bogie system (Li et al. 2020e). The roller bearing health state monitoring case is considered, and two rotating speeds are also used, which are 1590 and 1950 r/min. A total of ten health states are also simulated, with healthy, inner race fault, roller fault and outer race fault. For each fault, three severities are created, with incipient, medium and severe faults. Table 4.15 presents the details of the train bogie dataset. The test rig of the train bogie system with different fault modes are shown in Fig. 4.31. Based on the dataset, some open-set transfer diagnosis tasks are created as shown in Table 4.16.

(2) Diagnosis results

The experimental results of different open set tasks on train bogie data are shown in in Figs. 4.32 and 4.33. By comparing with different methods, ILWAL basically achieves the highest testing accuracy in different situations. The typical domain adaptive methods MMD-DA and DANN do not work well in challenging open set

4.5 Instance-Level Weighted Adversarial Learning for Open-Set Fault Diagnosis

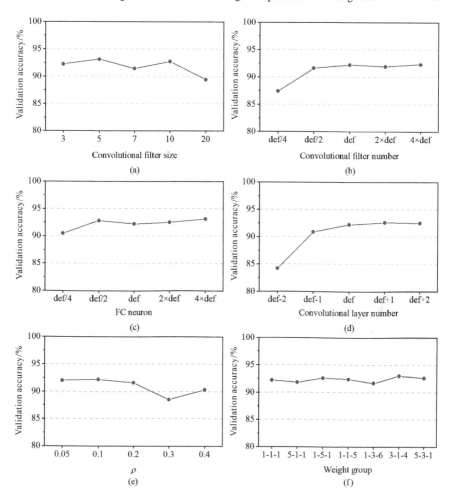

Fig. 4.30 The influence of different model parameters on the performance. "def" means the default values of the coefficients, and different times of the value are also examined. (a) the validation results with different convolutional filter sizes, (b) the validation results with different convolutional filter numbers, (c) the validation results with different FC neurons, (d) the validation results with different convolutional layer numbers, (e) the validation results with different threshold coefficients, and (f) the validation results with different weight groups. For example, the weight group 1–3–6 means $\alpha_d = 1$, $\alpha_e = 3$, and $\alpha_o = 6$

fault diagnosis problems, and no significant improvement is found compared to the basic methods. The presented instance-level weighted adversarial learning method can handle this problem well, and the benefits of entropy minimization techniques are also observed. The detailed results as shown in Fig. 4.34 by the confusion matrix also indicate the superiority of the presented method in different scenarios.

Table 4.15 Details of the train bogie dataset

Health states	Fault severity	Label
N	—	1
IF	Incipient	2
	Medium	3
	Severe	4
RF	Incipient	5
	Medium	6
	Severe	7
OF	Incipient	8
	Medium	9
	Severe	10

Fig. 4.31 Experimental setting in the train bogie dataset and the corresponding bearing faults (Li et al. 2019)

Table 4.16 Open-set transfer diagnosis tasks of the train bogie

Task	Source to target	Source classes
B_1	1590 r/min of the source to 1950 r/min of the target	All classes
B_2		1, 2, 3, 5, 6, 7, 8, 9, 10
B_3		1, 2, 4, 5, 6, 8, 9, 10
B_4		1, 2, 3, 7, 8, 9, 10
B_5		1, 4, 5, 6, 8, 9
B_6		1, 2, 7, 8, 9
B_7		1, 3, 5, 7, 10
B_8	1950 r/min of the source to 1590 r/min of the target	1, 2, 4, 5, 6, 7, 9, 10
B_9		1, 2, 3, 4, 6, 7, 8
B_{10}		1, 5, 7, 8, 9, 10

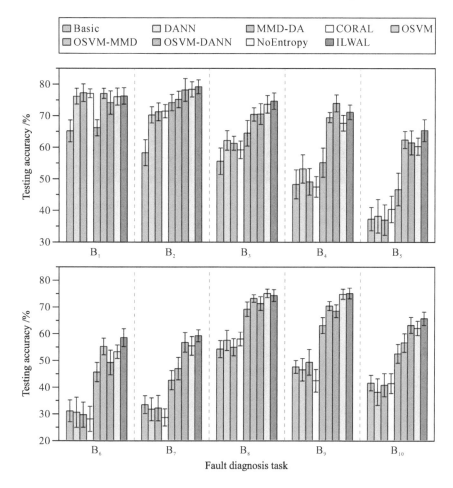

Fig. 4.32 Experimental results in the open set fault diagnosis problems using different methods on the train bogie data

4.5.4 Epilog

In this section, ILWAL is used for the mechanical system's open set transfer diagnosis. This method weights the target domain samples through the instance weighting mechanism, which can well reflect the similarity between the target sample and the source domain class. According to the results of the train bogie diagnosis case, ILWAL has obtained higher testing accuracy and can well identify undetected faults in the target domain. Although satisfactory results have been obtained by the presented method, the shortcomings and limitations of the method should be clarified. Current researches generally focus on the problem of open set domain adaptation, where the failure mode of the source domain is contained in the label set of the target domain.

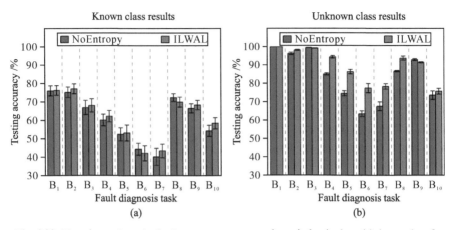

Fig. 4.33 Experimental results in the open-set cases on the train bogie data: (a) the results of the known classes, and (b) the results of the unknown classes

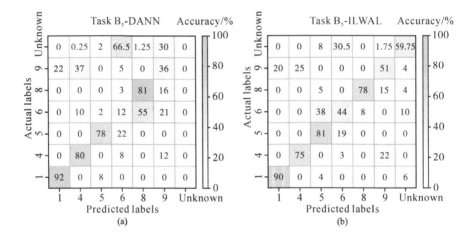

Fig. 4.34 Confusion matrix in task B_5 using: (a) DANN method, and (b) ILWAL method

However, for broader domain adaptation, partial domain adaptation methods are also common in practice, where the target failure mode is included in the source domain class, as described in the previous section. Therefore, there is a great need for a more general transfer learning framework that can automatically solve partial and open set problems. In addition, current research assumes that the target domain data can be used in advance for model training. Therefore, the presented method is more suitable for offline testing. In future work, more attention will be paid to the online domain adaptation problem instead of making assumptions about the target domain data. This will be more suitable for actual industrial scenarios.

4.6 Conclusions

This chapter systematically discusses the deep transfer learning in applications of intelligent fault diagnosis to engineering scenarios with respect to monitoring big data. Deep transfer learning is the combination of deep learning and transfer learning, in which deep learning automatically learns features from the monitoring data, and transfer learning achieves diagnosis tasks without any labeled data. The advantages of deep learning and transfer learning make deep transfer learning solve challenges of intelligent fault diagnosis in real-world applications, such as the huge labor in artificial feature extraction, the lack of labeled monitoring data, and insufficient fault types. However, there remains issues to implement diagnosis knowledge transfer in real cases.

(1) How to transfer when very limited data in the target are labeled. Section 4.2 presents a DBNCL algorithm for this issue. It uses continual learning to store multiple source knowledge of feature representation. The knowledge is further transferred to the target so that the diagnosis models trained with very limited target features can also work well with other unlabeled data.

(2) How to transfer when none of the data is labeled. The MLAN-based diagnosis model, as shown in Sect. 4.3, is constructed with the multi-layer adaptation module and the pseudo label learning algorithm for this issue. The strong capability of MLAN is to correct the distribution discrepancy between the source data and the target data. As a result, the diagnosis models fully trained with the laboratory mechanical systems can also work at related mechanical systems in real cases although none of the data is labeled.

(3) How to transfer when the required knowledge is less than the provided one from the source. In Sect. 4.4, a DPTLN explains that adversarial learning can assess the domain asymmetry of the source and the target, and adaptively learn the domain-asymmetry factors for source samples in the distribution adaptation. Based on the strategy, partial distribution across domain samples is adapted to make part of knowledge transfer from the source to the target.

(4) How to transfer when the provided knowledge cannot meet that is required by the target. Section 4.5 studies an ILWAL to solve such the open-set transfer diagnosis task. This method uses an instance weighted mechanism that targets the weights for the source according to the similarities across domain classes. By the weights, the outlier classes in the target are picked out and further blocked in the training of GAN-based transfer diagnosis methods. The readers may selectively be inspired by the sections based on the transfer diagnosis tasks they are concerning or experiencing.

Although deep transfer learning can promote applications of intelligent fault diagnosis to engineering scenarios, the transferability measure between the source and the target remains an open problem. The higher transferability the source and target domains are subject to, the better transfer performance the transfer diagnosis methods will achieve (Yang et al. 2021b). Based on the transferability analytics, an optimal source can be found or created to provide the most effective knowledge for the fault diagnosis tasks of the target. This will be an important research topic in our future works.

References

Arjovsky M, Chintala S, Bottou L (2017) Wasserstein generative adversarial networks. In: International Conference on Machine Learning. PMLR, pp 214–223
Cao Z, Long M, Wang J, Jordan MI (2018) Partial transfer learning with selective adversarial networks. In: Proceedings of the IEEE Conference on Computer Vision and Pattern Recognition. pp 2724–2732
Fawcett T (2006) An introduction to ROC analysis. Pattern Recogn Lett 27(8):861–874
Fink M (2005) Object classification from a single example utilizing class relevance metrics. Adv Neural Inf Process Syst 17:449–456
Fischer A, Igel C (2014) Training restricted Boltzmann machines: an introduction. Pattern Recogn 47(1):25–39
Ganin Y, Ustinova E, Ajakan H, Germain P, Larochelle H, Laviolette F, Marchand M, Lempitsky V (2016) Domain-adversarial training of neural networks. J Mach Learn Res 17(1):2096–2030
Garreau D, Jitkrittum W, Kanagawa M (2017) Large sample analysis of the median heuristic. arXiv preprint arXiv:170707269
Grandvalet Y, Bengio Y (2005) Semi-supervised learning by entropy minimization. CAP 367:281–296
Gretton A, Borgwardt KM, Rasch MJ, Schölkopf B, Smola A (2012) A kernel two-sample test. J Mach Learn Res 13(1):723–773
Guo L, Lei Y, Xing S, Yan T, Li N (2019) Deep convolutional transfer learning network: a new method for intelligent fault diagnosis of machines with unlabeled data. IEEE Trans Industr Electron 66(9):7316–7325
He K, Zhang X, Ren S, Sun J (2016) Deep residual learning for image recognition. In: Proceedings of the IEEE conference on computer vision and pattern recognition. pp 770–778
Hinton GE (2012) A practical guide to training restricted Boltzmann machines. Neural networks: tricks of the trade. Springer, pp 599–619
Ioffe S, Szegedy C (2015) Batch normalization: accelerating deep network training by reducing internal covariate shift. In: International conference on machine learning. PMLR, pp 448–456
Jain LP, Scheirer WJ, Boult TE (2014) Multi-class open set recognition using probability of inclusion. European conference on computer vision. Springer, pp 393–409
Jia F, Lei Y, Lin J, Zhou X, Lu N (2016) Deep neural networks: a promising tool for fault characteristic mining and intelligent diagnosis of rotating machinery with massive data. Mech Syst Signal Process 72:303–315
Jia F, Lei Y, Guo L, Lin J, Xing S (2018) A neural network constructed by deep learning technique and its application to intelligent fault diagnosis of machines. Neurocomputing 272:619–628
Kingma DP, Ba J (2014) Adam: a method for stochastic optimization. arXiv preprint. arXiv:14126980
LeCun Y, Bengio Y, Hinton G (2015) Deep learning. Nature 521(7553):436–444
Lee D-H (2013) Pseudo-label: the simple and efficient semi-supervised learning method for deep neural networks. In: Workshop on challenges in representation learning, international conference on machine learning, vol 2, p 896
Lei Y, Jia F, Lin J, Xing S, Ding SX (2016) An intelligent fault diagnosis method using unsupervised feature learning towards mechanical big data. IEEE Trans Industr Electron 63(5):3137–3147

References

Lei Y (2017) 3—Individual intelligent method-based fault diagnosis. In: Lei Y (ed) Intelligent fault diagnosis and remaining useful life prediction of rotating machinery. Butterworth-Heinemann, pp 67–174. https://doi.org/10.1016/B978-0-12-811534-3.00003-2

Lei Y, Yang B, Jiang X, Jia F, Li N, Nandi AK (2020) Applications of machine learning to machine fault diagnosis: a review and roadmap. Mech Syst Signal Process 138:106587

Li X, Zhang W, Ding Q (2019) Cross-domain fault diagnosis of rolling element bearings using deep generative neural networks. IEEE Trans Industr Electron 66(7):5525–5534. https://doi.org/10.1109/TIE.2018.2868023

Li X, Zhang W, Ding Q, Li X (2020a) Diagnosing rotating machines with weakly supervised data using deep transfer learning. IEEE Trans Industr Inf 16(3):1688–1697

Li X, Zhang W, Ding Q, Sun J-Q (2020b) Intelligent rotating machinery fault diagnosis based on deep learning using data augmentation. J Intell Manuf 31(2):433–452. https://doi.org/10.1007/s10845-018-1456-1

Li X, Zhang W, Ma H, Luo Z, Li X (2020c) Deep learning-based adversarial multi-classifier optimization for cross-domain machinery fault diagnostics. J Manuf Syst

Li X, Zhang W, Ma H, Luo Z, Li X (2020d) Partial transfer learning in machinery cross-domain fault diagnostics using class-weighted adversarial networks. Neural Netw 129:313–322

Li X, Zhang W, Xu N-X, Ding Q (2020e) Deep learning-based machinery fault diagnostics with domain adaptation across sensors at different places. IEEE Trans Industr Electron 67(8):6785–6794

Long M, Cao Y, Wang J, Jordan M (2015) Learning transferable features with deep adaptation networks. In: International conference on machine learning. PMLR, pp 97–105

Long M, Wang J, Cao Y, Sun J, Philip SY (2016) Deep learning of transferable representation for scalable domain adaptation. IEEE Trans Knowl Data Eng 28(8):2027–2040

Lu W, Liang B, Cheng Y, Meng D, Yang J, Zhang T (2017) Deep model based domain adaptation for fault diagnosis. IEEE Trans Industr Electron 64(3):2296–2305

Nair V, Hinton GE (2010) Rectified linear units improve restricted boltzmann machines. In: International conference on machine learning

Pan SJ, Yang Q (2010) A survey on transfer learning. IEEE Trans Knowl Data Eng 22(10):1345–1359

Pan SJ, Tsang IW, Kwok JT, Yang Q (2011) Domain adaptation via transfer component analysis. IEEE Trans Neural Netw 22(2):199–210

Panareda Busto P, Gall J (2017) Open set domain adaptation. In: Proceedings of the IEEE international conference on computer vision, pp 754–763

Parisi GI, Kemker R, Part JL, Kanan C, Wermter S (2019) Continual lifelong learning with neural networks: a review. Neural Netw 113:54–71

Ren Z, Zhu Y, Yan K, Chen K, Kang W, Yue Y, Gao D (2020) A novel model with the ability of few-shot learning and quick updating for intelligent fault diagnosis. Mech Syst Signal Process 138:106608. https://doi.org/10.1016/j.ymssp.2019.106608

Sun B, Saenko K (2016) Deep CORAL: Correlation alignment for deep domain adaptation. European conference on computer vision. Springer, pp 443–450

Tzeng E, Hoffman J, Zhang N, Saenko K, Darrell T (2014) Deep domain confusion: Maximizing for domain invariance. arXiv preprint. arXiv:14123474

Van der Maaten L, Hinton G (2008) Visualizing data using t-SNE. J Mach Learn Res 9(11)

Wang Y, Sun G, Jin Q (2020) Imbalanced sample fault diagnosis of rotating machinery using conditional variational auto-encoder generative adversarial network. Appl Soft Comput 92:106333. https://doi.org/10.1016/j.asoc.2020.106333

Xing S, Lei Y, Wang S, Jia F (2021a) Distribution-invariant deep belief network for intelligent fault diagnosis of machines under new working conditions. IEEE Trans Industr Electron 68(3):2617–2625

Xing S, Lei Y, Yang B, Lu N (2021b) Adaptive knowledge transfer by continual weighted updating of filter kernels for few-shot fault diagnosis of machines. IEEE Trans Ind Electron

Xu X, Lei Y, Li Z (2020) An incorrect data detection method for big data cleaning of machinery condition monitoring. IEEE Trans Industr Electron 67(3):2326–2336

Yang B, Lei Y, Jia F, Xing S (2019) An intelligent fault diagnosis approach based on transfer learning from laboratory bearings to locomotive bearings. Mech Syst Signal Process 122:692–706

Yang B, Lei Y, Jia F, Li N, Du Z (2020) A Polynomial kernel induced distance metric to improve deep transfer learning for fault diagnosis of machines. IEEE Trans Industr Electron 67(11):9747–9757. https://doi.org/10.1109/TIE.2019.2953010

Yang B, Lee C-G, Lei Y, Li N, Lu N (2021a) Deep partial transfer learning network: a method to selectively transfer diagnostic knowledge across related machines. Mech Syst Signal Process 156:107618

Yang B, Lei Y, Xu S, Lee C-G (2021b) An optimal transport-embedded similarity measure for diagnostic knowledge transferability analytics across machines. IEEE Trans Ind Electron

Yosinski J, Clune J, Bengio Y, Lipson H (2014) How transferable are features in deep neural networks? arXiv preprint. arXiv:14111792

Zellinger W, Grubinger T, Lughofer E, Natschläger T, Saminger-Platz S (2017) Central moment discrepancy (cmd) for domain-invariant representation learning. arXiv preprint. arXiv:170208811

Zhang J, Ding Z, Li W, Ogunbona P (2018) Importance weighted adversarial nets for partial domain adaptation. In: Proceedings of the IEEE conference on computer vision and pattern recognition, pp 8156–8164

Zhang W, Li X (2021) Federated transfer learning for intelligent fault diagnostics using deep adversarial networks with data privacy. IEEE/ASME Trans Mechatron 1–1. https://doi.org/10.1109/TMECH.2021.3065522

Zhang W, Li X, Ma H, Luo Z, Li X (2021a) Open set domain adaptation in machinery fault diagnostics using instance-level weighted adversarial learning. IEEE Trans Industr Inf 17(11):7445–7455

Zhang W, Li X, Ma H, Luo Z, Li X (2021b) Universal domain adaptation in fault diagnostics with hybrid weighted deep adversarial learning. IEEE Trans Ind Inform 1–1. https://doi.org/10.1109/TII.2021.3064377

Chapter 5
Data-Driven RUL Prediction

5.1 Introduction

With the development of capability and efficiency of industrial mechanical systems, their stable, safe, and reliable operations have received a lot of attention. Therefore, PHM strategy with the RUL prediction technique centered is gaining more and more attention and application, and has become an important foundation of maintenance support systems (Jardine et al. 2006; Kim et al. 2017; Javed et al. 2017; Rodrigues 2018). RUL prediction is a technique to predict the period of a mechanical system in which its performance degrades from current state to complete failure based on its historical maintenance/monitoring records, operating conditions, service environment, material properties and other information (Park et al. 2020). Compared with fault diagnosis, RUL prediction is able to evaluate the current health state of mechanical systems and provide an estimate of the time before maintenance (Elsheikh et al. 2019). Consequently, it provides the basis for optimal maintenance decisions, sufficient time for the schedule of spare parts inventory, and greatly reduces downtime (Lei et al. 2018).

There are mainly three categories of RUL prediction methods, i.e., model-based methods, data-driven methods and data-model fusion methods (Khan and Yairi 2018). Model-based methods mainly rely on full knowledge of mechanical system degradation mechanisms, which are difficult to perform especially in the case of complex mechanical system structures or working conditions. On the contrary, data-driven RUL prediction methods gain popularity owing to their ability to model relationships between monitoring data and degradation representation of mechanical systems automatically (Rigamonti et al. 2018). Therefore, it effectively reduces the dependence on a priori knowledge and expert experience, and becomes an important choice and powerful support for the RUL prediction of mechanical systems. This chapter will mainly focus on this kind of methods. Data-model fusion methods aim to integrate the expert knowledge of degradation models and the degradation information in monitoring data. Thus, they are expected to leverage the advantages of both model-based methods and data-driven methods. This is a kind of new methods emerging in recent years. Chapter 6 will address the data-model fusion methods.

As is mentioned in Chapter 4, the development of cyber-physical systems has made a variety of sensors mounted on mechanical systems to monitor their health states. Combined with the implementation of the acquisition of industrial data and interconnection of mechanical systems, data collected from manufacturing systems are exploding. The tremendous increase of monitoring data brings new opportunities and challenges to the data-driven based RUL prediction. On the one hand, the massive amount of monitoring data contains abundant mechanical system degradation information, which helps to analyze the degradation law of mechanical systems comprehensively and construct more accurate prediction models. On the other hand, the huge monitoring data put forward higher requirements for data analysis methods.

In order to deal with increasingly complex mechanical systems and massive monitoring data, machine learning algorithms are introduced into the field of RUL prediction (Babu et al. 2016; Deutsch and He 2017; Caceres et al. 2021; Li et al. 2021). With the support of the powerful information mining and representation learning capabilities of deep learning, a comprehensive characterization of mechanical system degradation information can be provided and the prediction accuracy can be effectively improved (Listou Ellefsen et al. 2019). However, most existing deep learning-based data-driven RUL prediction methods suffer from the following limitations.

- When facing high-dimensional raw monitoring data such as force signals, vibration signals and acoustic emission signals, existing predictions based on deep learning still need to use advanced signal processing technology or expert experience to construct health indicators (HI) of mechanical systems and the prediction performance largely depends on the hand-crafted HI, which is time-consuming and laborious and easily misses essential information.
- The uncertainty quantification of the prediction result is of great significance to facilitating maintenance decision making, but the existing deep learning-based structure can only provide the point estimate of the prediction result rather than the probability distribution.
- To obtain comprehensive information on the degradation of mechanical systems, deep prediction networks usually use multi-sensor monitoring data as model inputs. However, these methods usually lack an explicit learning mechanism when multi-sensor data are used. Thus, the model cannot effectively identify the distinctions of different sensors and highlight the effective degradation information.

To deal with the above-mentioned limitations, three new deep prognosis networks are introduced in the following part of this chapter. Experimental and industrial case studies are also conducted to illustrate the implementation process of these methods.

5.2 Deep Separable Convolutional Neural Network-Based RUL Prediction

5.2.1 Motivation

Since deep learning techniques can automatically estimate catastrophic events hidden in monitoring data and sufficiently mine the mechanical system degradation information, it has been proved as a promising method to deal with the massive amount of condition monitoring data and accurately obtain RUL prediction results (Yang et al. 2019). Although some existing deep learning-based prognosis methods have achieved promising results, they suffer from the following shortcomings:

- The prediction accuracy of existing methods is highly dependent on delicately designed features, which requests researchers and practitioners to obtain abundant prior knowledge and consumes massive human labor. Moreover, the handcrafted features are normally unable to be directly utilized in practical cases due to different operating environments.
- Existing methods have not exhaustively investigated the correlations amongst monitoring data from multiple sensors. However, different sensors usually capture different degradation information and are sensitive in fault reflection with respect to multiple aspects.

To deal with the above two limitations, a new framework named deep separable convolutional network (DSCN) is presented to take the monitoring data of different sensors as inputs and achieve the RUL prediction for mechanical systems. In DSCN, the separable convolutions are introduced to directly model the interrelationships of different sensor data, and the squeeze and excitation (SE) unit is constructed after the separable convolutional layer to adaptively acquire the importance of features, which increases the sensitivity of DSCN to effective feature maps. Based on the separable convolutions layer and the SE unit, the core block of DSCN is built. After that, multiple separable convolutional building blocks are stacked to learn the high-level representations from the input data, and the RUL can finally be accurately estimated by forwarding the learned features to the fully-connected layer. Finally, the prediction performance is verified by the vibration data from the accelerated degradation tests of rolling element bearings.

5.2.2 Deep Separable Convolutional Network

The architecture of DSCN is shown in Fig. 5.1. It maps the relationship between the obtained monitoring data from multiple types of sensors and their corresponding RUL results. DSCN includes two sub-networks for representation learning and RUL estimation. Each representation learning sub-network is stacked by the separable convolutional building blocks and is used to discover the discriminative information

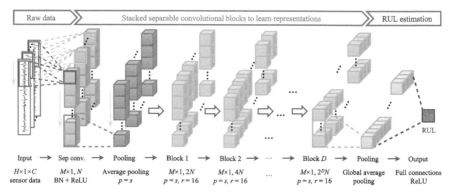

Fig. 5.1 An illustration of the DSCN architecture

automatically and learn the high-level representations from the input data. In detail, the input data of size $H \times 1 \times C$ is convolved by a separable convolutional layer, the channel-wise convolution kernel is set to $M \times 1$, and the number of pointwise convolution kernels is set to N. Then, both batch normalization and ReLU are adopted after separable convolutions. The average pooling layer is used to achieve down-sampling after the separable convolutional layer, in which the pooling size p is equal to the stride size s. After that, the stacking of multiple separable convolutional building blocks allows high-level representations to be learned. The parameters in each separable convolutional building block are the same. In addition, the non-overlapping window is used in average pooling and the setting of dimensionality reduction ratio r in the SE unit takes into account the trade-off between prediction accuracy and computational load, which is set to 16.

Further, the high-level representations obtained from the separable convolutional building blocks are transmitted into the RUL estimation sub-network to achieve the RUL prediction based on a fully-connected layer. A global average pooling layer is employed to obtain the high-level representations from the separable convolutional building block, which aims to lower the total amount of parameters. Finally, a fully-connected layer with one neuron is attached at the end of DSCN to perform the RUL estimation and the ReLU is adopted to implement the ability of nonlinear activation.

5.2.3 Architecture of DSCN

5.2.3.1 Separable Convolutions

The deep prognosis models usually take multi-channel time-series data as inputs to integrate the degradation information, and each channel represents one data sequence from a specific sensor. For the multi-sensor time-series input data, each channel owns temporal correlations as the collected data reflect the failure progression over the

5.2 Deep Separable Convolutional Neural Network-Based RUL Prediction

operating time of one component. In addition, the data in different channels have mutual connections due to the evolution and interaction of faults amongst components. Existing deep prognostic models cannot chapter and model the dependencies of different sensor data effectively in the mechanical system degradation processes (Li et al. 2019), which limits the generalization and prediction accuracy of the prognosis models. Therefore, a separable convolution operation (Mamalet and Garcia 2012) is introduced in this section to decouple temporal and cross-channel correlations, which models interrelationships of different sensor data effectively.

The temporal correlations and cross-channel correlations are unraveled in the separable convolutions by factorizing the standard convolutions. As shown in Fig. 5.2, the standard convolutions are decomposed into two parts: channel-wise convolutions and pointwise convolutions. The channel-wise convolutions are used to map temporal correlations of every sensor sequence separately by a single convolution kernel in each of the input channels. The pointwise convolutions are used to create a linear combination of the channel-wise convolution outputs by utilizing 1×1 convolutions. Then, the temporal and cross-channel correlations are decoupled through the channel-wise and pointwise convolutions.

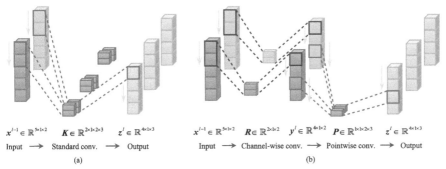

Fig. 5.2 An illustration for convolution operations: (a) standard convolutions, (b) separable convolutions

Let $x^{l-1} \in \mathbb{R}^{H \times W \times C}$ denote the input volume, where H is the height of the input, respectively, W is the width of the input and C is the number of input channels. Correspondingly, the convolution kernels in a standard convolutional layer are denoted as $K \in \mathbb{R}^{M \times 1 \times C \times N}$, where $M \times 1$ is the size of a convolution kernel, and N is the number of kernels. In the separable convolutions, K are factorized into the channel-wise convolution kernels $R \in \mathbb{R}^{M \times 1 \times C}$ and the pointwise convolution kernels $P \in \mathbb{R}^{C \times N}$. Therefore, the z_n^l with the nth output of separable convolutions is calculated by

$$y_c^l = R_c * x_c^{l-1} + b_c^l \tag{5.1}$$

$$z_n^l = \sum_{c=1}^{C} P_n * y_c^l + b_n^l \tag{5.2}$$

where y_c^l is the cth output of the channel-wise convolutions, R_c is the cth channel-wise convolution kernel that is applied to the cth channel of x^{l-1}, b_c^l is the bias vector corresponding to R_c, P_n is the nth pointwise convolution kernel of size 1×1, and b_n^l is the bias vector corresponding to P_n.

5.2.3.2 Feature Response Recalibrations

In general, different separable convolutional layers contain different degrees of degradation information. Therefore, an SE unit is constructed after the separable convolutional layer to highlight informative feature maps and suppress useless ones. As shown in Fig. 5.3, the SE unit is utilized to evaluate the informativeness of each feature map, which consists of two steps, the squeeze operation and excitation operation.

Firstly, in the squeeze operation, the overall information acquired from the separable convolutional layer is embedded into a channel descriptor $u^l \in \mathbb{R}^C$. The global average pooling is utilized to shrink the convolutional output z^l by

$$u_n^l = \frac{1}{H} \sum_{h=1}^{H} z_{n,h}^l \tag{5.3}$$

where u_n^l is the nth element of u^l. Then, the excitation operation utilizes a self-gating mechanism (Hu et al. 2018) to generate the corresponding channel weights ω^l and represent the informativeness of every channel. The ω^l is calculated by

$$\omega^l = \sigma\left(W_2^l \delta\left(W_1^l u^l\right)\right) \tag{5.4}$$

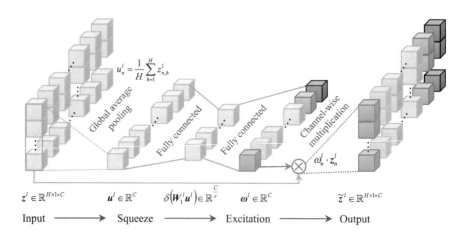

Fig. 5.3 An illustration for feature response recalibrations in an SE unit

where $\sigma(\cdot)$ and $\delta(\cdot)$ are sigmoid and ReLU activation functions, respectively, $W_1^l \in \mathbb{R}^{\frac{C}{r} \times C}$ and $W_2^l \in \mathbb{R}^{C \times \frac{C}{r}}$ are the weights, and r is the ratio of dimensionality reduction. Finally, the feature map z_n^l and the corresponding channel weight ω_n^l are multiplied to obtain the recalibrated feature maps.

5.2.3.3 Separable Convolutional Building Blocks

In this section, the residual connections are employed in DSCN to ease the training burden and accuracy saturation or accuracy reduction problems, and the pre-activation strategy in (He et al. 2016) is utilized to relieve the problem of overfitting and improve the regularization of the network. As shown in Fig. 5.4, the separable convolutional building block with a residual connection is constructed by adding an identity skip connection between the input x^{l-1} and the output x^l. Two separable convolutional layers, an average pooling layer and an SE unit are contained in the separable convolutional building block, the batch normalization (BN) (Ioffe and Szegedy 2015) is used as pre-activation and ReLU (Nair and Hinton 2010) is added before the separable convolution.

In this introduced framework, we utilize the full pre-activation strategy, which conducts the BN and ReLU before each separable convolution. In addition, an average pooling layer is inserted before the SE unit to reduce the dimension of the representation. Subsequently, the output of the separable convolutional building block x^l is generated by the recalibrated feature maps and the identity mapping x^{l-1}. Then, the output is fed into the next separable convolutional building blocks or subsequent processing layers. After that, the final output representations are input into the fully-connected layer to realize the RUL prediction.

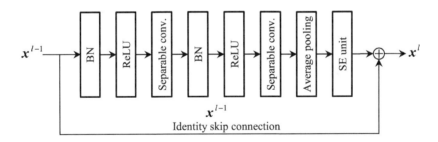

Fig. 5.4 A separable convolutional building block with a residual connection in DSCN

5.2.4 RUL Prediction Case of Accelerated Degradation Experiments of Rolling Element Bearings

For the effectiveness and superiority verification of the introduced DSCN in RUL prediction, the accelerated degradation tests of bearings are used hereafter. As shown

Fig. 5.5 Testbed of rolling element bearings

in Fig. 5.5, the testbed of rolling element bearings (Wang et al. 2018) is composed of an alternating current (AC) motor, a motor speed controller, a support shaft, two support bearings (heavy-duty roller bearings), a hydraulic loading system, and so on. The experimental platform is able to perform the accelerated degradation tests of bearings under different operating conditions. As shown in Table 5.1, the complete run-to-failure data of fifteen LDK UER204 ball bearings are acquired under three kinds of operating conditions. The radial force of the tested bearing is provided by the hydraulic loading system and the rotating speed is controlled by the AC induction motor. Figure 5.6 shows the pictures of normal and four types of failed bearings. As shown in Fig. 5.5, two accelerometers are positioned at 90° on the housing of the tested bearings to collect the vibration signals. Besides, the sampling frequency is set to 25.6 kHz, and 32768 data points are recorded each 1 min. The horizontal and vertical vibration signals under three operating conditions are shown in Fig. 5.7. In this section, the first four datasets under each working condition are used as the training dataset, and the remaining one is used as the testing dataset.

5.2.4.1 Evaluation Metrics

In this section, the prediction performance of DSCN is quantitatively evaluated by scoring function and root mean square error.

(1) **Scoring function**: The scoring function is first proposed by the 2008 Prognostics and Health Management Data Challenge (Saxena et al. 2008), which is calculated as

5.2 Deep Separable Convolutional Neural Network-Based RUL Prediction

Table 5.1 Operating conditions of XJTU-SY bearing datasets

Operating condition	Radial force	Rotating speed /(r/min)	Bearing dataset	
			Training dataset	Testing dataset
Condition 1	12 kN	2100	Bearing 1_1 Bearing 1_2 Bearing 1_3 Bearing 1_4	Bearing 1_5
Condition 2	11 kN	2250	Bearing 2_1 Bearing 2_2 Bearing 2_3 Bearing 2_4	Bearing 2_5
Condition 3	10 kN	2400	Bearing 3_1 Bearing 3_2 Bearing 3_3 Bearing 3_4	Bearing 3_5

Fig. 5.6 Pictures of the bearings in different fault degrees: (a) normal, (b) inner race wear, (c) cage fracture, (d) outer race wear, and (e) outer race fracture

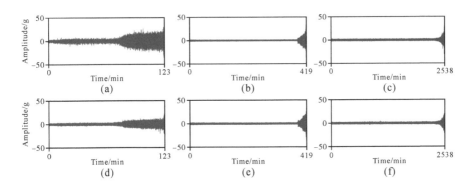

Fig. 5.7 Vibration signals of bearings: (a) Bearing 1_1 in he orizontal direction, (b) Bearing 2_1 in the horizontal direction, (c) Bearing 3_1 in the horizontal direction, (d) Bearing 1_1 in the vertical direction, (e) Bearing 2_1 in the vertical direction, and (f) Bearing 3_1 in the vertical direction

$$\text{score} = \begin{cases} \sum_{i=1}^{Q}(e^{-\frac{d_i}{13}} - 1) & \text{for} \quad d_i < 0 \\ \sum_{i=1}^{Q}(e^{\frac{d_i}{10}} - 1) & \text{for} \quad d_i \geq 0 \end{cases} \quad (5.5)$$

where Q is the total number of samples, and d_i is the error between the actual RUL and the predicted RUL of the ith sample. It needs to be mentioned that the score is lower when the prediction error is smaller, and when the estimated RUL is greater than the actual RUL, the scoring function will penalize it more heavily.

(2) **RMSE**: The score function will be seriously affected by a few outliers with large prediction errors. Therefore, RMSE is also introduced as the evaluation metric. RMSE is commonly used as the evaluation metric of RUL prediction, which is defined as

$$\text{RMSE} = \sqrt{\frac{1}{Q}\sum_{i=1}^{Q} d_i^2} \quad (5.6)$$

Figure 5.8 shows RMSE and the scoring function. From the depicted scoring function, we can observe that it is asymmetric, but RMSE assigns equal weights to the same error, thereby reducing the interference of outliers.

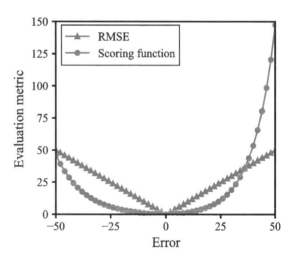

Fig. 5.8 Comparison between scoring function and RMSE

5.2.4.2 Data Preprocessing

Before the monitoring data are input into DSCN, the data needs to be preprocessed appropriately.

(1) **Normalization**: Firstly, vibration signals from both horizontal and vertical directions are normalized using z-score to improve network convergence. In addition, because of the large lifetime distribution range (from 33 to 2538 min), the lifetime difference may incur overfitting problems. Therefore, for each bearing in the training dataset, it is necessary to normalize the actual RUL value to the range of [0, 1] through dividing by their maximum lifetime. Hence, for the bearings in the testing dataset, the predicted RUL can be calculated by

$$\text{PreRUL}_t = \frac{\text{NormRUL}_t}{1 - \text{NormRUL}_t} \times P_t \tag{5.7}$$

where P_t is the time of the tth sampling time step, NormRUL_t is the output value of the DSCN at P_t, and PreRUL_t is the final estimated RUL value.

(2) **Time window embedding**: In order to fully utilize the temporal information, this section adopts a time window embedding strategy (Babu et al. 2016) after normalization, which utilizes a time window to splice the monitoring data at consecutive sampling time steps. Therefore, a high-level vector can be obtained through time window embedding, which is used as the input of the DSCN. The process of time window embedding is demonstrated in Fig. 5.9.

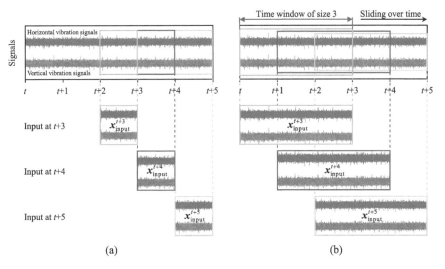

Fig. 5.9 Time window embedding illustration: (a) without a time window, and (b) time window of size 3

5.2.4.3 Configurations of DSCN

In the DSCN, there are many parameters that need to be predefined which contain the channel-wise convolution kernel size $M \times 1$, the number of point-wise convolution kernels N and separable convolutional building blocks D, the dimensionality reduction ratio of SE units r and the pooling size p. The fourfold cross-validation is performed on the training datasets to determine these hyperparameters. Considering the trade-off between accuracy and calculation loss, the final hyperparameters are determined as shown in Table 5.2. The mean square error is used as the loss function of DSCN, and the Adam optimizer (Kingma and Ba 2014) is adopted to minimize the MSE loss function by iteratively updating the network weights and biases. The weights and biases are initialized by the strategy described in (He et al. 2016), and the model is trained for 100 epochs from scratch.

Table 5.2 Configurations of DSCN in RUL prediction of bearings

Hyperparameter	Size	Hyperparameter	size
Kernel size $M \times 1$	8×1	Number of kernels N	16
Pooling size p	4	Reduction ration r	16
Number of blocks D	3	Time window size S	5
Mini-batch size	128	Number of epochs	100

5.2.4.4 Experimental Results

In this case study, we firstly discuss and analyze the advantages of utilizing separable convolutions as well as feature response recalibrations. Then, the DSCN is compared with four state-of-the-art approaches for demonstrating its effectiveness and superiority. In order to remove the random impact, we repeat each experiment 20 times. Particularly, the score and RMSE are computed since the half and till the end of a lifetime for each tested bearing, because prediction results at this stage are more reliable and meaningful (Singleton et al. 2014).

In order to verify the benefits of separable convolutions and feature response recalibrations, this section uses two prediction networks to perform RUL prediction for comparison, which is the network with standard convolutions (denoted as standard ConvNet) and the network with separable convolutions (denoted as separable ConvNet). These two prediction networks do not contain SE units, and have the same structure and parameters as the DSCN. Table 5.3 lists the total model parameters and training time of three different prognosis networks. It can be seen that the separable convolutions have greatly reduced the model size (82.07% reduction) compared with standard convolutions, which results in a lower computational cost. As for DSCN, the total model parameter has only increased by 5.50% due to the introduction of SE compared to the separable ConvNet. Figure 5.10 shows the RUL estimation performance of the three prognostic networks. It can be seen that compared with the standard ConvNet, separable ConvNet obtains a lower score

5.2 Deep Separable Convolutional Neural Network-Based RUL Prediction

Table 5.3 Comparison of total model parameters and training time for three different prognosis networks

Prognosis network	Total model parameters	Training time/s
Standard ConvNet	272171	4607.36
Separable ConvNet	48865	1940.12
DSCN	51553	2023.13

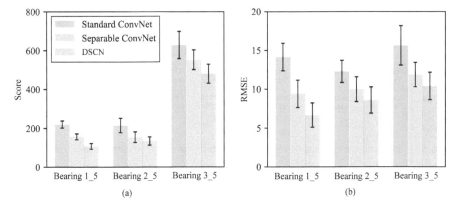

Fig. 5.10 Performance estimation results of three different prognosis networks: (a) scoring function values, and (b) RMSE values

and RMSE value, which means the separable convolutions can effectively improve the performance of the prognosis network through decoupling temporal correlations and cross-channel correlations sufficiently. Further, the DSCN has achieved better prediction performance compared with the other two prognosis networks due to the adaptive recalibration capacity of the SE unit to the feature responses from the separable convolutional layers. In summary, the DSCN has a significant advantage in computational complexity and accuracy when compared with the standard convolutional networks.

To further demonstrate the superiority of the DSCN, we implement four existing RUL prognosis approaches for comparison, including SVM (Babu et al. 2016), DBN (Deutsch and He 2017), multi-scale CNN (MCNN) (Zhu et al. 2018) and convolutional LSTM (CLSTM) (Zhao et al. 2017). For the SVM, DBN and MCNN methods, the features are pre-extracted and selected as inputs to train the corresponding models. While for CLSTM, a CNN layer is used for automatic feature extraction from the raw bearing monitoring signals, and then the features are inputted into the bidirectional LSTM to achieve the RUL estimation. For each prediction model, the same fourfold cross-validation operation is adjusted for hyperparameter tuning to obtain more accurate RUL prediction performance.

Figure 5.11 shows the performance difference of the five different prognosis approaches. It can be seen that compared with the traditional shallow models (i.e., SVM), the deep learning models (i.e., DBN, MCNN, CLSTM and DSCN) have achieved better performance for the three tested bearings regarding both score and

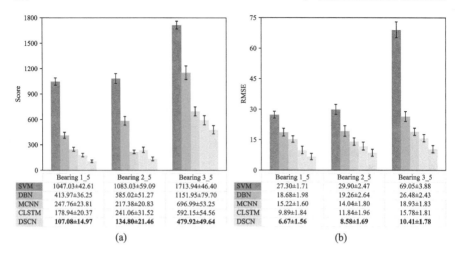

Fig. 5.11 Performance comparison regarding different metrics: (a) scoring function values, and (b) RMSE values

RMSE value. Such a phenomenon is caused by the fact that the traditional machine learning model is built upon shallow architectures, which limits the representation ability of bearing degradation behavior, whereas the deep learning models have the outstanding representation learning ability to mine valuable bearing degradation information for more accurate RUL estimation. In addition, it can also be observed from Fig. 5.11 that, the DSCN obtains the lowest RMSE and score value among all the compared deep prognosis models, which indicates that the DSCN has the highest prediction accuracy for each tested bearing. In summary, compared with other prognosis algorithms, the DSCN method has superiority in the RUL prediction of bearings.

5.2.5 Epilog

In this section, the DSCN is used for mechanical system RUL prediction. In DSCN, raw multi-sensor data are used directly as the input of the prognosis model. Next, the separable convolution operation is utilized to model the interrelationships of different sensor data effectively. In the meantime, to enable the neural network to effectively utilize the degradation-sensitive information, an SE unit is constructed behind the separable convolutional layer to recalibrate the adaptive feature response. Then, the separable convolution and SE units are employed to construct separable convolution blocks. In particular, to reduce network training and alleviate network overfitting, residual connection and pre-activation are used in separable convolutional building blocks. The high-level representations are learned through the separable convolutional building blocks stacking and the fully-connected layer is employed to estimate the RUL based on the learned representation. The DSCN is validated using the full-life vibration data of rolling element bearings, and also makes a comparison with some state-of-art prognosis approaches. The results show that the DSCN is superior to traditional data-driven approaches and has achieved high RUL prediction accuracy.

5.3 Recurrent Convolutional Neural Network-Based RUL Prediction

5.3.1 Motivation

Deep learning has been quite effective in analyzing the condition monitoring data of mechanical systems (Tsui et al. 2015). Based on the deep learning framework, CNN has gained special attention due to its powerful ability to process time-series signals, and has obtained satisfactory outputs in current research (Ambadekar and Choudhari 2020; Zhang et al. 2021). However, there are still two limitations in those studies:

(1) The time dependencies amongst different degradation states have not yet been considered in network construction.
(2) The uncertainty of the RUL prediction results cannot be quantified.

To tackle those limitations, a new framework named recurrent convolutional neural network (RCNN) is presented in this chapter for mechanical system RUL prediction (Wang et al. 2020). In RCNN, recurrent convolutional layers are first constructed for temporal dependency modeling among degradation states. Then, the variational inference is utilized for RUL uncertainty quantification. The performance is verified by the vibration data from the public FEMTO-ST bearing datasets. More importantly, the RUL prediction yielded by RCNN is a probabilistic estimation rather than point estimation, which facilitates maintenance decision making.

5.3.2 Recurrent Convolutional Neural Network

In this section, the RCNN architecture is provided in Fig. 5.12. The RCNN consists of a stack of recurrent convolutional layers, pooling layers and fully-connected layers. In RCNN, the multi-channel time-series sensor data of size $H \times 1 \times C$ are fed into the prognosis network to integrate the degradation information from different sensors, where H is the length of the sensor sequence, and C is the number of sensors. Then, the multiple-level representations are learned automatically by the recurrent convolutional layers and pooling layers from the sensor data. After that, we further modeled the temporal dependencies amongst different degradation states. In the RCNN architecture, there are N convolutional layers, and for the ith recurrent

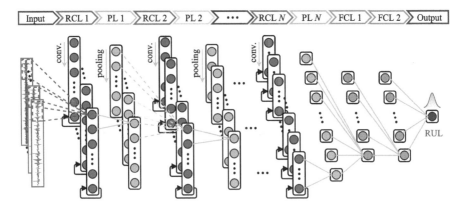

Fig. 5.12 Illustration of the RCNN architecture

convolutional layer, the convolution kernels are set the same, i.e., the number of kernels is $2^{(i-1)}M$ and the kernel size is $k \times 1$. As for the $N-1$ pooling layers, the non-overlapping window is used in the pooling operation and the max pooling is utilized as the function for down-sampling. The Nth pooling layer utilizes the global max pooling for downsampling. Then the high-level representation is embedded into a vector after the N convolutional layers and pooling layers. Furthermore, L fully-connected layers are further linked with the vector for the RUL prediction. In this section, there are three fully-connected layers in the RCNN. The first two fully-connected layers have F neurons and adopt ReLU to implement nonlinear activation. The last fully-connected layer has only one neuron which is used as the output layer to map the RUL value.

5.3.3 Architecture of RCNN

5.3.3.1 Recurrent Convolutional Layer

The convolution layers are the core building blocks in CNNs which are able to extract discriminative features automatically from the sensor data. However, the output of a layer cannot feedback into the input in a convolutional layer due to the lack of a cyclic feedback mechanism. That means the information in CNNs can only flow in the forward direction and CNNs only consider the current input information at each time step, so the precious degradation information is neglected. Therefore, CNNs lack the ability to model the contextual dependencies of different degradation states. In existing CNN-based prognosis approaches, this weakness is not considered and addressed, which affects the prediction accuracy and generalization of the constructed neural network. The RCNN architecture can solve such an issue and improve the prediction performance relies on a new core building block, i.e., recurrent convolutional layer.

5.3 Recurrent Convolutional Neural Network-Based RUL Prediction

The recurrent convolutional layer is able to establish a recurrent connection between the output and the input to conduct the looping of information, which is different from the convolutional layers in that the information is passed only in the forward direction. According to that, the output of the recurrent convolutional layer can be dependent on both the current input and the previous stored states, which gives the recurrent convolutional layer the information memory ability. Formally, the ith recurrent convolutional layer state x_t^i at time step t can be written as

$$x_t^i = f\left(x_t^{i-1}, h_{t-1}^i\right) \tag{5.8}$$

where $f(\cdot)$ is the nonlinear activation function, x_t^{i-1} is the input volume, i.e., the raw sensor data or the feature maps obtained in its previous layer i-1, and $h_{t-1}^i = x_{t-1}^i$ is the state at time step $t-1$ that is fed back through the recurrent connection.

Theoretically speaking, the recurrent convolutional layers own the capability of learning arbitrary long-term dependencies by directly utilizing the input data. However, this capability is limited to only learning a few time steps backwards due to its intrinsic problem of gradient vanishing or gradient exploding gradient that may happen in practical training.

To alleviate the problem mentioned above and capture long-term dependencies, the gating mechanism (Cho et al. 2014) is introduced into the recurrent convolutional layer. The recurrent convolutional layer is shown in Fig. 5.13, the reset gate r_t^i and update gate u_t^i are included in the recurrent convolutional layer, and they can be calculated by

$$r_t^i = \sigma\left(K_r^i * x_t^{i-1} + W_r^i * h_{t-1}^i + b_r^i\right) \tag{5.9}$$

$$u_t^i = \sigma\left(K_u^i * x_t^{i-1} + W_u^i * h_{t-1}^i + b_u^i\right) \tag{5.10}$$

where $\sigma(\cdot)$ is the logistic sigmoid function, $*$ denotes the convolution operator, K_r^i, W_r^i, K_u^i and W_u^i are the convolution kernels, b_r^i and b_u^i are the bias terms. At each time step t, the gated recurrent convolutional layer state x_t^i can be obtained by

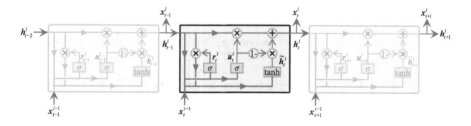

Fig. 5.13 The gating mechanism in the recurrent convolutional layer

$$x_t^i = u_t^i \circ h_{t-1}^i + (1 - u_t^i) \circ \tilde{h}_t^i \tag{5.11}$$

$$\tilde{h}_t^i = \tanh\left(K_h^i * x_t^{i-1} + W_h^i * (r_t^i \circ h_{t-1}^i) + b_h^i\right) \tag{5.12}$$

where ∘ denotes the Hadamard product, $\tilde{h}_t^i$ is the newly generated state, tanh(·) is the tanh activation function, K_h^i and W_h^i are the convolution kernels, and b_h^i is the bias term. From Eqs. (5.11) and (5.12), the state x_t^i is a linear combination of the previous state h_{t-1}^i and the current candidate state $\tilde{h}_t^i$, and is controlled by the reset and update gates.

According to the gating mechanism, the recurrent convolutional layer has obtained the capability to forget or emphasize the previous and current information. On the other hand, the reset gate decides how much the past information will be forgotten. In the meantime, the update gate is able to control the amount of the previous information that will flow to the current state. Such function improves the ability to remember long-term information and eliminates the issue of vanishing gradients. In addition, it should be noted that the recurrent convolutional layers can capture the dependencies in various time scales adaptively due to the separate reset and update gates in it. If we activate the reset gates frequently, the short-term dependencies will be captured and the corresponding feature maps will focus on the current input only. Contrarily, the long-term dependencies will be captured.

5.3.3.2 Pooling Layer and Fully-Connected Layer

Apart from the above described recurrent convolutional layers, there are also pooling layers and fully-connected layers in the RCNN. The pooling layers are used for representation dimensionality reduction and extracted feature compacting. The pooling layer follows the recurrent convolutional layer and is independently conducted on every input feature map, and then outputs a summary statistic of the previous feature map. Especially, the pooling operation greatly improves the statistical efficiency of the network since it creates an invariance to small shifts and distortions. In the ith pooling layer, the state of the jth feature map at time step t which is denoted as y_t^{ij} can be obtained by

$$y_t^{ij} = \text{pool}\left(x_t^{ij}, p, s\right) \tag{5.13}$$

where pool(·) is the down-sampling function, p is the pooling size, and s is the stride size.

The fully-connected layer is placed at the end of RCNN. The role of the fully connected layer is to perform the high-level representation and regression analysis, which is used as the output layer for the RUL prediction. The neurons in the fully-connected layer are fully connected to all neurons in the previous layer. For the ith fully-connected layer, the output at the time step t can be computed by

$$z_t^i = f(z_t^{i-1}) = f(W_f^i z_t^{i-1} + b_f^i) \tag{5.14}$$

where z_t^{i-1} is the input of the ith fully-connected layer, W_f^i is the weight matrix, and b_f^i is the bias vector.

5.3.3.3 Uncertainty Quantification

The uncertainty quantification of prognostic networks has played an important role in the decision making of mechanical system maintenance. In RCNN, the variational inference is utilized to quantify the uncertainty of RUL prediction results. The RCNN is viewed as a probabilistic model with random variables ω that obey the Gaussian prior distribution. The random variable ω includes all learned network parameters, i.e., the convolution kernels and biases in the recurrent convolutional layers $\omega_{\text{RCL}} = \{K_r^i, W_r^i, b_r^i, K_u^i, W_u^i, b_u^i, K_h^i, W_h^i, b_h^i\}_{i=1}^N$ and the weights and biases in the fully-connected layers $\omega_{\text{FCL}} = \{W_f^i, b_f^i\}_{i=1}^L$. Given a training dataset $X = \{x_t\}_{t=1}^T$ and the corresponding outputs $O = \{o_t\}_{t=1}^T$, the posterior distribution of ω can be calculated by the Bayes' theorem:

$$p(\omega|X, O) = \frac{p(O|X, \omega)p(\omega)}{p(O|X)} \tag{5.15}$$

According to the above equation, the predicted distribution for a new input x^* can be calculated by

$$p(o^*|x^*, X, O) = \int p(o^*|x^*, \omega) p(\omega|X, O) d\omega \tag{5.16}$$

The posterior distribution $p(\omega|X, O)$ in Eq. (5.16) cannot be obtained, because of the difficulty in analytically marginalizing the likelihood over ω. Therefore, the variational inference is introduced for the approximation of the posterior distribution $p(\omega|X, O)$.

First of all, the approximating variational distribution $q(\omega)$ is defined to factorize over the weights and biases (Hinton et al.) as follows:

$$\begin{aligned} q(\omega) &= q(\omega_{\text{RCL}}) q(\omega_{\text{FCL}}) \\ &= \prod_{i=1}^N q(K_r^i) q(W_r^i) q(b_r^i) q(K_u^i) q(W_u^i) q(b_u^i) q(K_h^i) \\ &\quad q(W_h^i) q(b_h^i) \prod_{i=1}^L q(W_f^i) q(b_f^i) \end{aligned} \tag{5.17}$$

where $\boldsymbol{W}^l$ and $\boldsymbol{b}^l$ are the weight and bias in the weight layer l. Each weight is defined as a Gaussian mixture distribution, and each bias is considered to follow a simple Gaussian distribution in the variational distribution. Correspondingly, $q(\boldsymbol{W}^l)$ and $q(\boldsymbol{b}^l)$ can be expressed as

$$q(\boldsymbol{W}^l) = \pi^l N(\boldsymbol{\mu}_W^l, \tau^{-1}\boldsymbol{I}) + (1 - \pi^l) N(0, \tau^{-1}\boldsymbol{I}) \tag{5.18}$$

$$q(\boldsymbol{b}^l) = N(\boldsymbol{\mu}_b^l, \tau^{-1}\boldsymbol{I}) \tag{5.19}$$

where $\pi^l \in [0, 1]$ is the probability, $\boldsymbol{\mu}_W^l$ and $\boldsymbol{\mu}_b^l$ are the variational parameters of weight and bias, respectively, and τ is the model precision.

Next, the approximate predicted distribution is derived by minimizing the Kullback–Leibler (KL) divergence between the posterior distribution and approximating variational distribution, which is written as

$$p(o^*|\boldsymbol{x}^*, \boldsymbol{X}, \boldsymbol{O}) \approx \int p(o^*|\boldsymbol{x}^*, \boldsymbol{\omega}) q^*(\boldsymbol{\omega}) d\boldsymbol{\omega} \tag{5.20}$$

where $q^*(\boldsymbol{\omega})$ is the value that is used to minimize the KL divergence. As the optimization objective of RCNN, minimizing KL divergence is equivalent to maximizing the evidence lower bound (Bishop 2006). Therefore, the objective function of RCNN can be written as

$$\begin{aligned} \Gamma &= \mathrm{KL}(q(\boldsymbol{\omega}) \| p(\boldsymbol{\omega}|\boldsymbol{X}, \boldsymbol{O})) \\ &= \int q(\boldsymbol{\omega}) \log \frac{q(\boldsymbol{\omega})}{p(\boldsymbol{\omega}|\boldsymbol{X}, \boldsymbol{O})} d\boldsymbol{\omega} \\ &\propto -\int q(\boldsymbol{\omega}) \log p(\boldsymbol{O}|\boldsymbol{X}, \boldsymbol{\omega}) d\boldsymbol{\omega} + \mathrm{KL}(q(\boldsymbol{\omega}) \| p(\boldsymbol{\omega})) \\ &= -\sum_{t=1}^{T} \int q(\boldsymbol{\omega}) \log p(o_t|\boldsymbol{x}_t, \boldsymbol{\omega}) d\boldsymbol{\omega} + \mathrm{KL}(q(\boldsymbol{\omega}) \| p(\boldsymbol{\omega})) \end{aligned} \tag{5.21}$$

In Eq. (5.21), the first term is evaluated by Monte Carlo integration, and the detailed procedure is as follows. First of all, the integrands are re-parameterized in each sum term by the standard normal distribution $q(\boldsymbol{\alpha}) = N(0, \boldsymbol{I})$ and the Bernoulli distribution $q(\boldsymbol{\beta}) = \mathrm{Bernoulli}(\pi)$. The $\boldsymbol{W}^l$ and $\boldsymbol{b}^l$ can be re-written as

$$\boldsymbol{W}^l = \boldsymbol{\beta}^l \left(\boldsymbol{\mu}_W^l + \tau^{-\frac{1}{2}} \boldsymbol{\alpha} \right) + (1 - \boldsymbol{\beta}^l) \tau^{-\frac{1}{2}} \boldsymbol{\alpha} \tag{5.22}$$

and

$$\boldsymbol{b}^l = \boldsymbol{\mu}_b^l + \tau^{-\frac{1}{2}} \boldsymbol{\alpha} \tag{5.23}$$

5.3 Recurrent Convolutional Neural Network-Based RUL Prediction

Secondly, the objective function can be reformulated based on Eqs. (5.22) and (5.23), which is written as

$$\Gamma = -\sum_{t=1}^{T} \int q(\alpha, \beta) \log p(o_t|x_t, \omega(\alpha, \beta)) d\alpha d\beta + \mathrm{KL}(q(\omega) \| p(\omega)) \quad (5.24)$$

Thirdly, the integral in the first term of Eq. (5.24) is estimated by Monte Carlo integration with a single sample $\hat{\omega}_t\ q(\omega)$ and the unbiased estimator $\log p(o_t|x_t, \hat{\omega}_t)$ is derived. Finally, the objective function is obtained as

$$\Gamma = -\sum_{t=1}^{T} \log p(o_t|x_t, \hat{\omega}_t) + \mathrm{KL}(q(\omega) \| p(\omega)) \quad (5.25)$$

Further, the second term in Eq. (5.21) is approximated as L^2 regularization based on (Gal and Ghahramani 2016), which is written as

$$\mathrm{KL}(q(\omega) \| p(\omega)) = \sum_{l=1}^{M+L} \left(\frac{\pi^l c^2}{2} \| \mu_W^l \|_2^2 + \frac{c^2}{2} \| \mu_b^l \|_2^2 \right) \quad (5.26)$$

where c is the prior length-scale. The objective function in Eq. (5.25) is written as

$$\begin{aligned}
\Gamma &= -\sum_{t=1}^{T} \log p(o_t|x_t, \hat{\omega}_t) + \sum_{l=1}^{M+L} \left(\frac{\pi^l c^2}{2} \| \mu_W^l \|_2^2 + \frac{c^2}{2} \| \mu_b^l \|_2^2 \right) \\
&\propto \frac{1}{T} \sum_{t=1}^{T} \frac{-\log p(o_t|x_t, \hat{\omega}_t)}{\tau} + \sum_{l=1}^{M+L} \left(\frac{\pi^l c^2}{2\tau T} \| \mu_W^l \|_2^2 + \frac{c^2}{2\tau T} \| \mu_b^l \|_2^2 \right) \quad (5.27) \\
&= \frac{1}{T} \sum_{t=1}^{T} E(o_t, \hat{o}(x_t, \hat{\omega}_t)) + \sum_{l=1}^{M+L} \left(\frac{\pi^l c^2}{2\tau T} \| \mu_W^l \|_2^2 + \frac{c^2}{2\tau T} \| \mu_b^l \|_2^2 \right)
\end{aligned}$$

where $E(\cdot, \cdot)$ is the prior length-scale, i.e., mean square error, mean absolute error and Huber loss.

For the probabilistic RCNN, the network output $\hat{o}(\cdot)$ with a sample $\hat{\omega}$ is estimated by randomly masking rows in each weight matrix during the forward pass, which is similar to dropout (Hinton et al. 2012) in each weight layer of RCNN. And the second term in Eq. (5.27) is equivalent to the L^2 regularization term of each weight and bias during model optimization. Consequently, the uncertainty of RCNN is quantified by introduced dropout with probability π and L^2 regularization with weight decay coefficient λ into the recurrent convolutional layer and fully-connected layer. Correspondingly, the objective function in Eq. (5.27) is proportional to the following objective function:

$$\Gamma \propto \Gamma_{\text{dropout}} = \frac{1}{T} \sum_{t=1}^{T} E(o_t, \hat{o}_t) + \lambda \sum_{l=1}^{M+L} \left(\left\| \boldsymbol{W}^l \right\|_2^2 + \left\| \boldsymbol{b}^l \right\|_2^2 \right) \quad (5.28)$$

where Γ_{dropout} is an objective function that is minimized by some optimizers, i.e., stochastic gradient descent and Adam optimizer. Then, the Monte Carlo estimation is utilized to achieve the prediction as:

$$p(o^* | \boldsymbol{x}^*, \boldsymbol{X}, \boldsymbol{O}) \approx \int p(o^* | \boldsymbol{x}^*, \boldsymbol{\omega}) q(\boldsymbol{\omega}) \mathrm{d}\boldsymbol{\omega} \approx \frac{1}{V} \sum_{v=1}^{V} p(o^* | \boldsymbol{x}^*, \hat{\boldsymbol{\omega}}_v) \quad (5.29)$$

where the $\boldsymbol{x}^*$ is a new input, V is the number of performing random forward passes through the network, and the $\hat{\boldsymbol{\omega}}_v \ q(\boldsymbol{\omega})$.

5.3.4 RUL Prediction Case Study of FEMTO-ST Accelerated Degradation Tests of Rolling Element Bearings

5.3.4.1 FEMTO-ST Bearing Datasets

The public FEMTO-ST bearing datasets are available on the website of NASA Prognostics Data Repository (Nectoux et al. 2012). The structure of the experimental platform, which is referred to as PRONOSTIA, is shown in Fig. 5.14. As shown in Table 5.4, three operating conditions are set for accelerated degradation tests, and a total number of seventeen bearings are tested. The bearing model is NSK 6804RS. In order to obtain the full-life vibration signal of each bearing, two acceleration sensors are installed in the horizontal and vertical directions of the tested bearings outer race respectively. The last two bearing datasets of Condition 1 and Condition 2 are selected as the testing dataset to verify the RCNN. In the experiments, the sampling frequency is 25.6 kHz, and the sample duration is 0.1 s, the data points are recorded every 10 s. Figure 5.15 shows the pictures before and after the tested bearing failure. It can be seen from the figure that the failure may occur on a certain component or multiple components during the bearing degradation process. Further, the horizontal and vertical full-life vibration signals of Ber 1_1 are shown in Fig. 5.16. The input size of RCNN is 2560 × 1 × 2 since both horizontal and vertical vibration signals are used.

5.3.4.2 Configuration of RCNN

The parameters of RCNN mainly include the number of recurrent convolutional layers N, the number of convolution kernels M, the kernel size, the pooling size

5.3 Recurrent Convolutional Neural Network-Based RUL Prediction

Fig. 5.14 The PRONOSTIA experimental platform

Table 5.4 Descriptions about FEMTO-ST bearing datasets

Operating condition	Radial force /N	Rotating speed /(r/min)	Bearing dataset	
			Training dataset	Testing dataset
Condition 1	4000	1800	Ber 1_1 Ber 1_2 Ber 1_3 Ber 1_4 Ber 1_5	Ber 1_6 Ber 1_7
Condition 2	4200	1650	Ber 2_1 Ber 2_2 Ber 2_3 Ber 2_4 Ber 2_5	Ber 2_6 Ber 2_7
Condition 3	5000	1500	Ber 3_1 Ber 3_2 Ber 3_3	—

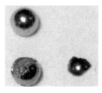

Fig. 5.15 Pictures of the normal and failed bearings in the tests

p, and the number of neurons F. In order to realize the parameter configuration of RCNN, 5-fold cross-validation on the training datasets is employed to determine the structural parameters. Meanwhile, the dropout and L^2 regularization is introduced to the recurrent convolutional layer and fully-connected layer, and V stochastic forward passes are performed to acquire the predicted mean and variance. In addition, the root mean square error is used as the loss function, Adam optimizer is used with a mini-batch size of 128, and the network parameters are updated through iterative training to optimize the objective function Γ_{dropout} in Eq. (5.28). The RCNN is trained for 200 epochs, and its detailed configuration is summarized in Table 5.5.

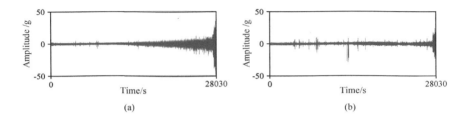

Fig. 5.16 Vibration signals of Ber 1_1 in a lifecycle: (a) horizontal, and (b) vertical

Table 5.5 Configuration of RCNN

Hyperparameter	Size	Hyperparameter	Size
Number of kernels M	16	Kernel size $k \times 1$	8×1
Number of layers L	4	Pooling size p	8
Number of neurons F	100	Weight decay coefficient λ	10^{-5}
Number of forward passes V	1000	Dropout probability π	0.15
Mini-batch size	128	Number of epochs	200

5.3.4.3 Effect of Dropout Probability on Uncertainty Quantification

The performance of the RCNN for bearing RUL prediction is investigated and discussed. It is mainly divided into two aspects. Firstly, we discussed the influence of network depth and dropout probability on prediction accuracy and uncertainty quantification. Then, RCNN is compared with the state-of-the-art prognosis approaches to verify its superiority.

For deep prognosis networks, the depth of the network has an important impact on the representation learning ability. In the RCNN architecture, the network depth increases with the periodical stacking of recurrent convolutional layers and pooling layers. As shown in Table 5.6, we use the RCNN to predict the life of bearings in the case of five network depths, and evaluate the prediction performance of the network according to CRA and C_{PE}. CRA value is computed by

$$\text{CRA} = \frac{1}{S} \sum_{s=1}^{S} \left(1 - \frac{|\text{ActRUL}_s - \text{PreRUL}_s|}{\text{ActRUL}_s} \right) \tag{5.30}$$

5.3 Recurrent Convolutional Neural Network-Based RUL Prediction

Table 5.6 The effect of the network depth on the prediction performance of RCNN

Number of layers N	Total model parameters	CRA					C_{PE}			
		Ber 1_6	Ber 1_7	Ber 2_6	Ber 2_7		Ber 1_6	Ber 1_7	Ber 2_6	Ber 2_7
2	57421	0.5623	0.4102	0.4696	0.6829		6307	5886	1796	588
3	208269	0.6981	0.6372	0.7395	0.7967		6214	5681	1790	573
4	804877	0.7781	0.7419	0.8116	0.8453		6164	5619	1758	564
5	3177741	0.7717	0.7533	0.8250	0.8637		6187	5613	1760	564
6	12642061	0.7320	0.7795	0.8320	0.8439		6180	5617	1765	568

where S is the number of testing samples, ActRUL_s and PreRUL_s are the actual RUL and the predicted RUL values corresponding to the sth testing sample, respectively. C_{PE} is used to measure convergence and is defined by the Euclidean distance between the origin and the centroid of the area under the prediction error curve (Saxena et al. 2010). C_{PE} value is calculated by

$$C_{\text{PE}} = \sqrt{(C_x - t_p)^2 + C_y^2} \tag{5.31}$$

where t_p is the first prediction time instance, and (C_x, C_y) is the centroid of the area under the prediction error curve. The lower the C_{PE} value is, the faster the predicted RUL converges to the actual RUL.

As shown in Table 5.3, the value of CRA increases as the increment of network depth, while the value of C_{PE} changes in the opposite direction. This demonstrates that RCNN has a much stronger capability in representation learning along with the increase of network depth, which results in a higher accuracy and faster convergence of the prediction results. Nevertheless, limited by computational burden and accuracy saturation or even overfitting problem due to sample limitations, the number of recurrent convolutional layers L is set to be 4 in this chapter.

The uncertainty of RCNN can be quantified by Monte Carlo dropout and the dropout probability can largely affect the uncertainty quantification of RCNN. Figure 5.17 is the prediction results of Ber 2_7 with different values of π. In particular, Fig. 5.18 shows the RUL distributions at 1700s (three-quarters of lifetime), and the corresponding mean and standard deviation (SD) are calculated. As shown in Figs. 5.17 and 5.18, RCNN with a larger dropout probability results in a wider uncertainty interval. So, during the forward pass, a larger dropout probability can mask more units in the weight layers and then increase the modeling uncertainty with a larger SD of RUL. Therefore, in order to provide more accurate maintenance decisions, it is more inclined to find a small π value. However, it should be noted that it is difficult to capture the uncertainty effectively from the measurement noise if the dropout probability is too small, and at the same time, it is not favorable to the improvement of network generalization and avoiding overfitting problems. In summary, the dropout probability π is set to be 0.15 in RUL prediction.

5.3.4.4 Comparison with State-Of-The-Art Prognosis Approaches

In this section, four state-of-the-art prognosis approaches presented in (Ren et al. 2018), (Zhu et al. 2018), (Babu et al. 2016) and (Hinchi and Tkiouat 2018) are implemented to demonstrate the superiority of RCNN, and are denoted as M1, M2, M3 and M4, respectively. Among them, the first three methods are developed on the standard CNN, and the fourth method is a combination of CNN and LSTM. Specifically, M1 takes 64 time–frequency domain features extracted by fast Fourier

5.3 Recurrent Convolutional Neural Network-Based RUL Prediction

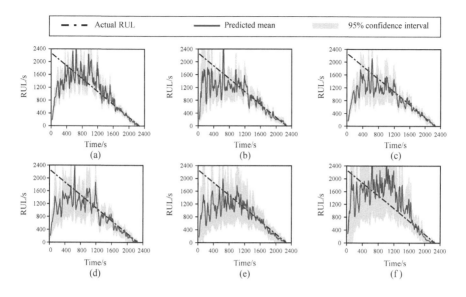

Fig. 5.17 RUL prediction results of Ber 2_7 using different dropout probability: (a) $\pi=0.05$, (b) $\pi=0.1$, (c) $\pi=0.15$, (d) $\pi=0.2$, (e) $\pi=0.25$, and (f) $\pi=0.3$

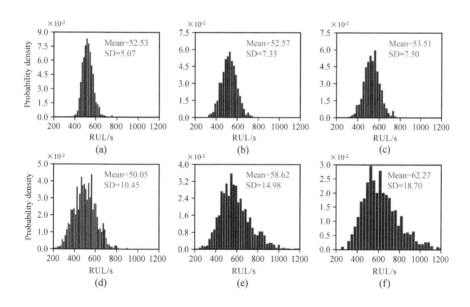

Fig. 5.18 Predictions for Ber 2_7 at time instance of 1700s using different dropout probability: (a) $\pi=0.05$, (b) $\pi=0.1$, (c) $\pi=0.15$, (d) $\pi=0.2$, (e) $\pi=0.25$, and (f) $\pi=0.3$

transform as input, and predicts the RUL of the bearings through stacked 3 convolution layers, 3 average pooling layers, and 7 fully connected layers. M2 uses wavelet transform to extract time–frequency domain features, and uses 8-layer CNN for RUL prediction. M3 takes the raw vibration signals as inputs and utilizes a stack of 2 convolutional layers, 2 average pooling layers and 1 fully-connected layer to achieve the prediction of RUL. M4 uses the convolutional layer and the global pooling layer to extract features from vibration signals, and utilizes the convolutional layer and the fully connected layer for RUL prediction. From Table 5.7, it can be seen that RCNN has the highest CRA values and lowest C_{PE} values, which indicates that RCNN has the highest prediction accuracy and the fastest convergence speed on the FEMTO-ST bearing datasets. More importantly, RCNN is useful for the maintenance decisions making of mechanical systems due to its ability of the distribution of probability prediction rather than point estimation. In summary, RCNN is more superior in rolling element bearing RUL prediction compared to the other four prognosis approaches.

Table 5.7 Performance comparison of five different prognosis approaches on FEMTO-ST bearing datasets

Bearing dataset		M1	M2	M3	M4	RCNN
Ber 1_6	CRA	0.5550	0.7227	0.6123	0.7410	**0.7781**
	C_{PE}	6313.0213	6214.1844	6324.7402	6167.4558	**6164.8550**
Ber 1_7	CRA	0.4789	0.7231	0.4592	0.6721	**0.7419**
	C_{PE}	5910.2072	5647.1975	5941.6631	5654.8450	**5619.0031**
Ber 2_6	CRA	0.6731	0.7590	0.6455	0.7206	**0.8116**
	C_{PE}	1758.2144	1759.7486	1835.5745	1772.8769	**1758.4386**
Ber 2_7	CRA	0.4305	0.7889	0.5809	0.8183	**0.8453**
	C_{PE}	666.6638	583.9817	613.4448	571.1441	**564.5396**

5.3.5 Epilog

In this section, an end-to-end framework named RCNN is constructed to provide the probabilistic estimation RUL of machinery. The time-series data acquired by different sensors are served as its inputs. Then, a recurrent convolutional layer is built to model the temporal dependencies of different degradation states, combined with a max pooling layer to reduce the dimensionality of representations. Through multiple stacking recurrent convolutional layers and max pooling layers, the high-level representations can be learned automatically from the input sensor data. After that, the fully-connected layers are added to map the learned representations to the preliminary RUL. Especially, to quantify the uncertainty of predicted RUL, the variational inference combined with the Monte Carlo dropout are utilized. The run-to-failure data of rolling element bearings are used to evaluate the prediction performance of RCNN. From the experimental results, it can be concluded that RCNN has advantages in accuracy compared with other common prognostics approaches. More importantly, RCNN has the ability to provide probabilistic RUL prediction results, which is meaningful for maintenance decision-making in practical engineering.

5.4 Multi-scale Convolutional Attention Network-Based RUL Prediction

5.4.1 Motivation

With the rapid development of measurement and instrumentation technology, information technology and intelligent control technology, cyber-physical systems have been widely applied in industrial mechanical systems, which contributes to the explosive growth of monitoring data acquired by various sensors. In order to exhaustively utilize the complete degradation information of mechanical systems, existing deep learning-based RUL prediction methods commonly utilize multi-sensor data as the inputs of constructed networks. However, these methods lack a particular mechanism to effectively integrate these multi-sensor data. To be specific, during their construction of the networks, they all assume that the different sensor data make the same contribution to the RUL prediction. While in practical cases, monitoring data collected from various sensors are more likely to possess different degrees of degradation information. In other words, monitoring data collected from some sensors may be informative, while the rest might be non-informative. If no certain mechanism is adopted for the clear differentiation of different sensors and emphasis of sensor data with critical information, the prediction accuracy of constructed networks would be compromised due to the existence of irrelevant and redundant information, which will further jeopardize the generalization in RUL prediction of mechanical systems. Targeted at the above weaknesses, a new deep prognosis network called multi-scale convolutional attention network (MSCAN) is presented in this section for predicting RUL of mechanical systems (Wang et al. 2021).

5.4.2 Multi-scale Convolutional Attention Network

The architecture of the MSCAN is provided in Fig. 5.19, which is composed of multi-scale representation learning subnetworks and an RUL prediction module. In MSCAN, self-attention modules are first constructed to highlight the discriminative information from the input multi-sensor data. Then, the multi-scale mechanism is

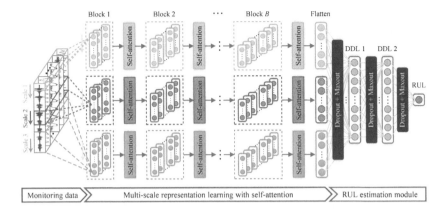

Fig. 5.19 Illustration of the MSCAN architecture

developed to learn representations from multiple temporal scales. Finally, the learned high-dimensional representations are integrated by dynamic dense layers (DDLs) to execute regression reasoning and RUL estimation.

5.4.3 Architecture of MSCAN

As shown in Fig. 5.19, MSCAN is composed of representation learning subnetworks and the RUL estimation module. The representation learning network is designed by stacking multiple convolution building blocks, and each of them is followed by a self-attention module. Further, a multi-scale mechanism is employed in the subnetwork to fully derive representation to guarantee the integrity of learned representations. In the RUL estimation module, instead of the traditional fully-connected layer (FCL), a new layer, DDL, is built to execute regression analysis and reasoning. The details of MSCAN are presented as follows.

5.4.3.1 Convolutional Building Blocks

In MSCAN, each convolutional building block consists of 2 convolutional layers and 1 pooling layer. The convolutional layers are used to learn multiple levels of representations from the monitoring data. For each convolutional layer, a set of learnable kernels are firstly utilized to convolve the inputs, and then an element-wise nonlinear activation function is employed on the outputs of convolution operations. Through these two operations, a convolutional layer can extract different feature maps. Mathematically, $x^{l-1} \in \mathbb{R}^{H \times 1 \times C}$ and $k^l \in \mathbb{R}^{F \times 1 \times C \times N}$ respectively denote the input data and the convolution kernels, where H is the length of the input data, C is the number of input channels, $F \times 1$ is the size of convolution kernels, and N is the

5.4 Multi-scale Convolutional Attention Network-Based RUL Prediction

number of convolution kernels. Then, the nth feature map of the lth convolutional layer, i.e., x_n^l, can be obtained by:

$$x_n^l = \sigma_{\mathrm{r}}(u_n^l) \tag{5.32}$$

$$u_n^l = k_n^l * x^{l-1} + b_n^l = \sum_{c=1}^{C} k_{n,c}^l * x_c^{l-1} + b_n^l \tag{5.33}$$

where $\sigma_{\mathrm{r}}(\cdot)$ is the ReLU activation function, u_n^l is the output of convolution operations, $*$ denotes the convolution operator, k_n^l is the nth convolution kernel, and b_n^l is the bias term.

In a convolutional building block, the pooling layer is followed after the second convolutional layer. Formally, the nth feature map of the lth pooling layer, y_n^l, can be computed by:

$$y_n^l = \mathrm{pool}(y_n^{l-1}, p, s) \tag{5.34}$$

where y_n^{l-1} is the nth input feature map, pool($\cdot$) is the max pooling function, p denotes the pooling size, and s denotes the stride size. In MSCAN, the max pooling operations are conducted with a non overlapping window, i.e., $p = s$.

5.4.3.2 Self-Attention Modules

In MSCAN, multi-channel time-series monitoring sensor data are fed into the network, where each channel corresponds to a sensor sequence. These input data have two noticeable properties.

- The data acquired via different sensors possess different degrees of mechanical system degradation information. In other words, some may contain well informative degradation features, but others have limited ones or are even full of measurement noise.
- The essential degradation features, such as shocks, only occur in partial positions in one senor sequence.

Hence, to differentiate and emphasize different sensor data, it is necessary to model interchannel relationships along the channel dimension explicitly. Moreover, to capture the pivotal degradation features, it is essential to independently encode the contextual relationships of each channel along the temporal dimension (TD). Accordingly, as shown in Fig. 5.20, the self-attention module is designed into two parts: channel-wise attention and temporal attention, which enables MSCAN to learn "what" and "where" to attend in the channel and temporal dimensions.

As shown in Fig. 5.20, $z^{l-1} \in \mathbb{R}^{I \times 1 \times J}$ denotes the feature maps extracted from different input sensor sequences. The self-attention module sequentially presumes

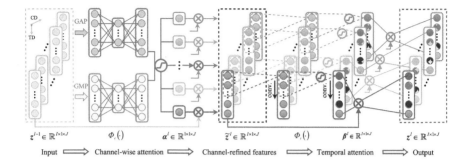

Fig. 5.20 The illustration for self-attention modules in MSCAN

channel-wise attention weights $\boldsymbol{\alpha}^l \in \mathbb{R}^{1 \times 1 \times J}$ and temporal attention weights $\boldsymbol{\beta}^l \in \mathbb{R}^{I \times 1 \times J}$, where I is the length of each feature map, $J = N \times S$ is the number of the feature maps, and S is the channel number of the input sensor sequences. The integral attention process can be expressed as:

$$\tilde{z}^l = \boldsymbol{\alpha}^l \otimes z^{l-1} = \Phi_c(z^{l-1}) \otimes z^{l-1} \tag{5.35}$$

$$z^l = \boldsymbol{\beta}^l \otimes \tilde{z}^l = \Phi_t(\tilde{z}^l) \otimes \tilde{z}^l \tag{5.36}$$

where $\otimes$ denotes the element-wise product, $\tilde{z}^l \in \mathbb{R}^{I \times 1 \times J}$ denotes the channel-refined feature maps, i.e., the outputs of the channel-wise attention, $z^l \in \mathbb{R}^{I \times 1 \times J}$ denotes the terminal refined output feature maps, i.e., the outputs of the temporal attention, $\Phi_c(\cdot)$ and $\Phi_t(\cdot)$ respectively denote the channel-wise and temporal attention functions, which are detailed as follows.

(1) Channel-wise attention

Through the modeling of the interrelationships amongst channels, channel-wise attention can be built. To be specific, the global average pooling (GAP) and the global max pooling (GMP) are firstly utilized to integrate the global information of each channel. Therefore, the channel descriptors $\boldsymbol{v}^l \in \mathbb{R}^J$ and $\boldsymbol{m}^l \in \mathbb{R}^J$ can be obtained. It should be noted that $\boldsymbol{v}^l$ and $\boldsymbol{m}^l$ separately contain J channel-wise statistics that are acquired as follows:

$$v_j^l = \frac{1}{I} \sum_{i=1}^{I} z_{j,i}^{l-1} \tag{5.37}$$

$$m_j^l = \max\left(z_j^{l-1}\right) \tag{5.38}$$

5.4 Multi-scale Convolutional Attention Network-Based RUL Prediction

Then, a MLP with one hidden layer follows to capture the interchannel relationships and assess the informativeness of each channel. After that, the element-wise summation is conducted to fuse the outputs of two MLPs. Mathematically, the channel-wise attention weights α^l can be obtained by:

$$\alpha^l = \sigma_{h\_s}\left(W^l_{12}\left(W^l_{11}v^l\right) \oplus W^l_{22}\left(W^l_{21}m^l\right)\right) \quad (5.39)$$

where $\oplus$ denotes the element-wise summation, $\sigma_{h\_s}(\cdot)$ denotes the hard sigmoid activation function (Nwankpa et al. 2018), $W^l_{11} \in \mathbb{R}^{\frac{J}{r} \times J}$, $W^l_{12} \in \mathbb{R}^{J \times \frac{J}{r}}$, $W^l_{21} \in \mathbb{R}^{\frac{J}{r} \times J}$ and $W^l_{22} \in \mathbb{R}^{J \times \frac{J}{r}}$ denote the weight matrices in the MLPs, the biases are omitted. Finally, by conducting channel-wise multiplication operation between the inputs z^{l-1} and the channel-wise attention weights α^l along the channel dimension, the channel-refined feature maps $\tilde{z}^l$ can be obtained. It is worth mentioning that each feature map obtained by the convolutional building block can be regarded as a pattern detector (Zeiler and Fergus 2014). During representation learning, these feature maps are able to automatically extract various patterns from the input multi-sensor data. And the channel-wise attention can recognize important degradation ones from the extracted patterns. In summary, the channel-wise attention focuses on "what" is valuable among the multi-sensor data.

(2) Temporal attention

By encoding the contextual relationships of each channel, temporal attention can be built to capture informative locations. As illustrated in Fig. 5.20, considering that dilated convolutions (Yu et al. 2017) are beneficial to enlarge the receptive field, stacked dilated convolutions are firstly employed to handle the channel-refined feature maps $\tilde{z}^l$. The stacked depth wise dilated convolutions apply a single dilated convolution kernel into each channel of $\tilde{z}^l$ and can independently map the context of each channel. To be specific, the jth weight vector of the temporal attention weights β^l can be calculated by:

$$\beta^l_j = \sigma_{h\_s}\left(D^l_{2,j} *_d \left(D^l_{1,j} *_d \tilde{z}^l_j + b^l_{1,j}\right) + b^l_{2,j}\right) \quad (5.40)$$

where $*_d$ denotes the dilated convolution operation with a dilation rate of size d, $D^l_{1,j} \in \mathbb{R}^{F' \times 1}$ and $D^l_{2,j} \in \mathbb{R}^{F' \times 1}$ denote the dilated convolution kernels corresponding to the jth channel, $F' \times 1$ denotes the size of dilated convolution kernels, $b^l_{1,j}$ and $b^l_{2,j}$ are the bias terms. Finally, by conducting element-wise multiplication between $\tilde{z}^l$ and β^l, the refined feature maps z^l can be obtained and fed into the following convolutional building blocks. Throughout the above computational process, instead of considering each temporal location equally, the temporal attention is able to pay more attention to the locations that are relative to mechanical system degradation. In a word, temporal attention focuses on "where" is salient given one sensor sequence, which is complementary to channel-wise attention.

5.4.3.3 Multi-scale Representation Learning via Parallel Convolutional Pathways

In convolutional networks, different sizes enable the convolution kernels to extract information from different temporal scales (Szegedy et al. 2015). A large convolution kernel specializes in the information that is distributed more globally, whereas a small convolution kernel specializes in the information that is distributed more locally. For the multi-sensor monitoring data, the distribution of decisive degradation features changes over operation time. Specifically, there exist only few degradation features in the monitoring data in the initial operating time, but as time passes, the degradation features accumulate in the monitoring data. Besides, for different kinds of monitoring data, such as vibration signals and force signals, the scales of degradation features are distinct. Consequently, if only a single convolution kernel size is used in the networks, it is unavoidable for the loss of degradation information, resulting in the reduction of prediction accuracy. To overcome this shortcoming, a multi-scale learning strategy is designed in MSCAN.

As shown in Fig. 5.19, three different convolutional pathways are designed in parallel to extract representations from the initial multi-sensor monitoring data. To ensure the completeness of representations, the convolutional pathways, i.e., $F \times 1$, $2F \times 1$ and $4F \times 1$, have no interaction among the different convolutional pathways during representation extracting, and the degradation information is independently extracted from different temporal scales. It should be noted that the network structure and hyperparameter settings of the three convolutional pathways are the same except for the convolution kernel sizes. For each convolutional pathway, the high-dimensional representations can be extracted by B stacked convolutional building blocks and self-attention modules. After that, these obtained representations are flattened as a 1-dimensional vector and concatenated together to input into the RUL estimation subnetwork to perform RUL estimation.

5.4.3.4 Dynamic Dense Layers

In the RUL estimation module, rather than the traditional FCL, a new layer, DDL, is constructed to execute regression reasoning and RUL estimation. As presented in Fig. 5.21, two pivotal techniques are included in the DDL. One is the dynamic Gaussian dropout which is employed to diminish the overfitting problem and strengthen the generalization of the constructed network. Instead of a standard dropout that randomly masks neurons with a constant probability of Bernoulli distribution (Srivastava et al. 2014), the dynamic Gaussian dropout introduces multiplicative noise which is sampled from $N(1, var(t))$ into each neuron of the inputs. Additionally, the variance of Gaussian noise is fluctuating over the training epoch to avoid co-adaptations between neurons. The other one is the maxout activation (Goodfellow et al. 2013) that is utilized to augment the reasoning ability of DDL and strengthen the prediction accuracy. The maxout activation outputs the maximum of G candidate linear units and thus makes a piece-wise linear approximation to an arbitrary convex function.

5.4 Multi-scale Convolutional Attention Network-Based RUL Prediction

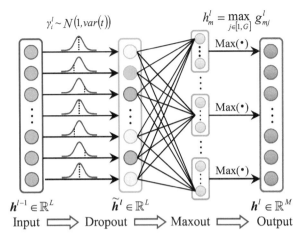

Fig. 5.21 Dynamic Gaussian dropout and maxout activation in the DDL

In MSCAN, the RUL estimation subnetwork is formed by three stacked DDLs. The neuron number of the first two DDLs (that is set to be M) is the same. Whereas the third DDL has only one neuron and is set as the output layer of MSCAN to give the final RUL. Mathematically, for the mth neuron of the lth DDL, the output, h_m^l, can be computed by:

$$\gamma_i^l \sim N(1, var(t)) \tag{5.41}$$

$$\tilde{\boldsymbol{h}}^{l-1} = \boldsymbol{\gamma}^l \otimes \boldsymbol{h}^{l-1} \tag{5.42}$$

$$g_{mj}^l = \boldsymbol{W}_{mj}^l \tilde{\boldsymbol{h}}^{l-1} + b_{mj}^l \tag{5.43}$$

$$\boldsymbol{h}_m^l = \max_{j \in [1,G]} g_{mj}^l \tag{5.44}$$

where $\boldsymbol{h}^{l-1}$ denotes the inputs of the lth DDL, i.e., the high-dimensional representations from the representation learning subnetwork, γ_i^l denotes the Gaussian noise sampled from $N(1, var(t))$, $i = 1, 2, \cdots, L$, L denotes the number of the inputs, $\boldsymbol{W}_{mj}^l$ denotes the weight vector, and b_{mj}^l is the bias term. During training, the variance of Gaussian noise is increased linearly over the training time, i.e., $var(t) = t/T$, where t is the current training epoch and T is the total number of training epochs.

5.4.4 RUL Prediction Case of a Life Testing of Milling Cutters

For the performance evaluation of MSCAN, multi-sensor monitoring data collected from life testing of milling cutters are used. In addition, the RUL prediction results of MSCAN are compared with several state-of-the-art prognosis models.

5.4.4.1 Data Description

As presented in Fig. 5.22, the lifetime tests of milling cutters are carried out on a computer numerical control (CNC) milling machine. During each test, a C45 steel workpiece is machined on an APMT 1135 milling machine. As tabulated in Table 5.8, three operating conditions are set, and a total of 18 milling cutters are tested in accordance with international standard ISO 8688–1. To monitor the degradation processes of milling cutters roundly, five types of sensors are mounted in the experiment rig, i.e., vibration accelerometer (PCB 356A15), triaxial dynamometer (Kistler 9257B), microphone (Brüel & Kjær 4966), current sensor (Anyway CSA201-P030T01) and acoustic emission (AE) sensor (PAC WD). The sampling frequency is set to be 10 kHz for the first four sensors, whereas, for the last acoustic emission sensor, the sampling frequency is 1 MHz.

The degradation of milling cutters mainly contains two categories: wear and brittle fracture. Figure 5.23 presents two deterioration pictures of the cutters after reaching their end of life. For the wear condition evaluation of cutters, a GP-300C industrial microscope is used for measuring the width of the flank wear land after each cutting. Figure 5.24 plots the collected monitoring data of C1_1 during the entire operating life.

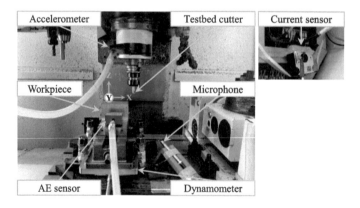

Fig. 5.22 CNC milling machine and sensor placement

5.4 Multi-scale Convolutional Attention Network-Based RUL Prediction

Table 5.8 Cutting conditions of milling cutters

Cutting condition	Spindle speed /(r/min)	Feed rate /(mm/min)	Depth of cut /mm	Training dataset	Testing dataset
Condition 1	2500	200	2	C1_1 C1_2 C1_3 C1_4	C1_5 C1_6
Condition 2	3500	300	1.75	C2_1 C2_2 C2_3 C2_4	C2_5 C2_6
Condition 3	4500	400	1.5	C3_1 C3_2 C3_3 C3_4	C3_5 C3_6

 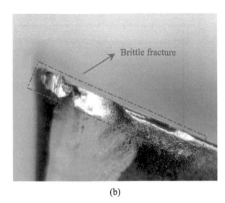

Fig. 5.23 Pictures of the cutter deteriorations: (a) flank wear, and (b) brittle fracture

5.4.4.2 Configuration of MSCAN

During the RUL prediction, the hyperparameters of MSCAN are tuned by fourfold cross-validation, and eventual selected hyperparameters are listed in Table 5.9. In the training phase, the mean square error is used as the loss function and the Adam optimizer is used with a mini-batch size of 128 to iteratively update. After 100 training epochs, the trained model is tested to predict the RUL of milling cutters. Specifically, each constructed network will run 10 times for reducing the influence of randomness. In addition, two prognosis metrics, i.e., scoring function (Saxena et al. 2008) and RMSE (Wang et al. 2019) are used to quantitatively assess the prediction performance. For each cutter in the testing datasets, the values of metrics are computed from half to the end of life.

5.4.4.3 Advantages of Self-Attention Modules and Multi-scale Learning

Figure 5.25 presents the RUL prediction results of C1_5 and C1_6, in which the mean value of prediction results is marked as the blue solid line, whereas the SD of prediction results is depicted in the orange shaded region. From Fig. 5.25, it can be distinctly seen that as operating time passes, the predicted mean of MSCAN converges to

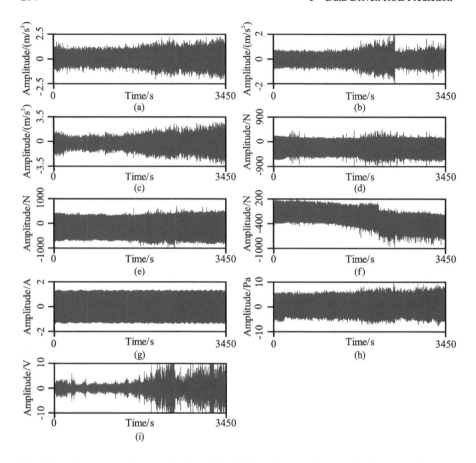

Fig. 5.24 Lifetime monitoring signals of C1_1: (a) x-axis vibration signals, (b) y-axis vibration signals, (c) z-axis vibration signals, (d) x-axis force signals, (e) y-axis force signals, (f) z-axis force signals, (g) current signals, (h) sound signals, and (i) AE signals

Table 5.9 Configuration of MSCAN in RUL prediction of milling cutters

Hyperparameter	Value	Hyperparameter	Value
Number of convolution kernels N	8	Size of convolution kernels $F \times 1$	4×1
Pooling size p	8	Number of blocks B	3
Ratio of dimension reduction r	8	Size of dilated convolution kernels $F' \times 1$	4×1
Number of candidate linear units G	4	Dilation rate d	2
Number of neurons M	25	Number of training epochs T	100

5.4 Multi-scale Convolutional Attention Network-Based RUL Prediction

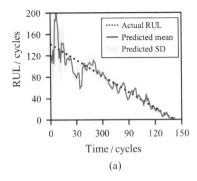 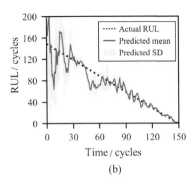

Fig. 5.25 RUL prediction results of milling cutters: (a) C1_5, and (b) C1_6

the actual RUL gradually, meanwhile, the SD interval is narrow. To further explicate the benefits from the designed self-attention modules and multi-scale learning, the other three prognosis networks are designed to perform RUL prediction tasks, which are denoted as Network-1, Network-2 and Network-3, respectively. The basic frameworks of these networks are identical with those of MSCAN, except that.

(1) Network-1 does not employ self-attention modules and multi-scale learning mechanisms.
(2) Network-2 only employs self-attention modules.
(3) Network-3 only employs multi-scale learning mechanisms.

Moreover, the hyperparameter settings of these three networks are inherited from the setting of MSCAN. Table 5.10 illustrates the performance results of these four networks. From the comparison results, it can be noticed that MSCAN receives lower score values and smaller RMSE values compared with the classic convolutional network (i.e., Network-1), which indicates that the use of self-attention modules or multi-scale learning improves the prediction performance of the network. For Network-2, by incorporating self-attention modules into the network, it is able to adaptively spot multi-sensor degradation information from both the channel and the temporal dimensions. For Network-3, with the utilization of multiple convolutional pathways, the network can learn more complete representations, and the prediction accuracy is thus improved. Besides, with the help of the comprehensive combination of self-attention modules with multi-scale learning, the MSCAN obtains the lowest RMSE and score values for all cutters, which confirms again the benefits of utilizing self-attention modules and multi-scale learning.

5.4.4.4 Benefits from Dynamic Gaussian Dropout and Maxout Activation

In MSCAN, a new layer called DDL is built to substitute the traditional FCLs to execute regression reasoning and RUL estimation. To illustrate the benefits of DDL,

Table 5.10 Performance estimation results of four different networks

Testing dataset		Network-1	Network-2	Network-3	MSCAN
C1_5	Score	126.83 ± 8.53	30.72 ± 3.36	44.00 ± 6.25	**22.27 ± 2.91**
	RMSE	10.15 ± 0.25	4.73 ± 0.56	5.78 ± 0.82	**3.82 ± 0.44**
C1_6	Score	65.39 ± 11.68	33.90 ± 11.97	48.04 ± 4.09	**20.61 ± 5.28**
	RMSE	7.26 ± 0.16	4.63 ± 0.99	6.09 ± 0.73	**3.03 ± 0.76**
C2_5	Score	58.28 ± 6.65	29.02 ± 6.04	42.51 ± 5.02	**19.91 ± 2.02**
	RMSE	7.31 ± 0.31	4.35 ± 0.95	5.64 ± 0.22	**3.20 ± 0.07**
C2_6	Score	193.50 ± 24.69	111.57 ± 8.74	137.55 ± 12.15	**70.36 ± 8.38**
	RMSE	12.55 ± 0.84	7.66 ± 0.99	9.86 ± 0.83	**6.05 ± 0.32**
C3_5	Score	100.13 ± 8.53	62.08 ± 6.30	74.21 ± 8.72	**53.77 ± 6.10**
	RMSE	8.24 ± 0.77	5.44 ± 0.69	6.30 ± 0.49	**4.85 ± 0.20**
C3_6	Score	96.03 ± 7.57	58.03 ± 7.13	79.05 ± 8.18	**49.92 ± 5.65**
	RMSE	9.38 ± 0.59	6.55 ± 0.75	8.01 ± 0.57	**5.47 ± 0.86**

as presented in Fig. 5.26, a comparison is carried out and detailed in this section. Three compared prognosis networks, i.e., Network-4, Network-5 and Network-6, have the identical representation learning subnetwork as MSCAN except for RUL estimation modules. Specifically, Network-4 employs three FCLs to construct the RUL estimation module, in which each FCL uses the ReLU activation function and the neuron number of the first two FCLs is assigned to be 100. Similarly, the RUL estimation modules of Network-5 and Network-6 are also composed of three FCLs. The differences are that Network-5 employs the standard dropout, Network-6 utilizes the designed dynamic Gaussian dropout. From Fig. 5.26, it can be observed that the networks using dropout surpass Network-1 in terms of both the score and RMSE, which indicates that the application of the dropout technique in the RUL estimation subnetwork is necessary. The reasons are that, on the one hand, dropout is able to lower the redundancy of the high-dimensional representations and relieve the overfitting problem. On the other hand, dropout promotes the generalization ability of the networks. Especially, owing to that dynamic Gaussian dropout provides a time-varying co-adaptation prevention strategy, it receives a lower score value and a smaller RMSE value for each cutter as compared to the standard dropout. In consequence, the dynamic Gaussian dropout outperforms the standard dropout in prognosis. Furthermore, the MSCAN achieves higher RUL prediction accuracy than Network-6 on account of introducing maxout activation to substitute ReLU, which implies the maxout activation contributes to improving the prediction performance of the network.

5.4 Multi-scale Convolutional Attention Network-Based RUL Prediction

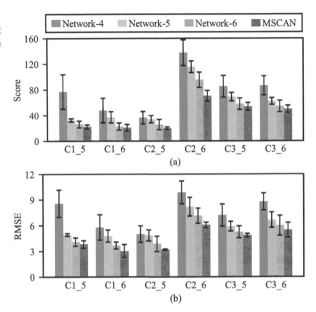

Fig. 5.26 Performance comparison of four different prognosis networks: (a) score metric, and (b) RMSE metric

5.4.4.5 Comparison with State-Of-The-Art Prognosis Approaches

To further evidence the advantage of the MSCAN, four state-of-the-art prognosis approaches are also tested under the milling cutters dataset, including RF (Wu et al. 2017), DBN (Chen et al. 2018), residual dense network (RDN) (Li et al. 2019) and CLSTM (Hinchi and Tkiouat 2018). For RF and DBN, features listed in (Wu et al. 2017) and (Chen et al. 2018) are first extracted from vibration, AE and force signals. As for RDN and CLSTM, the raw monitoring signals are used as inputs directly. Moreover, hyperparameter tuning is also carried out accordingly by using cross-validation for each compared approach. Table 5.11 reports the performance comparison results of these prognosis approaches in RUL prediction of milling cutters. From Table 5.11, it can be observed that the MSCAN receives a lower score value and a smaller RMSE value than the other four prognosis approaches for each cutter, which indicates that RUL prediction results of MSCAN are more accurate and robust. From the aforementioned analysis, there is no doubt that the MSCAN surpasses the other four existing prognosis approaches in RUL prediction of milling cutters, and the superiority of self-attention mechanisms, multi-scale learning and DDLs is further validated.

5.4.5 Epilog

In this section, a new prediction network called MSCAN is presented to fuse multi-sensor monitoring data. The network is composed of multi-scale representation

Table 5.11 Performance evaluation of four different networks

Testing dataset		RF (Wu et al. 2017)	DBN (Chen et al. 2018)	RDN (Li et al. 2019)	CLSTM (Hinchi and Tkiouat 2018)	MSCAN
C1_5	Score	235.82 ± 15.08	146.58 ± 7.15	111.65 ± 11.88	41.47 ± 5.07	**22.27 ± 2.91**
	RMSE	17.32 ± 1.12	13.96 ± 0.98	10.79 ± 0.59	5.10 ± 0.42	**3.82 ± 0.44**
C1_6	Score	145.25 ± 9.49	77.06 ± 9.55	53.71 ± 7.04	35.21 ± 8.99	**20.61 ± 5.28**
	RMSE	9.94 ± 1.06	7.42 ± 0.42	6.19 ± 0.87	4.58 ± 0.50	**3.03 ± 0.76**
C2_5	Score	117.89 ± 14.98	50.25 ± 10.11	63.08 ± 7.11	31.74 ± 7.04	**19.91 ± 2.02**
	RMSE	8.48 ± 0.90	6.02 ± 0.97	7.64 ± 0.38	4.63 ± 0.46	**3.20 ± 0.07**
C2_6	Score	302.83 ± 35.69	218.19 ± 27.74	168.82 ± 22.65	120.31 ± 10.19	**70.36 ± 8.38**
	RMSE	16.37 ± 1.59	13.39 ± 0.91	11.71 ± 1.25	8.21 ± 0.65	**6.05 ± 0.32**
C3_5	Score	160.28 ± 15.14	100.54 ± 11.81	108.66 ± 9.26	65.60 ± 8.18	**53.77 ± 6.10**
	RMSE	11.01 ± 1.02	8.87 ± 1.29	9.01 ± 0.72	5.62 ± 0.68	**4.85 ± 0.20**
C3_6	Score	128.19 ± 11.45	88.45 ± 10.57	103.60 ± 10.88	61.32 ± 7.43	**49.92 ± 5.65**
	RMSE	12.12 ± 0.89	8.17 ± 0.62	9.98 ± 0.99	7.15 ± 0.51	**5.47 ± 0.86**

learning subnetworks with self-attention and an RUL estimation module. The multi-sensor data are input into the network to incorporate the full degradation information. After that, several stacked convolutional blocks and self-attention modules form the representation learning subnetwork to discriminate representations from the inputs under different scales. Finally, the RUL estimation module which is composed of a new layer named DDL is constructed to perform regression analysis and provide the RUL prediction results. To evaluate the prediction performance of MSCAN, multi-sensor monitoring data collected from life testing of milling cutters are used. Experimental results prove that the RUL prediction of the MSCAN is more accurate than the compared state-of-the-art prognosis approaches.

5.5 Conclusions

This chapter addresses the key issues of deep learning in the data-driven remaining useful life prediction of mechanical systems. Three research studies are introduced totally, and three types of deep RUL prediction networks are presented accordingly.

For the problem that existing deep prognosis networks need to manually construct degradation indicators from high-dimensional monitoring data, a deep separable convolutional network is developed for RUL prediction by introducing separable convolutions and building squeeze and excitation units. In DSCN, raw monitoring data are directly used as the inputs of the network, and multiple levels of representations are automatically learned from input data by stacking separable convolutional building blocks, which reduces the requirements of domain expertise and improves the accuracy of RUL prediction. The effectiveness of DSCN is validated using the vibration data from accelerated degradation tests of rolling element bearings.

For the problem that existing convolutional prognosis networks cannot provide probabilistic RUL prediction results, a recurrent convolutional neural network is presented. In RCNN, recurrent connections and gating mechanisms are first employed, which is able to effectively memorize useful degradation information over time. Then, the variational inference is used to quantify the uncertainty of RCNN in RUL prediction, resulting in probabilistic RUL prediction results. The RCNN is evaluated using vibration data from accelerated degradation tests of rolling element bearings. Experimental results show that RCNN is able to handle the uncertainty in prognosis fairly well and is superior to compared existing convolutional prognosis networks.

For the problem that existing deep prognosis networks lack an explicit learning mechanism to effectively fuse multi-sensor data, the MSCAN method is introduced. Through analyzing the characteristics of multi-sensor data, two self-attention modules, i.e., channel-wise attention and temporal attention, are first built in MSCAN. Then, a multi-scale learning strategy is developed to automatically learn representations from different temporal scales, ensuring the completeness of representations. Finally, a new layer, i.e., dynamic dense layer, is built to replace the traditional fully-connected layer, which is able to relieve the overfitting problem and improve the accuracy and generalization of MSCAN. The effectiveness of MSCAN is verified using multi-sensor monitoring data from life testing of milling cutters.

References

Ambadekar P, Choudhari C (2020) CNN based tool monitoring system to predict life of cutting tool. SN Applied Sciences 2(5):1–11

Babu GS, Zhao P, Li X-L (2016) Deep convolutional neural network based regression approach for estimation of remaining useful life. International conference on database systems for advanced applications. Springer, pp 214–228

Bishop CM (2006) Pattern recognition. Machine learning 128(9)

Caceres J, Gonzalez D, Zhou T, Droguett EL (2021) A probabilistic Bayesian recurrent neural network for remaining useful life prognostics considering epistemic and aleatory uncertainties. Structural Control and Health Monitoring 28(10). https://doi.org/10.1002/stc.2811

Chen Y, Jin Y, Jiri G (2018) Predicting tool wear with multi-sensor data using deep belief networks. Int J Adv Manuf Technol 99(5):1917–1926

Cho K, Van Merriënboer B, Gulcehre C, Bahdanau D, Bougares F, Schwenk H, Bengio Y (2014) Learning phrase representations using RNN encoder-decoder for statistical machine translation. arXiv preprint arXiv:14061078

Deutsch J, He D (2017) Using deep learning-based approach to predict remaining useful life of rotating components. IEEE Transactions on Systems, Man, and Cybernetics: Systems 48(1):11–20

Elsheikh A, Yacout S, Ouali M-S (2019) Bidirectional handshaking LSTM for remaining useful life prediction. Neurocomputing 323:148–156. https://doi.org/10.1016/j.neucom.2018.09.076

Gal Y, Ghahramani Z (2016) Dropout as a Bayesian approximation: Representing model uncertainty in deep learning. In: International conference on machine learning, 2016. PMLR, pp 1050–1059

Goodfellow I, Warde-Farley D, Mirza M, Courville A, Bengio Y (2013) Maxout networks. In: International conference on machine learning, PMLR, pp 1319–1327

He K, Zhang X, Ren S, Sun J (2016) Deep residual learning for image recognition. In: Proceedings of the IEEE conference on computer vision and pattern recognition, pp 770–778

Hinchi AZ, Tkiouat M (2018) Rolling element bearing remaining useful life estimation based on a convolutional long-short-term memory network. Procedia Comput Sci 127:123–132

Hinton GE, Srivastava N, Krizhevsky A, Sutskever I, Salakhutdinov RR (2012) Improving neural networks by preventing co-adaptation of feature detectors. arXiv preprint arXiv:12070580

Hu J, Shen L, Sun G (2018) Squeeze-and-excitation networks. In: Proceedings of the IEEE conference on computer vision and pattern recognition, pp 7132–7141

Ioffe S, Szegedy C (2015) Batch normalization: Accelerating deep network training by reducing internal covariate shift. In: International conference on machine learning. PMLR, pp 448–456

Jardine AKS, Lin D, Banjevic D (2006) A review on machinery diagnostics and prognostics implementing condition-based maintenance. Mech Syst Signal Process 20(7):1483–1510. https://doi.org/10.1016/j.ymssp.2005.09.012

Javed K, Gouriveau R, Zerhouni N (2017) State of the art and taxonomy of prognostics approaches, trends of prognostics applications and open issues towards maturity at different technology readiness levels. Mech Syst Signal Process 94:214–236

Khan S, Yairi T (2018) A review on the application of deep learning in system health management. Mech Syst Signal Process 107:241–265. https://doi.org/10.1016/j.ymssp.2017.11.024

Kim N-H, An D, Choi J-H (2017) Prognostics and health management of engineering systems. Springer International Publishing, Switzerland

Kingma DP, Ba J (2014) Adam: A method for stochastic optimization. arXiv preprint arXiv:14126980

Lei Y, Li N, Guo L, Li N, Yan T, Lin J (2018) Machinery health prognostics: a systematic review from data acquisition to RUL prediction. Mech Syst Signal Process 104:799–834

Li X, Zhang W, Ma H, Luo Z, Li X (2021) Degradation alignment in remaining useful life prediction using deep cycle-consistent learning. IEEE Transactions on Neural Networks and Learning Systems:1–12. https://doi.org/10.1109/TNNLS.2021.3070840

Li Y, Xie Q, Huang H, Chen Q (2019) Research on a tool wear monitoring algorithm based on residual dense network. Symmetry 11(6):809

Listou Ellefsen A, Bjørlykhaug E, Æsøy V, Ushakov S, Zhang H (2019) Remaining useful life predictions for turbofan engine degradation using semi-supervised deep architecture. Reliab Eng Syst Saf 183:240–251. https://doi.org/10.1016/j.ress.2018.11.027

Mamalet F, Garcia C (2012) Simplifying convnets for fast learning. International Conference on Artificial Neural Networks. Springer, pp 58–65

Nair V, Hinton GE (2010) Rectified linear units improve restricted Boltzmann machines. In: Icml

Nectoux P, Gouriveau R, Medjaher K, Ramasso E, Chebel-Morello B, Zerhouni N, Varnier C (2012) PRONOSTIA: an experimental platform for bearings accelerated degradation tests. In: IEEE International Conference on Prognostics and Health Management, PHM'12. IEEE Catalog Number: CPF12PHM-CDR, pp 1–8

Nwankpa C, Ijomah W, Gachagan A, Marshall S (2018) Activation functions: comparison of trends in practice and research for deep learning. arXiv preprint arXiv:181103378

Park K, Choi Y, Choi WJ, Ryu H, Kim H (2020) LSTM-based battery remaining useful life prediction with multi-channel charging profiles. IEEE Access 8:20786–20798. https://doi.org/10.1109/ACCESS.2020.2968939

Ren L, Sun Y, Wang H, Zhang L (2018) Prediction of bearing remaining useful life with deep convolution neural network. IEEE Access 6:13041–13049

Rigamonti M, Baraldi P, Zio E, Roychoudhury I, Goebel K, Poll S (2018) Ensemble of optimized echo state networks for remaining useful life prediction. Neurocomputing 281:121–138. https://doi.org/10.1016/j.neucom.2017.11.062

References

Rodrigues LR (2018) Remaining useful life prediction for multiple-component systems based on a system-level performance indicator. IEEE/ASME Trans Mechatron 23(1):141–150. https://doi.org/10.1109/TMECH.2017.2713722

Saxena A, Celaya J, Saha B, Saha S, Goebel K (2010) Metrics for offline evaluation of prognostic performance. Int J Prognostics Health Manage 1(1):4–23

Saxena A, Goebel K, Simon D, Eklund N (2008) Damage propagation modeling for aircraft engine run-to-failure simulation. In: 2008 international conference on prognostics and health management. IEEE, pp 1–9

Singleton RK, Strangas EG, Aviyente S (2014) Extended Kalman filtering for remaining-useful-life estimation of bearings. IEEE Trans Industr Electron 62(3):1781–1790

Srivastava N, Hinton G, Krizhevsky A, Sutskever I, Salakhutdinov R (2014) Dropout: a simple way to prevent neural networks from overfitting. J Mach Learn Res 15(1):1929–1958

Szegedy C, Liu W, Jia Y, Sermanet P, Reed S, Anguelov D, Erhan D, Vanhoucke V, Rabinovich A (2015) Going deeper with convolutions. In: Proceedings of the IEEE conference on computer vision and pattern recognition, pp 1–9

Tsui KL, Chen N, Zhou Q, Hai Y, Wang W (2015) Prognostics and health management: a review on data driven approaches. Mathematical Problems in Engineering 2015

Wang B, Lei Y, Li N, Li N (2018) A hybrid prognostics approach for estimating remaining useful life of rolling element bearings. IEEE Trans Reliab 69(1):401–412

Wang B, Lei Y, Li N, Wang W (2021) Multiscale convolutional attention network for predicting remaining useful life of machinery. IEEE Trans Industr Electron 68(8):7496–7504. https://doi.org/10.1109/TIE.2020.3003649

Wang B, Lei Y, Li N, Yan T (2019) Deep separable convolutional network for remaining useful life prediction of machinery. Mech Syst Signal Proc 134:106330

Wang B, Lei Y, Yan T, Li N, Guo L (2020) Recurrent convolutional neural network: a new framework for remaining useful life prediction of machinery. Neurocomputing 379:117–129. https://doi.org/10.1016/j.neucom.2019.10.064

Wu D, Jennings C, Terpenny J, Gao RX, Kumara S (2017) A comparative study on machine learning algorithms for smart manufacturing: tool wear prediction using random forests. J Manuf Sci Eng 139(7)

Yang B, Liu R, Zio E (2019) Remaining useful life prediction based on a double-convolutional neural network architecture. IEEE Trans Industr Electron 66(12):9521–9530

Yu F, Koltun V, Funkhouser T (2017) Dilated residual networks. In: Proceedings of the IEEE conference on computer vision and pattern recognition, pp 472–480

Zeiler MD, Fergus R (2014) Visualizing and understanding convolutional networks. European Conference on Computer Vision. Springer, pp 818–833

Zhang W, Li X, Ma H, Luo Z, Li X (2021) Transfer learning using deep representation regularization in remaining useful life prediction across operating conditions. Reliab Eng Syst Saf 211, 107556. https://doi.org/10.1016/j.ress.2021.107556

Zhao R, Yan R, Wang J, Mao K (2017) Learning to monitor machine health with convolutional bi-directional LSTM networks. Sensors (Basel):17(2). https://doi.org/10.3390/s17020273

Zhu J, Chen N, Peng W (2018) Estimation of bearing remaining useful life based on multiscale convolutional neural network. IEEE Trans Industr Electron 66(4):3208–3216

Chapter 6
Data-Model Fusion RUL Prediction

6.1 Introduction

In industrial applications, it is generally feasible to obtain the mechanical system prior degradation information, such as degradation mechanisms, degradation trends of similar equipment, and empirical knowledge (Hanachi et al. 2019), from experts. Meanwhile, with the advancement of measurement and instrumentation technology, the dynamic acquisition of mechanical system condition monitoring data is no longer a difficult issue. In this context, if the prior information can be effectively integrated with condition monitoring data, it is possible to obtain more accurate and industrial application-oriented RUL prediction methods. To this end, data-model fusion RUL prediction has been developed and is becoming a vibrant field in PHM of mechanical systems.

In general, data-model fusion RUL prediction first endeavors to formulate suitable mathematical models to empirically describe the degradation trend of the considered mechanical system. Then based on the continuously obtained real-time condition monitoring data, model parameters can be dynamically updated, and the failure time when mechanical system exceeds the failure threshold (FT) can be finally obtained accurately (Elattar et al. 2016). Compared with the commonly used model-based and data-driven RUL prediction methods, data-model fusion RUL prediction is able to integrate both the empirical knowledge of mechanical system degradation process and the dynamic information obtained from condition monitoring data. Therefore, it has become a promising tool in the mechanical system RUL prediction realm.

In data-model fusion RUL prediction, degradation model formulation is the key procedure in applying RUL prediction methods effectively. The degradation models used in literature include auto-regressive models, random coefficient models, Wiener process models, Gamma process models, inverse Gaussian process models, Markov models, proportional hazard models, etc. (Lei et al. 2018). Among them, Wiener process models are the most commonly used model category owing to their capability of describing the overall degradation trend and instantaneous fluctuation of mechanical system simultaneously (Si et al. 2011). In existing literature, the most generally adopted Wiener process model describes the overall degradation trend as a linear drift function (Liao and Tian 2013), and utilize a Brownian diffusion term to reflect the instantaneous fluctuation. Considering the fact that this model is only able to accurately describe degradation processes with linear characteristics, researchers further dedicated to formulating models capable of describing more complex degradation processes. Gebraeel et al. (Gebraeel et al. 2005) formulated an exponential Wiener process model, of which the improvements are further conducted by several researchers regarding multiple aspects in RUL prediction (Li et al. 2015; Liu et al. 2013). Lei et al. (Lei et al. 2016) formulated a polynomial model and applied it in the RUL prediction of rolling element bearings. Si et al. (2012) developed a more general Wiener process model, in which the drift term is described as a time-dependent function. The analytical solution of RUL is correspondingly derived. Based on this work, researchers made various extensions considering different aspects in RUL prediction (Le Son et al. 2013; Si et al. 2014). As the degradation process of mechanical system can be both time-dependent and state-dependent, Li et al. (Li et al. 2017a) further developed a time- and state-dependent Wiener process model, and provided the analytical solution-based computation algorithm for RUL prediction.

In the context of Industry 4.0, the higher demand of mechanical system operation reliability and efficiency promotes its PHM towards a more intelligent direction, which in turn requests RUL prediction to possess higher accuracy and stability, and the internal as well as external variabilities influencing the degradation process of mechanical system must be taken into consideration (Lee et al. 2013; Li et al. 2017b). For internal variabilities, they are generally generated via various sorts of intrinsic factors within the mechanical system, such as material microdefects, manufacturing and assembling errors, component quality variation, etc., and would impact the RUL prediction accuracy substantially if not appropriately dealt with (Zheng et al. 2016). As for external variabilities like time-varying operating conditions, they would incur different impact levels, and further change the stress level and degradation rate of mechanical system. Without full consideration of external factors, RUL prediction methods would be unable to meet the stability requirement in industrial applications. Moreover, the combination of both internal and external factors would further contribute to more complicated failure phenomenon, such as competing failure (Yan et al. 2021), multiple degradation phases (Zhang et al. 2018), etc. Therefore, the joint impact of internal and external variabilities should also be reflected in RUL prediction methods to fertilize their effective applications in industries.

In the rest of this chapter, four major types of variabilities, i.e., random fluctuation variability, unit-to-unit variability, time-varying operational conditions, and competing failure processes, are introduced. Then, the data-model fusion RUL prediction methods concentrating on these variabilities will be elaborated in detail. Furthermore, abundant experimental or industrial cases in RUL prediction of mechanical systems will be provided as a guidance for the implementation of these methods.

6.2 RUL Prediction with Random Fluctuation Variability

6.2.1 Motivation

The health degradation of mechanical systems is generally incurred by incipient faults of materials such as microfractures, pitting, wearing, cracks. Under specific stress level and operating environment, those microdefects will be propagated, and mechanical systems eventually fail by the propagated faults. Generally speaking, the overall degradation trend of mechanical systems is subject to certain statistical distributions, which provides the opportunity to empirically formulate a degradation model to describe the trend. However, due to the existence of random factors such as material microdefects, manufacturing error, operating load fluctuation, measurement error, etc., there could be numerous temporal random fluctuations during the entire degradation trend of mechanical system condition monitoring data. These random fluctuations can be observed during the operation cycle of mechanical systems and would impact the RUL prediction of mechanical systems substantially. For a clear description of this impact, the lifecycle condition monitoring data and its corresponding RUL prediction results of an illustrative rolling element bearing are given in Fig. 6.1.

Similar as illustrated in Fig. 6.1, it can be observed that during the lifecycle of most mechanical systems, there usually exists a long healthy operating period before degradation phenomenon occurs. During this healthy operating period, the amplitude of condition monitoring data does not reflect any obvious degradation trend. With the increase of operating time, mechanical system further enters the degrading period, during which the condition monitoring data amplitude will gradually increase and eventually exceed the FT. Such a lifecycle is usually referred as two-stage degradation processes (Li et al. 2015). For two-stage degradation processes, it is widely adopted that condition monitoring is deployed during the entire lifetime, but RUL prediction is triggered only when obvious degradation trend occurs. To this end, the

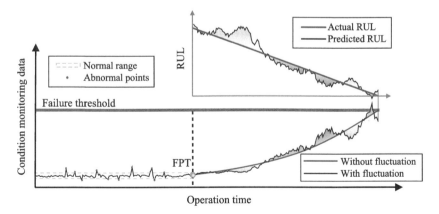

Fig. 6.1 Illustrative example of rolling element bearing degradation process

selection of first predicting time (FPT) is of vital importance for the accuracy of RUL prediction: if FPT is selected too early, i.e., mechanical system is still operating during its normal stage, degradation trend information will be contaminated by useless healthy condition monitoring data, thus compromising the RUL prediction accuracy. Contrariwise, if FPT is selected too late, useful degradation data will be omitted, which leads to the late response of RUL prediction to degradation trend. However, due to the existence of random fluctuation, abnormal information will be introduced into condition monitoring data, causing the sudden change of data amplitude and eventually interfering the selection of FPT.

Apart from its influence on FPT selection, random fluctuation will also impact the degradation state estimation and cause corresponding RUL prediction errors. As can be seen in Fig. 6.1, the random fluctuation of monitoring data has an obviously negative correlation with RUL prediction error. To be specific, when the negative random fluctuation occurs, signal amplitude will be lower than the ideal amplitude of actual degradation state, causing underestimation of degradation state and eventually overestimation of RUL. On the contrary, if random fluctuation is positive, signal amplitude will be higher than actual degradation state, which will further lead to the predicted RUL lower than actual RUL.

In conclusion, random fluctuation of mechanical system condition monitoring data is able to compromise the RUL prediction in two aspects, i.e., interfering the selection of FPT, and introducing RUL prediction errors. In view of these two aspects, this section will provide a RUL prediction method considering random fluctuation variability (Li et al. 2015). First, an abnormal data identification mechanism, which jointly considers the statistical characteristics and degradation trend information, will be introduced for FPT selection. Then, the RUL prediction based on joint filtering of parameter and state will be addressed. With this method, both the FPT selection and degradation state estimation accuracies will be enhanced, and the RUL prediction accuracy will correspondingly be increased in the context of random fluctuation variability.

6.2.2 RUL Prediction Considering Random Fluctuation Variability

As can be seen in Fig. 6.2, the flowchart of the presented method mainly contains two modules, i.e., FPT selection, and RUL prediction based on joint parameter and state filtering.

In the FPT selection module, Kurtosis, the health indicator which is sensitive to incipient faults is firstly extracted from both the online acquired condition monitoring data and historical data of the same equipment. Then, the historical data in the normal operating period will be utilized to determine a 3σ interval for abnormal detection. An FPT selection method based on consecutive triggering logic is utilized, which will eventually yield an accurate FPT selection result.

6.2 RUL Prediction with Random Fluctuation Variability

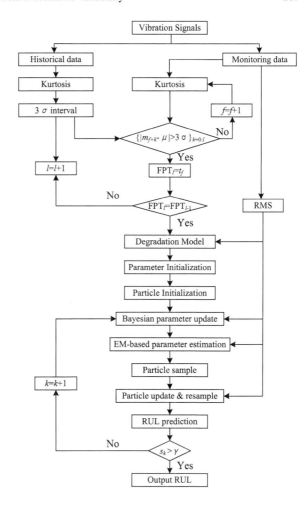

Fig. 6.2 Flowchart of the RUL prediction method considering random fluctuation variability

As for the RUL prediction module, an exponential model is firstly formulated to empirically describe the propagation trend of RMS, which is an effective indicator for describing the long-term degradation process of mechanical system. Then, the analytical RUL probability density function (PDF) is correspondingly derived. To eliminate the prediction errors incurred by random fluctuation, Bayesian online update as well as expectation maximization (EM) algorithm will be used for online parameter update. Based on the accurate prior distribution provided by Bayesian update and EM algorithm, particle filtering is further executed to dynamically update the degradation state. With this procedure, online condition monitoring data can be dually utilized for parameter and state update, which is able to reduce the influence of random fluctuation on RUL prediction results. More details of each step are explained below.

6.2.2.1 First Predicting Time Selection

The essential difference between the abnormal data in the healthy operating period and the degradation data are that the amplitude of abnormal data exhibits random pattern, while the amplitude of degradation data owns obvious tendency, i.e., the distribution of data amplitude will consecutively exceed normally operating data distribution interval. According to this statistical characteristic, a consecutive triggering mechanism is utilized to determine whether condition monitoring signal is exceeding healthy interval or not, so as to determine accurate FPT. However, in consecutive triggering logic, a vital parameter, i.e., consecutive triggering number, could have significant influence on the FPT selection accuracy. To be specific, if consecutive triggering number is set too small, it is prone to selecting a too-early FPT. Contrariwise, a delayed FPT will be selected. In a word, pre-specified consecutive number cannot guarantee an accurate FPT selection. Thus, an adaptive consecutive triggering number setting logic is utilized, so as to obtain a reasonable parameter specification result. Details of the consecutive triggering mechanism based on the above-mentioned adaptive setting logic are as follows:

(1) Use historical data to compute the mean μ and standard deviation σ of healthy operating distribution interval parameters, and denote the healthy interval as $[\mu + 3\sigma, \mu - 3\sigma]$. Set consecutive triggering number and maximum triggering number as $l = 0$.

(2) Let $l = l + 1$, locate the time t_f when consecutive $l + 1$ values $\{m_{f+k}\}_{k=0:l}$ meets $\{|m_{f+k} - \mu| > 3\sigma\}_{k=0:l}$, and determine the time t_f as FPT_l.

(3) Increase l from 1 until l whilst $\text{FPT}_l = \text{FPT}_{l-1}$ is held, and output FPT_l as the optimal FPT.

The basis behind the above triggering mechanism is that abnormal states caused by random fluctuation are extremely unlikely to appear $l + 1$ times consecutively during the healthy operation stage. On the contrary, if the selected FPT starts to keep stable with the increase of l, it generally implies the occurrence of an incipient fault.

6.2.2.2 RUL Prediction Based on Joint Filtering of Parameter and State

As shown in Fig. 6.1, the random fluctuation during degradation process will introduce negatively correlated prediction errors, and it is thus important to eliminate the impact of random fluctuation on state and parameter estimation so as to ensure accurate RUL prediction results. To this end, Bayesian update is used in combination with EM algorithm for parameter update, and particle filtering for state estimation. Based on the accurately estimated state and parameter, RUL can be recursively predicted with online condition monitoring data. Details of this procedure are provided as follows.

(1) Degradation Modeling

6.2 RUL Prediction with Random Fluctuation Variability

To describe the degradation process empirically, the exponential Wiener process is employed to formulate the degradation trend as

$$x_k = \phi + \theta' \exp\left(\beta' t_k + \sigma B(t_k) - \frac{\sigma^2}{2} t_k\right) \tag{6.1}$$

where ϕ is a known constant, θ' and β' are random variables characterizing the stochastic part, σ is a constant representing the deterministic part, and $\sigma B(t_k)$ is a Brownian motion (BM) following $N(0, \sigma^2 t_k)$, which represents the randomness of degradation process.

For convenience, the exponential model is transformed into the logged form as

$$s_k = \ln[x_i - \phi] = \ln(\theta') + \left(\beta' - \frac{\sigma^2}{2}\right) t_k + \sigma B(t_k) = \theta + \beta t_k + \sigma B(t_k) \tag{6.2}$$

where $\theta = \ln(\theta')$ follows $N(\mu_0, \sigma_0^2)$ and $\beta = \beta' - \sigma^2/2$ follows $N(\mu_1, \sigma_1^2)$. Note that for the convenience of analysis, it is assumed that $\phi = 0$.

Given the pre-specified FT γ, the RUL L_k at time t_k can be correspondingly defined as

$$L_k = \inf\{l_k : s(l_k + t_k) \geq \gamma | S_{1:k}\} \tag{6.3}$$

According to the properties of Brownian motion (Si et al. 2013), the PDF of RUL can be expressed as

$$f_{L_k|S_{1:k}}(l_k|S_{1:k}) = \frac{\gamma - s_k}{\sqrt{2\pi l_k^3 \left(\sigma_{\beta,k}^2 l_k + \hat{\sigma}_k^2\right)}} \exp\left[-\frac{(\gamma - s_k - \mu_{\beta,k} l_k)^2}{2 l_k \left(\sigma_{\beta,k}^2 l_k + \hat{\sigma}_k^2\right)}\right], l_k \geq 0 \tag{6.4}$$

(2) Parameter Estimation

Assuming that a sequence of measurement data $S_{1:k} = \{s_1, s_2, \cdots, s_k\}$ is available. As previously defined, random errors $\sigma B(t_k)$ are independent identically distributed (i.i.d.) normal random variables, the likelihood function of θ and β can be given according to the measurement data $S_{1:k}$ as

$$p(S_{1:k}|\theta, \beta) = \left(\frac{1}{\sqrt{2\pi\sigma^2 \Delta t}}\right)^k \exp\left[-\frac{(s_1 - \theta - \beta t_1)^2}{2\sigma^2 t_1} - \sum_{j=2}^{k} \frac{(s_j - s_{j-1} - \beta \Delta t)^2}{2\sigma^2 \Delta t}\right] \tag{6.5}$$

where $\Delta t = t_j - t_{j-1}$ is the fixed time interval. Then the likelihood function of θ and β conditioned on $S_{1:k}$ is still healthy resulted from the normal distribution of θ and β, i.e. $\theta, \beta | S_{1:k} \sim N\left(\mu_{\theta,k}, \sigma_{\theta,k}^2, \mu_{\beta,k}, \sigma_{\beta,k}^2, \rho_k\right)$, with

$$\mu_{\theta,k} = \frac{\left(s_1\sigma_0^2 + \mu_0\sigma^2 t_1\right)\left(\sigma^2 + \sigma_1^2 t_k\right) - \sigma_0^2 t_1\left(s_k\sigma_1^2 + \mu_1\sigma^2 - 0.5\sigma^4\right)}{\left(\sigma_0^2 + \sigma^2 t_1\right)\left(\sigma_1^2 t_k + \sigma^2\right) - \sigma_0^2\sigma_1^2 t_1}$$

$$\sigma_{\theta,k}^2 = \frac{\sigma_0^2\sigma^2 t_1\left(\sigma^2 + \sigma_1^2 t_k\right)}{\left(\sigma_0^2 + \sigma^2 t_1\right)\left(\sigma_1^2 t_k + \sigma^2\right) - \sigma_0^2\sigma_1^2 t_1}$$

$$\mu_{\beta,k} = \frac{\left(s_k\sigma_1^2 + \mu_1\sigma^2 - 0.5\sigma^4\right)\left(\sigma_0^2 + \sigma^2 t_1\right) - \sigma_1^2\left(s_1\sigma_0^2 + \mu_0\sigma^2 t_1\right)}{\left(\sigma_0^2 + \sigma^2 t_1\right)\left(\sigma_1^2 t_k + \sigma^2\right) - \sigma_0^2\sigma_1^2 t_1} \quad (6.6)$$

$$\sigma_{\beta,k}^2 = \frac{\sigma_1^2\sigma^2 t_1\left(\sigma_0^2 + \sigma^2 t_1\right)}{\left(\sigma_0^2 + \sigma^2 t_1\right)\left(\sigma_1^2 t_k + \sigma^2\right) - \sigma_0^2\sigma_1^2 t_1}$$

$$\rho_k = \frac{-\sigma_0\sigma_1\sqrt{t_1}}{\sqrt{\left(\sigma_0^2 + \sigma^2 t_1\right)\left(\sigma_1^2 t_k + \sigma^2\right)}}$$

From the above equations, it can be easily known that the posterior estimations of θ and β can be dynamically updated along with the online obtained new measurements. However, before updating, there are still several unknown parameters needed to be estimated. For convenience, we denote those parameters as $\Theta = [\sigma^2, \mu_0, \sigma_0^2, \mu_1, \sigma_1^2]$. In this part, we utilize EM algorithm to estimate unknown parameters. Denote $\hat{\Theta}_k^{(i)} = \left[\hat{\sigma}_k^{2(i)}, \hat{\mu}_{0,k}^{(i)}, \hat{\sigma}_{0,k}^{2(i)}, \hat{\mu}_{1,k}^{(i)}, \hat{\sigma}_{1,k}^{2(i)}\right]$ as the estimation during the ith step, let $\partial\left[\ell\left(\Theta_k | \hat{\Theta}_k^{(i)}\right)\right] / \partial\Theta_k = 0$, then the parameter estimation in the next step $\hat{\Theta}_k^{(i+1)}$ can be obtained as (Si et al. 2013)

$$\hat{\sigma}_k^{2(i+1)} = \frac{1}{k}\left(\frac{s_1^2 - 2s_1(\mu_{\theta,k} + \mu_{\beta,k} t_1) + \sigma_{\theta,k}^2 + \sigma_{\beta,k}^2}{t_1} + \right.$$

$$2\left(\rho_k\sigma_{\theta,k}\sigma_{\beta,k} + \mu_{\theta,k}\mu_{\beta,k}\right) + t_1\left(\mu_{\beta,k}^2 + \sigma_{\beta,k}^2\right) +$$

$$\left. \sum_{j=2}^{k} \frac{\left(s_j - s_{j-1}\right)^2 - \left(s_j - s_{j-1}\right)\Delta t \mu_{\beta,k} + (\Delta t)^2\left(\mu_{\beta,k}^2 + \sigma_{\beta,k}^2\right)}{\Delta t}\right) \quad (6.7)$$

$$\hat{\mu}_{0,k}^{(i+1)} = \mu_{\theta,k}, \quad \hat{\sigma}_{0,k}^{2(i+1)} = \sigma_{\theta,k}^2$$
$$\hat{\mu}_{1,k}^{(i+1)} = \mu_{\beta,k}, \quad \hat{\sigma}_{1,k}^{2(i+1)} = \sigma_{\beta,k}^2$$

(3) State Estimation

After the estimation of model parameters, degradation states are further estimated. In order to address the stochasticity of degradation process, particle filtering is adopted.

6.2 RUL Prediction with Random Fluctuation Variability

Based on the degradation model provided in (6.1), state estimation is conducted as follows.

First, initialize degradation model parameters as $\sigma^2 = \hat{\sigma}_0^2$, $\mu_0 = \mu_{\theta,0}$, $\sigma_0^2 = \sigma_{\theta,0}^2$, $\mu_1 = \mu_{\beta,0}$ and $\sigma_1^2 = \sigma_{\beta,0}^2$. Then initial particles $\{s_0^i\}_{i=1:N_s}$ are sampled from $p(\tilde{s}_0|s_0, \hat{\Theta}_0) \sim N(s_0, \hat{\sigma}_0^2 \Delta t)$. At each condition monitoring moment, once the measurement s_k at t_k is obtained, model parameters are updated as previously elaborated. According to the estimated parameters and the following variant of the degradation model

$$s_k = s_{k-1} + \beta \Delta t + \sigma W(\Delta t) \tag{6.8}$$

The PDF of the degradation state is given as

$$p(\tilde{s}_k|\hat{s}_{k-1}, \hat{\Theta}_k) = \frac{1}{\sqrt{2\pi \Delta t \left(\sigma_{\beta,k}^2 \Delta t + \hat{\sigma}_k^2\right)}} \exp\left[-\frac{\left(\tilde{s}_k - \hat{s}_{k-1} - \mu_{\beta,k}\Delta t\right)^2}{2\Delta t \left(\sigma_{\beta,k}^2 \Delta t + \hat{\sigma}_k^2\right)}\right] \tag{6.9}$$

where $\tilde{s}_k$ is the actual degradation state at t_k, $\hat{\Theta}_k = \left[\hat{\sigma}_k^2, \mu_{\theta,k}, \sigma_{\theta,k}^2, \mu_{\beta,k}, \sigma_{\beta,k}^2\right]$ are the estimated parameters, $\hat{s}_{k-1}$ is the estimated state at t_{k-1}. Note that $\hat{s}_{k-1}$ is different from the measurement s_{k-1}, and if $\hat{s}_{k-1}$ is substituted with s_{k-1}, the above equation will be transformed to $p(\tilde{s}_k|s_{k-1}, \hat{\Theta}_k)$, which is the PDF of the state acquired using the original exponential model.

Now, the prominent task is to estimate the actual degradation state $\tilde{s}_k$ at t_k based on the measurement s_k. Here we adopt $p(\tilde{s}_k|\hat{s}_{k-1}, \hat{\Theta}_k)$ as the importance function which contains the prior information of the state estimation at t_{k-1}. Then, sample particle sets $\{s_k^i\}_{i=1,2,\cdots,N_s}$ from Eq.(6.9). Note that the sampled particles can only approximate the apriori PDF of the state, which is merely a prediction rather than an estimation. In order to approximate the aposteriori PDF, particle weights are further updated according to the measurement s_k as follows (Chen et al. 2010):

$$w_k^i = w_{k-1}^i \frac{1}{\sqrt{2\pi \hat{\sigma}_k^2 t_k}} \exp\left[-\frac{\left(s_k - s_k^i\right)^2}{2\hat{\sigma}_k^2 t_k}\right], \quad w_k^i = w_k^i \Bigg/ \sum_{i=1}^{N_s} w_k^i \tag{6.10}$$

With the updated particles and their corresponding weights, the *posteriori* PDF of the degradation state can be approximated in discrete density form:

$$p(\tilde{s}_k|s_k, \hat{\Theta}_k) \approx \sum_{i=1}^{N_s} w_k^i \delta(\tilde{s}_k - s_k^i) \tag{6.11}$$

By iteratively performing the above procedure, the particles possessing higher similarity to the actual degradation states will acquire higher weights, and the dissimilar particles will be eliminated. Finally, the degradation state can be estimated in the form of

$$\hat{s}_k = \sum_{i=1}^{N_s} \left(w_k^i s_k^i \right) \tag{6.12}$$

The above weighted average form of updated particles guarantees that the estimated state $\hat{s}_k$ can be closer the actual degradation state than the measurement s_k.

After the update of particle weights, resampling is carried out to generate a new set of particles $\{s_k^{i*}\}_{i=1:N_s}$, and the particle weights are reset to $w_k^i = 1/N_s$. By doing so, the problem of particle degeneracy can be tackled to some extent.

Now that we have obtained the estimated degradation state, the RUL PDF at time t_k can eventually be obtained. According to the expression of RUL PDF in Eq. (6.4), we further compute the RUL PDF with estimated state as

$$f_{L_k|S_{1:k}}(l_k|S_{1:k}) = \frac{\gamma - \hat{s}_k}{\sqrt{2\pi l_k^3 \left(\sigma_{\beta,k}^2 l_k + \hat{\sigma}_k^2 \right)}} \exp\left[-\frac{\left(\gamma - \hat{s}_k - \mu_{\beta,k} l_k \right)^2}{2 l_k \left(\sigma_{\beta,k}^2 l_k + \hat{\sigma}_k^2 \right)} \right], l_k \geq 0 \tag{6.13}$$

For a good overall understanding, the RUL prediction process is summarized in Table 6.1.

6.2.3 RUL Prediction Case of FEMTO-ST Accelerated Degradation Tests of Rolling Element Bearings

6.2.3.1 Experimental Data Introduction

In this case study, vibration data from the FEMTO-ST dataset (Nectoux et al. 2012) is utilized. This experimental system is the same as the one described in Sect. 5.3.4 and will thus not be elaborated here. We hereafter select four tested bearings for method verification. For each selected bearing, its failure is considered to occur when the amplitude of the vibration signal equals or exceeds 20 g (Patil et al. 2018). During each test, the bearing degraded naturally from an as-good-as-new state without seeded incipient fault. Depending on the unit-to-unit variability among bearings and the randomness of degradation processes, the fault modes may be slightly different for different bearings. Consequently, the vibration signal pattern is different for each

6.2 RUL Prediction with Random Fluctuation Variability

Table 6.1 RUL prediction process considering random fluctuation variability

Step	Action
1	At time t_0, initialize parameters $\sigma^2 = \hat{\sigma}_0^2$, $\mu_0 = \mu_{\theta,0}$, $\sigma_0^2 = \sigma_{\theta,0}^2$, $\mu_1 = \mu_{\beta,0}$ and $\sigma_1^2 = \sigma_{\beta,0}^2$. Randomly sample particles $\{s_0^i\}_{i=1:N_s}$ from $p(\tilde{s}_0 \mid s_0, \hat{\Theta}_0) \sim N(s_0, \hat{\sigma}_0^2 \Delta t)$ with the weights $w_0^i = 1/N_s$
2	Once a measurement s_k is obtained at t_k for $k \geq 1$, set $\sigma^2 = \hat{\sigma}_{k-1}^2$, $\mu_0 = \mu_{\theta,k-1}$, $\sigma_0^2 = \sigma_{\theta,k-1}^2$, $\mu_1 = \mu_{\beta,k-1}$ and $\sigma_1^2 = \sigma_{\beta,k-1}^2$. Then $\mu_{\theta,k}$, $\sigma_{\theta,k}^2$, $\mu_{\beta,k}$ and $\sigma_{\beta,k}^2$ are estimated by (6.6)
3	Based on $\mu_{\theta,k}$, $\sigma_{\theta,k}^2$, $\mu_{\beta,k}$ and $\sigma_{\beta,k}^2$, $\sigma^2 = \hat{\sigma}_k^2$, $\mu_0 = \mu_{\theta,k}$, $\sigma_0^2 = \sigma_{\theta,k}^2$, $\mu_1 = \mu_{\beta,k}$ and $\sigma_1^2 = \sigma_{\beta,k}^2$ are calculated by (6.7)
4	Sample particles $\{s_k^i\}_{i=1:N_s}$ from $p(\tilde{s}_k \mid \hat{s}_{k-1}, \hat{\Theta}_k)$ using (6.9)
5	Update particle weights w_k^i via (6.10)
6	Resample new particles $\{s_k^{i*}\}_{i=1:N_s}$ and reset their weights as $w_k^i = 1/N_s$
7	Based on $\mu_{\theta,k}$, $\sigma_{\theta,k}^2$, $\mu_{\beta,k}$ and $\sigma_{\beta,k}^2$ in Step 2, $\hat{\sigma}_k^2$ from Step 3 and particles $\{s_k^i\}_{i=1:N_s}$ in Step 6, predict the RUL by (6.13)
8	Let $k = k + 1$, repeat Step 2–7 until $s_k > \gamma$

tested bearing. In Fig. 6.3, the vibration signals during the entire lifespans of four tested bearings are provided. It can be observed that the vibration amplitudes of bearings 1 and 2 have gradually increasing trends, which indicates that their faults propagate in a gradual manner. While the vibration amplitudes of bearings 3 and 4 exhibit rapid increases at the end of lifespans, which represents abrupt degradation processes. Take bearing 4 as an example, its detailed information of the vibration signals during the healthy operation stage and the failure stage are presented in Fig. 6.3. It is known that obvious impacts are more intense in the vibration signals of the failure stage than those of the healthy operation stage.

6.2.3.2 FPT Selection

In the FPT selection module, Kurtosis, the health indicator sensitive to incipient fault, is firstly extracted from vibration signals. Based on Kurtosis, the FPT selection approach is used to determine the FPTs of the four tested bearings. For comparison, another FPT selection approach proposed by (Ginart et al. 2006) is also utilized to select the FPT. The comparison results are shown in Fig. 6.4, and all FPT values are tabulated in Table 6.2, where "N/A" means that no FPT is detected during the entire lifetime. It is seen from Fig. 6.4 that the alarm set by Ginart's approach is much higher than the 3σ interval and thus unable to distinguish the healthy and the abnormal states. On the contrary, the 3σ interval is able to detect the abnormal states. However, as can be observed during the healthy operation stage that some abnormal

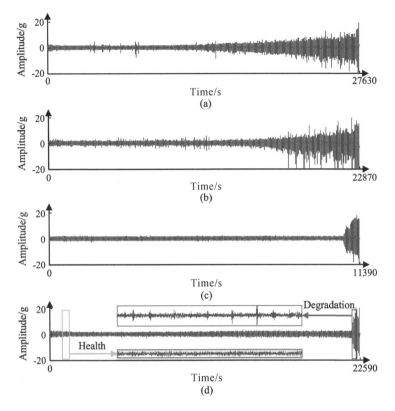

Fig. 6.3 Vibration signals of the four tested bearings: (a) bearing 1, (b) bearing 2, (c) bearing 3, and (d) bearing 4

Table 6.2 Selected FPTs for the four bearings

Selection Method	Bearing 1	Bearing 2	Bearing 3	Bearing 4
Ginart's approach	21670 s	18270 s	N/A	22580 s
Presented approach	14630 s	16440 s	10910 s	22130 s

states exist due to random noises rather than actual faults, which are marked by "∗" in Fig. 6.4. Fortunately, with the assistance of consecutive triggering mechanism, the actual time when faults occur is appropriately selected as the FPT.

In Fig. 6.4, all RMS of the four bearings and their corresponding FPT selected by the two approaches are provided. It can be seen that compared with Kurtosis, RMS has a more obvious degradation trend with the propagation of faults. However, since bearings 1 and 2 both exhibit gradual degradation pattern, it is therefore difficult to yield exact time of the fault occurrence. As a consequence, the FPTs selected by our approach would also have a little time delay as shown in Fig. 6.4. Although the FPTs

6.2 RUL Prediction with Random Fluctuation Variability

are delayed, the time difference is still much smaller than that of Ginart's approach, especially in the case of bearing 1, which is acceptable for the RUL prediction of bearings. As for bearings 3 and 4, they behave abrupt degradation trends. Therefore, it is not as difficult as bearings 1 and 2 to accurately select the FPTs. In these two cases, the introduced FPT selection approach still outperforms Ginart's approach.

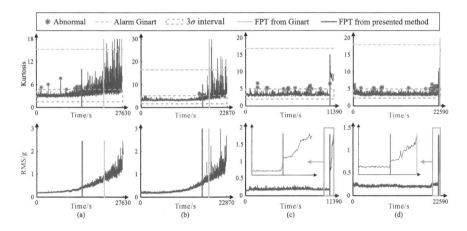

Fig. 6.4 FPT selection results for the tested bearings: (a) bearing 1, (b) bearing 2, (c) bearing 3, and (d) bearing 4

6.2.3.3 RUL Prediction

After the FPT selection module, RMS is further extracted from vibration signals, and the RULs of the above four tested bearings are predicted using the improved exponential model. In the joint filtering step, the particle number is set to be 1000. The state estimation results are provided in Fig. 6.5, and the RUL prediction results are presented in Fig. 6.6. To achieve an exhaustive comparison, the original exponential model (Si et al. 2013) and the Paris model (An et al. 2013) are used to predict the RUL of the four bearings as well. The same FPTs and particle numbers are also applied for them, and the estimation as well as prediction results are also presented in Figs. 6.5 and 6.6, respectively. Figure 6.5 indicates that the estimation results of the original model almost overlap with the actual RMS, while two PF-based methods, i.e. the Pairs model and the improved exponential model, are able to reduce random errors of the degradation processes and therefore reflect the overall degradation trends of the bearings. In Fig. 6.6, all of these methods output inaccurate results at the beginning because of the lack of online monitoring data. However, when enough online monitoring data are available for utilization, the original exponential model and the Paris model still output RUL prediction results with large prediction errors, especially for bearing 2. The improved exponential model, however, yields the prediction results gradually converging to the actual RUL fastest and exhibits the highest prediction accuracy among all methods. Such a phenomenon can be explained as follows.

For the original exponential model, it neglects the state estimation procedure, thus obtained the prediction results are impacted by the random errors of the degradation process. With the help of PF, the Paris model outperforms the original exponential method, but it is still weak in the aspect of model parameter estimation. As a consequence, the RUL prediction results of the Paris model still exhibit large random errors. Compared with these two methods, the improved exponential model takes the advantages of the superiority of PF in the state estimation and the Bayesian updating and EM algorithm in model parameter estimation. Therefore, it presents the best prediction results among the three models.

To quantitively compare the prediction performances of the three models, CRA and Convergence (Saxena et al. 2010) are utilized as performance metrics and computed at the times indexed from the half to the end of the lifetime. The metric scores are provided in Fig. 6.7. It can be noticed that the improved exponential model has the highest CRA and the lowest Convergence values among all three models for all tested bearings. Such a result indicates that the improved exponential model owns the superiority of providing the prediction results with the highest accuracy and the convergence with the fastest speed in all three methods. In a word, the improved exponential model performs best in the RUL prediction of the bearings.

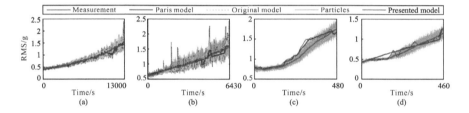

Fig. 6.5 State estimation results of the tested bearings: (a) bearing 1, (b) bearing 2, (c) bearing 3, and (d) bearing 4

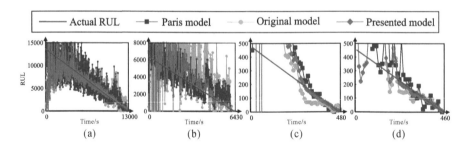

Fig. 6.6 RUL prediction results of the tested bearings: (a) bearing 1, (b) bearing 2, (c) bearing 3, and (d) bearing 4

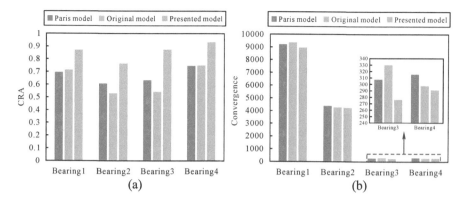

Fig. 6.7 Scores of two performance of the tested bearings: (a) CRA, and (b) Convergence

6.2.4 Epilog

Considering the influence of random fluctuation variability on RUL prediction of mechanical systems, this section introduces a RUL prediction method based on joint filtering of parameter and state. First, an FPT selection technique based on continuous triggering mechanism is elaborated, which provides an appropriate initialization of RUL prediction. Then based on online condition monitoring data, joint filtering of parameter and state is executed to reduce the disturbance of random fluctuations on RUL prediction. Through the utilization of PHM 2012 bearing accelerated degradation testing data, the application capability of this method in practical cases is demonstrated.

6.3 RUL Prediction with Unit-to-Unit Variability

6.3.1 Motivation

Due to the existence of individual uncertain factors such as product quality, operating environment, etc., the degradation processes of the same type mechanical systems usually exhibit distinctive difference. Even if they are working in the same operating environment, their degradation rates might still not be alike. Such a phenomenon is generally referred to as unit-to-unit variability (UtUV). The UtUV is defined as the diversity in degradation rates among different units from a population due to different operating conditions and health conditions, and can cause large-scale fluctuations in the degradation process of mechanical systems. Generally, the UtUV is restricted on a group of systems with the same type and specification, such as a group of gears produced from the same production and operation batch, and the term unit refers to a single object from the group, i.e., a specific gear in the previously mentioned gear group. Due to the sparsely distributed degradation trends of different units in the whole dataset, one model formulated regarding to the whole datasets is unable to be ensured to output accurate and converging RUL prediction results with respect to one specific monitored unit. Therefore, in order to enhance the RUL prediction accuracy and increase the effectiveness of predictive maintenance strategy, it is necessary to incorporate the influence of UtUV factor into RUL prediction.

Regarding the influence of UtUV factor on RUL prediction, most of the existing methods assumed that the degradation rates of similar mechanical equipment under the same working environment follow a specific probability distribution, and the most commonly used one is the normal distribution. In existing literature, the probability distribution parameters are estimated using the data from the failed units, and then the evaluation results of the degradation rate of the current samples are updated online based on the real-time data of the monitored units. In general, this type of methods is able to obtain the general distribution of the degradation rates with the help of the data from the failed units, then update the degradation rate dynamically, guaranteeing a more convergent and more accurate RUL prediction results. However, there are still two shortcomings in those methods. One shortcoming is that these studies assumed that the UtUV parameter follows a certain distribution. In practical cases, UtUV parameter varies with the changes of equipment types and operating environment, and it is thus unreasonable to restrict the UtUV parameter of different units to a certain distribution. The other shortcoming is that the parameter evaluation process of the monitored units is limited by the distribution divergence of the training units. If the degradation rates of the monitored units deviate from the distribution range of the training units, the parameters can only converge to the boundary of the range instead of to the true value.

To better explain the above two shortcomings, Fig. 6.8 shows the online evaluation process of the degradation rates of two typical units. The measurements of different units are also illustrated. For those two units, the degradation rate of unit 1 falls within the distribution range of the training units, and the degradation rate of unit 2 deviates from that of the training units. Regarding to shortcoming 1, as it is assumed that the degradation rate of the training units obeys a certain type of probability distribution, the initial update of the degradation rate of unit 1 is very slow, and a long period is required for the convergence to the actual value, which seriously affects the convergence speed of the parameter estimation. For shortcoming 2, it can be noticed that the parameter evaluation process for unit 2 is limited by the initial distribution of the training units. Since the actual value of the degradation rate is not within the distribution range of the training units, the online evaluation process is compromised, and the actual degradation rate of the monitoring unit cannot be achieved.

To overcome these two shortcomings, a Wiener process-based method (WPM) considering UtUV is proposed in literature (Li et al. 2019) for predicting the RUL. The following three aspects are mainly addressed:

- An age- and state-dependent WPM is specially designed to describe the UtUV.

6.3 RUL Prediction with Unit-to-Unit Variability

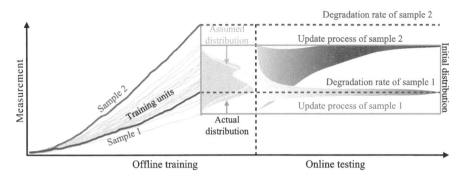

Fig. 6.8 The effect of the offline evaluation result on the online update process of the monitoring samples degradation rates

- An enhanced MLE algorithm called unit maximum likelihood estimation (UMLE) is proposed to estimate UtUV parameters without any limitation of the distribution mode.
- A fuzzy resampling algorithm is developed to deal with the sample dilution problem in the PF process.

The detailed process of the WPM-based method considering UtUV will be introduced in the following subsection.

6.3.2 RUL Prediction Model Considering Unit-to-Unit Variability

The flowchart of the presented method is shown in Fig. 6.9. Based on the developed WPM, the RUL prediction is carried out by performing offline training and online testing sequentially with respect to the measurements. In the offline training module, UtUV is introduced into the age- and state-dependent WPM, and the UMLE algorithm is utilized to estimate model parameters from historical data. In the online testing module, initial particles are generated based on the parameter estimation results, then the fuzzy resampling PF algorithm is used to update the UtUV parameter and the health state particle set online. Finally, the online parameter and state update results are substituted into the model to predict the RUL PDF. Detailed procedures are elaborated as follows.

6.3.2.1 Development of the WPM

1) Degradation Modeling and Analytical Solution of RUL

Assuming the degradation process of mechanical system is related to both the age and the health state. According to (Li et al. 2017a), the following age- and state-dependent

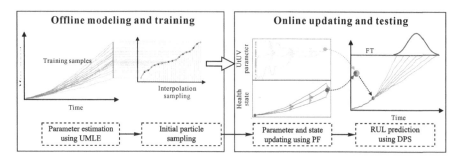

Fig. 6.9 Flowchart of the RUL prediction method considering UtUV

WPM expression can be used to describe such degradation process.

$$s(t) = s_k + \int_{t_k}^{t} \mu(s(\tau), \tau) d\tau + \int_{t_k}^{t} \sigma(\tau) dB(\tau) \tag{6.14}$$

where s_k is the state at time t_k; $s(t)$ is the state at a future time t, with $t \geqslant t_k$; $\mu(s(\tau), \tau)$ is the drift coefficient function; $\sigma(t)$ is the diffusion coefficient function; and $\{B(\tau), \tau \geqslant t_k\}$ denotes a standard BM process.

The RUL is defined as the remaining time until the state surpasses the FT

$$l_k = \inf\{l : s(l + t_k) \geqslant \lambda | s_k\} \tag{6.15}$$

where l_k is the RUL at t_k, $\inf\{\cdot\}$ denotes the inferior limit function, and λ is the FT.

It can be seen from Eq. (6.15) that the RUL prediction task is essentially the problem of determining the moment when a general Wiener process crossing a constant FT. To derive the analytical solution of the RUL PDF $f(l|s_k)$, the following three conversion steps can be performed.

(1) **Translation Transformation:** The degradation process in Eq. (6.14) can be transformed into a new degradation process starting from the coordinate origin through the following translation transformation as

$$\begin{cases} \tilde{s}(l) = s(t) - s_k \\ l = t - t_k \end{cases} \tag{6.16}$$

(2) **Standard Transformation:** The degradation process $\{\tilde{s}(l), l \geqslant 0\}$ is further transformed into a standard BM process using the following standard transformation (Ricciardi 1976):

$$\begin{cases} B(\tilde{l}) = \varphi(\tilde{s}, l) = s(l + t_k) - s_k - \int_0^l \mu(s(\tau + t_k), \tau + t_k) d\tau \\ \tilde{l} = \psi(l) = \int_0^l \sigma^2(\tau + t_k) d\tau \end{cases} \tag{6.17}$$

6.3 RUL Prediction with Unit-to-Unit Variability

The FT λ is changed correspondingly following the transformation step:

$$\varphi(\lambda,l) = \lambda - s_k - \int_0^l \mu(s(\tau + t_k), \tau + t_k) d\tau \tag{6.18}$$

(3) **Inverse Transformation:** Through the aforementioned transformations, a general Wiener process $\{s(t), t \geq t_k\}$ is transformed into a standard BM process $\{B(\tilde{l}), \tilde{l} \geq 0\}$. The PDF of a standard BM process crossing a time-varying FT has been derived in (Si et al. 2012), which is expressed as

$$f_B(\tilde{l}) \simeq (\frac{\varphi(\lambda,l)}{\psi(l)} + \frac{d\varphi(\lambda,l)}{d\psi(l)}) \frac{1}{\sqrt{2\pi\psi(l)}} \exp(-\frac{\varphi^2(\lambda,l)}{2\psi(l)}) \tag{6.19}$$

Since the standard BM process $\{B(\tilde{l}), \tilde{l} \geq 0\}$ is transformed from $\{s(t), t \geq t_k\}$, the probability of the standard BM process crossing the time-varying FT $\varphi(\lambda,l)$ equals the probability of the general Wiener process $\{s(t), t \geq t_k\}$ crossing the constant FT λ

$$F(l|s_k) = \Pr(s(l + t_k) \leq \lambda) = \Pr(\tilde{s}(l) \leq \lambda - s_k) = \Pr(B(\tilde{l}) \leq \varphi(\lambda, l))$$
$$= F_B(\tilde{l}) \tag{6.20}$$

By calculating the derivatives of $F(l|s_k)$ and $F_B(\tilde{l})$ with respect to l, the pdf of RUL is derived as follows:

$$f(l|s_k) \simeq (\frac{\lambda - s_k - \int_0^l \mu(s(\tau + t_k), \tau + t_k) d\tau}{\int_0^l \sigma^2(\tau + t_k) d\tau} + \frac{\mu(s(l + t_k), l + t_k)}{\sigma^2(l + t_k)}) \times$$
$$\frac{\sigma^2(l + t_k)}{\sqrt{2\pi \int_0^l \sigma^2(\tau + t_k) d\tau}} \exp(-\frac{(\lambda - s_k - \int_0^l \mu(s(\tau + t_k), \tau + t_k) d\tau)^2}{2 \int_0^l \sigma^2(\tau + t_k) d\tau})$$
$$\tag{6.21}$$

2) Degradation Process Simulation Algorithm-based RUL Prediction

As can be seen in Eq. (6.21), The future degradation trajectory $\{s(\tau + t_k)\}_{0 \leq \tau \leq l}$ involves the integration of future states, which are random and unknown in real application. In view of this problem, a degradation process simulation-based (DPS) RUL prediction algorithm is developed. The key idea of the DPS-based algorithm is to generate a set of possible degradation trajectories via Monte Carlo simulation, then the RUL PDF can be approximately obtained by inputting those trajectories into the above analytical solution. The procedures of the algorithm are as follows.

(1) Set the initial time interval as Δl_0 and generate an initial state set $\{s_n(t_k) = s_k\}_{n=1:N_x}$, where N_s is the number of simulated trajectories.

(2) Propagate each trajectory state stepwise via

$$s_n(l_i + t_k) = s_n(l_{i-1} + t_k) + \mu(s_n(l_{i-1} + t_k), l_{i-1} + t_k)\Delta l_{i-1} + V_{i-1} \quad (6.22)$$

where $s_n(l_i + t_k)$ is the state of the nth trajectory at $l_i + t_k$ ($i \in N^+$, $l_i = \sum_{j=0}^{i-1} \Delta l_j$) and V_{i-1} is the transition noise following a uniform distribution $U(-v_{i-1}, v_{i-1})$ with $v_{i-1} = 3\sigma(l_{i-1} + t_k)\sqrt{\Delta l_{i-1}}$.

(3) Adjust the time interval according to

$$\begin{cases} \Delta l_i = \Delta l_{i-1}, i = i+1, M_i = 0 \\ \Delta l_i = \min(M_{am}\Delta l_{i-1}/M_i, \Delta l_0), i = i+1, 0 < M_i \leqslant M_{am} \\ \Delta l_{i-1} = M_{am}\Delta l_{i-1}/M_i, i = i, M_i > M_{am} \end{cases} \quad (6.23)$$

where M_i is the number of the trajectories which surpass the FT during the ith transition, and M_{am} is the allowed maximum number of one-step failed trajectories.

(4) Repeat (2) and (3) until all trajectories have crossed the FT.

(5) Calculate the RUL probability of the nth trajectory by inputting its state values into the discrete version of analytical solution

$$p_n(l = l_{L_n}) \cong (\frac{\lambda - s_k - \sum_{i=0}^{L_n-1}\mu(s_n(l_i + t_k), l_i + t_k)\Delta l_i}{\int_0^{l_{L_n}} \sigma^2(\tau + t_k)d\tau} +$$

$$\frac{\mu(s_n(l_{L_n} + t_k), l_{L_n} + t_k)}{\sigma^2(l_{L_n} + t_k)}) \times$$

$$\frac{\sigma^2(l_{L_n} + t_k)}{\sqrt{2\pi \int_0^{l_{L_n}} \sigma^2(\tau + t_k)d\tau}} \exp(-\frac{(\lambda - s_k - \sum_{i=0}^{L_n-1}\mu(s_n(l_i + t_k), l_i + t_k)\Delta l_i)^2}{2\int_0^{l_{L_n}} \sigma^2(\tau + t_k)d\tau})$$

(6.24)

where L_n is the transition step of the nth trajectory, and $l_{L_n} = \sum_{i=0}^{L_n-1} \Delta l_i$ is its corresponding RUL.

(6) Approximate the RUL PDF by numerically integrating the probabilities of all trajectories.

$$\hat{f}(l|s_k) \cong \frac{\sum_{n=1}^{N_s} p_n(l = l_{L_n})\delta(l - l_{L_n})}{\sum_{n=1}^{N_s} \delta(l - l_{L_n})} \quad (6.25)$$

where $\delta(\cdot)$ is the Dirac delta function.

Compared with Monte Carlo simulation, the advantage of the DPS-based algorithm owns the superiority thanks to the exactness of the analytical solution.

6.3 RUL Prediction with Unit-to-Unit Variability

Therefore, it is able to output more accurate RUL prediction results in a flexible way.

6.3.2.2 Offline Training

In order to introduce the effect of UtUV into the degradation rate, a random parameter is added into the formulated WPM as

$$s(t) = s_k + a \int_{t_k}^{t} \omega(s(\tau), \tau, b) d\tau + \int_{t_k}^{t} \sigma_B dB(\tau) \qquad (6.26)$$

where a is the UtUV parameter that describes the changes in degradation rate among different units and is assumed to be a random variable. B and σ_B are two constant parameters governing the general degradation behaviors ubiquitously existed among all units.

In the offline training module, the distribution of the UtUV parameter a, and the values of b and σ_B are estimated based on the historical data from training units. In order to fulfill this task, a UMLE algorithm is presented, which is able to estimate the UtUV parameter of each unit individually unrestrictedly. Consider the case where the historical data S of N training units are available. Denote the UtUV parameter a of the nth training unit as a_n and transform Eq. (6.14) into the following discrete version:

$$s_{n,k} = s_{n,k-1} + a_n \omega(s_{n,k-1}, t_{k-1}, b) \Delta t_{n,k-1} + \sigma_B dB(\Delta t_{n,k-1}) \qquad (6.27)$$

where $s_{n,k}$ is the measurement of the nth unit at t_k and $\Delta t_{n,k-1} = t_{n,k} - t_{n,k-1}$ is the time interval.

Let $\Delta S_n = [s_{n,1} - s_{n,0}, \cdots, s_{n,K_n} - s_{n,K_n-1}]$, where K_n is the number of the measurements of the nth training unit. According to the i.i.d characteristic of Brownian motion, ΔS_n follows the normal distribution $N(a_n \omega_n, \sigma_B^2 \Sigma_n)$, with

$$\omega_n = [\omega(s_{n,0}, t_{n,0}, b) \Delta t_{n,0}, \omega(s_{n,1}, t_{n,1}, b) \Delta t_{n,1}, \cdots, \omega(s_{n,K_n-1}, t_{n,K_n-1}, b) \Delta t_{n,K_n-1}]^T \qquad (6.28)$$

$$\Sigma_n = \text{diag}([\Delta t_{n,0}, \cdots, \Delta t_{n,K_n-1}]) \qquad (6.29)$$

where Σ_n denotes a $K_n \times K_n$ diagonal matrix with $[\Delta t_{n,0}, \cdots, \Delta t_{n,k_n-1}]$ being the main diagonal elements.

The log-likelihood function over the unknown model parameter set $\Theta = [a_1, a_2, \cdots, a_N, b, \sigma_B^2]^T$ is formulated as

$$\ell(\Theta|S) = -\frac{\ln(2\pi)}{2}\sum_{n=1}^{N}(K_n - 1) - \frac{1}{2}\sum_{n=1}^{N}(K_n - 1)\ln\sigma_B^2 - \frac{1}{2}\sum_{n=1}^{N}\ln|\Sigma_n|$$
$$-\frac{\sum_{n=1}^{N}(\Delta S_n - a_n\omega_n)^T \Sigma_n^{-1}(\Delta S_n - a_n\omega_n)}{2\sigma_B^2} \quad (6.30)$$

Calculate the first partial derivatives of Eq. (6.30) with respect to a_n ($n = 1, \cdots, N$) and σ_B^2.

$$\frac{\partial l(\Theta|S)}{\partial a_n} = \frac{\omega_n^T \Sigma_n^{-1}(\Delta S_n - a_n\omega_n)}{2\sigma_B^2}, n = 1, \cdots, N \quad (6.31)$$

$$\frac{\partial l(\Theta|S)}{\partial \sigma_B^2} = -\frac{\sum_{n=1}^{N} K_n}{2\sigma_B^2} + \frac{\sum_{n=1}^{N}(\Delta S_n - a_n\omega_n)^T \Sigma_n^{-1}(\Delta S_n - a_n\omega_n)}{2\sigma_B^4} \quad (6.32)$$

Let the aforementioned derivatives equal zeros. The estimation results of a_n and σ_B^2 can be given as

$$\hat{a}_n = \frac{\omega_n^T \Sigma_n^{-1} \Delta S_n}{\omega_n^T \Sigma_n^{-1} \omega_n}, n = 1, \cdots, N \quad (6.33)$$

$$\hat{\sigma}_B^2 = \frac{\sum_{n=1}^{N}(\Delta S_n - a_n\omega_n)^T \Sigma_n^{-1}(\Delta S_n - a_n\omega_n)}{\sum_{n=1}^{N}(K_n - 1)} \quad (6.34)$$

Then substitute Eq. (6.33) and Eq. (6.34) into Eq. (6.30), the following function holds:

$$\ell(b|S) = -\frac{\ln(2\pi) + \ln\hat{\sigma}_B^2 + 1}{2}\sum_{n=1}^{N}(K_n - 1) - \frac{1}{2}\sum_{n=1}^{N}\ln|\Sigma_n| \quad (6.35)$$

As the above function is only dependent on one variable b, which can be optimized using one-dimension optimal search. Then, $\hat{a}_n$ ($n = 1, \cdots, N$) and $\hat{\sigma}_B^2$ can further be acquired by submitting $\hat{b}$ into Eq. (6.33) and Eq. (6.34), respectively. Now that the UtUV parameters of each unit are individually estimated, the initial distribution of UtUV parameters can be described using the discrete estimation results with no restriction on the distribution type.

6.3.2.3 Online Testing

In the previous subsection, the UtUV parameter of each training unit has been estimated. In the online testing module, the UtUV parameter and state are jointly updated by employing the PF algorithm (Zhang et al. 2010; Orchard et al. 2012) according

6.3 RUL Prediction with Unit-to-Unit Variability

to the real-time measurements of the testing unit. After that, the RUL is predicted through the DPS algorithm with the updated parameter and state.

To reproduce the number of the UtUV parameter samples from N to N_p under the same distribution, a linear interpolation method is used to initialize the UtUV parameter particles. At first, the estimated parameters $\{\hat{a}_n\}_{1:N}$ are sorted from small to large as a new parameter sequence $\{\hat{a}_n^*\}_{1:N}$, and denote their corresponding sequence numbers as $\{1, 2, \cdots, N\}$. To change the particle number from N to N_p, the interval between two resampled particles is adjusted as $\Delta = (N-1)/(N_p-1)$. If the ith resampled particle is within the $(n-1)$th and the nth original particles, then $n-2 < (i-1)\Delta \leq n-1$. From this inequality, it is derived that $n \in [1+(N-1)(i-1)/(N_p-1), 2+(N-1)(i-1)/(N_p-1)]$. With this linear interpolation, the following relationship between the interpolation value and the original value holds:

$$\hat{a}_n^* - \hat{a}_{n-1}^* = \frac{a_0^i - \hat{a}_{n-1}^*}{(i-1)\Delta - n + 2} \quad (6.36)$$

Therefore, the resampled ith particle is

$$a_0^i = \hat{a}_{n-1}^* + (2 - n + \frac{(N-1)(i-1)}{N_p - 1})(\hat{a}_n^* - \hat{a}_{n-1}^*), i = 1, 2, \cdots, N_p \quad (6.37)$$

A series of initial UtUV parameter particles $\{a_0^i\}_{i=1:N_p}$ is acquired by using Eq. (6.37). Let all of the initial state particles $\{s_0^i\}_{i=1:N_p}$ equal s_0, i.e., the actual measurement of the tested unit at the initial time, the state particles are transmitted following:

$$\tilde{s}_k^i = s_{k-1}^i + a_{k-1}^i \omega(s_{k-1}^i, t_{k-1}, \hat{b}) \Delta t_{k-1} + \hat{\sigma}_B \mathrm{d}B(\Delta t_{k-1}) \quad (6.38)$$

where s_{k-1}^i and a_{k-1}^i are the ith state particle and parameter particle at t_{k-1}, respectively. $\tilde{s}_k^i$ is the transition result of the ith state particle, and $\Delta t_{k-1} = t_k - t_{k-1}$ is the monitoring interval.

When the new measurement s_k at t_k is observed, the particle weights are updated and normalized as follows:

$$\tilde{\omega}_k^i = \omega_{k-1}^i \frac{1}{\sqrt{2\pi \hat{\sigma}_B^2 \Delta t_k}} \exp\left(-\frac{(s_k - \tilde{s}_k^i)^2}{2\hat{\sigma}_B^2 \Delta t_k}\right), \omega_k^i = \frac{\tilde{\omega}_k^i}{\sum_{i=1}^{N_p} \tilde{\omega}_k^i} \quad (6.39)$$

Note that the above PF algorithm updates the weight of each particle separately and describes the distribution of the particles using the updated weights. Therefore, it still has no restriction to the distribution of the UtUV parameter.

After that, the parameter and state particles are resampled according to their particle weights. The basic idea of resampling is to concentrate on particles with large weights and eliminate particles with small weights. The commonly used resampling algorithm in original PF is systematic resampling, which reproduces the particles with large weights (Arulampalam et al. 2002). However, this algorithm generally leads to the sample impoverishment problem, i.e., most particles are copied from the same one after a few iterations, leading to a loss of diversity among particles. To deal with this problem, a fuzzy resampling algorithm is presented. The principle of fuzzy resampling is to fuzz the particles sampled by adding random noise. A diagram of the fuzzy resampling algorithm is shown in Fig. 6.10, and its detailed procedure is presented as follows:

- Calculate the accumulated sequence of the particle weights $\{C_k^i = \Sigma_{j=1}^i \omega_k^j\}_{i=1:N_p}$, and let $C_k^0 = 0$. Each particle corresponds to an interval with the ith particle corresponding to $(C_k^{i-1}, C_k^i]$.
- Generate N_p-ordered random numbers following the uniform distribution $U(0,1)$, which are denoted as $\{r^d\}_{d=1:N_p}$.
- If the dth random sample falls into the ith interval, i.e., $r^d \in (C_k^{i-1}, C_k^i]$, then the resampled dth particle is $\{\tilde{a}_k^d = a_{k-1}^i, \tilde{s}_k^d = \tilde{s}_k^i\}$.
- Fuzz the resampled parameter particles using $a_k^d = \tilde{a}_k^d + \eta \xi_k^d$, where ξ_k^d is a normally distributed random noise $N(0, \sigma_{a_{k-1}}^2/N_p)$ with $\sigma_{a_{k-1}}^2$ denoting the variance of $\{a_{k-1}^i\}_{i=1:N_p}$, and η is a fuzzy coefficient which equals 1 if a_{k-1}^i is resampled more than once or equals 0 if a_{k-1}^i is resampled only once.
- Reset all particle weights to $1/N_p$.

After the resampling, new parameter and state particles $\{[a_k^i, s_k^i]\}_{i=1:N_p}$ are generated. The median of the parameter particles $\bar{a}_k$ and the median of the state particles $\bar{s}_k$ are used as the estimated UtUV parameter and the health state, respectively. The RUL PDF of the tested unit at t_k is calculated by inputting the estimated parameter and state set $[\bar{a}_k, \hat{b}, \hat{\sigma}_B^2, \bar{s}_k]$ into the DPS algorithm presented in Sect. 6.3.2.1. The RUL prediction will be iteratively carried out until the measurements exceed the FT.

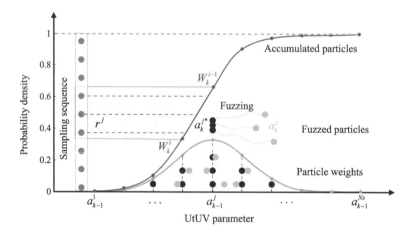

Fig. 6.10 Diagram of the fuzzy resampling algorithm

6.3.3 RUL Prediction Case of Turbofan Engine Degradation Dataset

6.3.3.1 Experimental Data Introduction

In this subsection, the above introduced method is verified by the simulation dataset of turbofan engine degradation published on the NASA Prognostics Data Repository website (Saxena and Goebel 2008). The dataset was simulated by a commercial software developed by NASA to investigate the degradation process of aeroengines.The dataset contains 21 sets of characteristic parameters of 100 aero engines during their full-service life. 100 engines operate at the same altitude, flight speed, and ambient temperature. The external operating environment of different samples is the same, and the difference in failure time is mainly incurred by the difference in the degradation rates of different samples. This dataset has been widely adopted in RUL prediction field due to its high authenticity.

6.3.3.2 Health Indicator Construction

To reflect the degradation processes of engines, a global HI estimation method (Ramasso 2014) is firstly used to fuse the multi-sensor data into a single HI. Here we choose 11 from the 21 features, which are recorded in the following columns [7, 8, 9, 10, 12, 14, 16, 17, 20, 25, 26] of the dataset. The feature vector for the ith unit at t_k is denoted as $x_k^i = [x_{k,1}^i, x_{k,2}^i, \ldots, x_{k,11}^i]$. These original features are fuzzed into a HI using the following linear weighting model:

$$\text{HI}_k^i(x_k^i, \theta) = \theta_0 + \sum_{n=1}^{11} \theta_{n, x_{k,n}^i} \tag{6.41}$$

where $\theta = [\theta_0, \theta_1, \cdots, \theta_{11}]$ is the weight vector. As the presented method is based on the premise that the degradation process has an increasing trend, the target value of the HI is defined as

$$\hat{\text{H}}\text{I}_k^i(x_k^i, \theta) = \begin{cases} 0, & t_k \leqslant T_i 5\% \\ \exp(\dfrac{\log(0.05)}{0.95 T_i}(T_i - t_k)), & T_i 5\% < t_k < T_i 95\% \\ 1, & t_k \geqslant T_i 95\% \end{cases} \tag{6.42}$$

where T_i is the length of the ith unit. It should be noticed that this HI function ranges between 0 and 1 during the lifetime, which has an opposite trend to the HI constructed in (Ramasso 2014).

The HIs of the 100 units are constructed as follows. At first, the HI target values of the 90 random training units are calculated using Eq. (6.42). Then, the feature values and the target values are input into Eq. (6.41), and the standard least-squares algorithm is utilized to estimate the weight vector θ. After that, the HIs of the 90 training units and the 10 tested units can further be obtained through inputting their features and the estimated weight vector into Eq. (6.41). By utilizing the moving average technique with the window length Δt_ω of 15, those HIs can further be smoothed. The FT is defined as the median of the final HI values of the 90 training units, i.e., $\lambda = 0.7873$. The HI of the 100 units and the FT are shown in Fig. 6.11. It can be seen from the HIs that the degradation trajectories of 100 samples exhibit a loose distribution pattern, and the failure time varies obviously. The shortest failure time is about 140 cycles, and the longest failure time reaches 360 cycles. Such a phenomenon indicates a nonnegligible variability among the degradation rates. Therefore, in the RUL prediction procedure, the influence of UtUV must be fully considered.

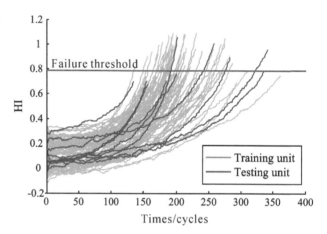

Fig. 6.11 HIs of the 100 turbofan engines

6.3.3.3 RUL Prediction of the Turbofan Engine Dataset

During the HI construction, Eq. (6.42) is used to describe the turbofan engines HI as an exponential function with respect to time. In order to be unified with the HI function form, the following exponent-type WPM is selected to describe the degradation of turbofan engines.

$$s(t) = s_k + a \int_0^t b \exp^{b\tau} d\tau + \int_0^t \sigma_B dB(\tau) \tag{6.43}$$

According to the above equation, the measurement data of 90 training units are used to estimate model parameters, and then 10 sets of testing units are chosen for RUL prediction. To illustrate the effectiveness of the presented method, two other methods are also used. Method 1 (M1) is a method based on the nonlinear diffusion model in (Si et al. 2012). It uses the original MLE algorithm to estimate model

6.3 RUL Prediction with Unit-to-Unit Variability

parameters and predicts RUL without updating UtUV parameters and state. Method 2 (M2) uses the UMLE algorithm to estimate model parameters, and uses the original PF algorithm to update UtUV parameters and state (An et al. 2013). For simplicity, the presented method is named Method 3 (M3).

The prediction results of the three methods are shown in Fig. 6.12. It can be seen from the RUL prediction curves that the results of M1 deviate seriously from the actual RUL. Such unsatisfactory results are caused by the fact that M1 does not update the degradation rate with the online measurement. For M2, most of the result curves converge to the actual RUL. However, it can be noted that there are three curves deviate far from the distribution of the training units, and the UtUV parameters and state of these three units cannot be accurately estimated. However, the results of M3 gradually converge to actual RUL, obtaining more accurate prediction results than M1 and M2. The reason is that M3 uses PF combined with fuzzy resampling algorithm to update UtUV parameters and state. With the introduction of fuzzing operation, the diversity of particles is substantially increased. Therefore, the parameters and states of testing units far from the distribution can also be accurately estimated.

For quantitative comparison, the CRA and Con scores of the three methods are shown in Figs. 6.13 and 6.14, respectively. It can be seen that the presented method M3 acquires the highest CRA scores in eight out of ten units, and the lowest Con scores in seven out of ten units. It shows that for most units, the result of M3 is more accurate than the results of M1 and M2, and it converges to the actual RUL faster. These results further verify the superiority of this method in mechanical system RUL prediction.

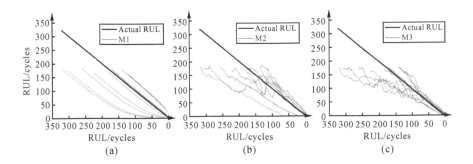

Fig. 6.12 RUL prediction results of the ten tested turbofan engines. (a) M1, (b) M2, and (c) M3

6.3.4 Epilog

This section mainly considers the influence of UtUV on the data-model fusion RUL prediction method. Through a comprehensive literature review, it is concluded that existing methods have two shortcomings when dealing with UtUV:

(1) Describe the degradation rate with a pre-specified probability distribution

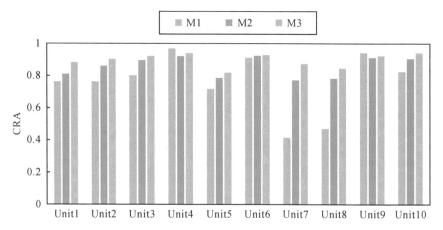

Fig. 6.13 CRA score of the three methods for the testing units

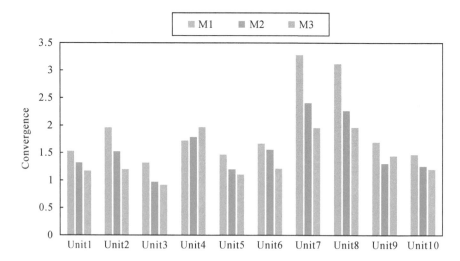

Fig. 6.14 Convergence score of the three methods for the testing units

(2) The parameter evaluation is limited by the initial distribution

In order to solve the first shortcoming, a WPM depending on both the age and the state was specially designed to describe degradation process considering UtUV; For the second shortcoming, the UtUV parameter and state were updated jointly using PF combined with the fuzzy resampling algorithm. With the parameter estimation and update procedure, the UtUV was considered and handled systematically during the RUL prediction. The effectiveness of the presented method was demonstrated using a turbofan engine degradation dataset.

6.4 RUL Prediction with Time-Varying Operational Conditions

6.4.1 Motivation

Most mechanical systems operate under time-varying conditions due to the change of functional requirements and working environment. For example, the power generation rate of wind turbine will change with the varying wind speed, and the operating status of aircraft engine will also change with the variation of aircraft altitude. Those changes of external working conditions will lead to different degrees of working stress for the mechanical system, and eventually changes its degradation rate. In the meantime, the change of external working conditions will also affect the amplitude of monitoring signal. For example, if bearings and gears, which are common rotating components used in engineering, are operating at high speed and high load, not only the degradation rates but also the amplitude of vibration signals will be stimulated to increase. In order to intuitively illustrate the influence of working condition change on the degradation rate of mechanical system and the amplitude of monitoring signal, Fig. 6.15 shows the health state degradation process and monitoring signal diagram of a rolling element bearing when operating alternately under two working conditions.

From the degradation process in Fig. 6.15, it can be seen that the degradation rate is changing under different operating conditions. Meanwhile, it can also be noticed that the change of operating condition also leads to the variation of signal amplitude. Such change of degradation rate increases the uncertainty of the lifespan of mechanical systems, while the change of monitoring signal amplitude interferes with the health state assessment. In a word, the joint interaction of the above two factors further increases the difficulty of mechanical system RUL prediction. In most related works, it is often defined that the failure of mechanical systems is the first hitting event of monitoring signal amplitude to the pre-specified FT. However, for the mechanical system operating under variable working conditions, the degradation trend is often compromised due to the amplitude fluctuation caused by the change of working conditions. As shown in Fig. 6.15, if the failure time of mechanical systems is still defined based on the monitoring signal amplitude under variable working conditions, false alarm accidents will occur, while the real health state has not reached the final failure yet.

Through the above analysis, it can be concluded that the change of working conditions will incur two effects on the RUL prediction of mechanical system, i.e., changing the rate of degradation process, and changing the amplitude of monitoring signal. The change of degradation rate will have a direct impact on the lifespan of mechanical system, while the change of monitoring signal amplitude will interfere with the health state evaluation and indirectly affect the RUL prediction results. Existing methods generally rely on the signal amplitude to judge whether the mechanical system is functional or not, thus cannot fully reveal the influence mechanism of working condition change on the RUL. How to distinguish the different effects of working condition changes on health states and monitoring signals, and clarify the

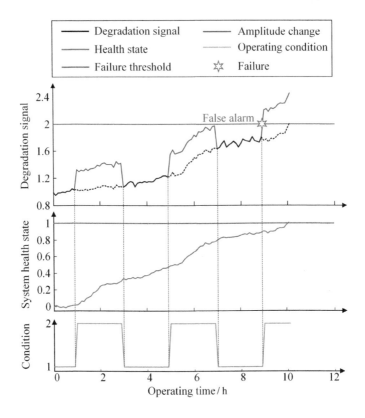

Fig. 6.15 Illustrative example of rolling element bearing degradation process under time-varying operating conditions

mapping relationship between internal health states, external response signals and time-varying working condition parameters has therefore become a critical problem to be solved.

In view of these problems, a two-factor state-space model based method (hereafter referred to as F2S2) for RUL prediction under time-varying operating conditions is introduced (Li et al. 2019): (1) The influence of operating conditions on the degradation rate is introduced into the state transfer function, and the degradation rate is expressed as a function of operating conditions. Then, the state transfer function is scaled on the time horizon accordingly. By introducing a certain operating condition as the reference condition, the time-varying degradation rate can be transformed into a constant decay rate under the reference condition. The state degradation process under variable working conditions is further transformed into the state degradation process under a constant working condition by time scale transformation. (2) The influence of working condition change on the amplitude of monitoring signal is introduced into the observation function. First, the working condition parameters are regarded as the signal amplitude modulation factor, and the amplitude of monitoring signal is then demodulated according to the operating condition parameters to convert the monitoring signal under variable working conditions into the monitoring signal under reference working condition. Through the above two-step transformation, the RUL prediction problem under variable working conditions is transformed into the residual life prediction problem under the constant working condition. As for failure definition, it is regarded as the moment when the health state first exceeds the specific FT, which is not based on the amplitude of monitoring signal. By doing so, the prediction error caused by amplitude change is avoided, so as to improve the RUL prediction accuracy.

6.4.2 RUL Prediction Model Considering Time-Varying Operational Conditions

The flowchart of the F2S2 method is shown in Fig. 6.16, which includes four steps: (1) state-space model development, (2) model parameter estimation, (3) system state estimation, and (4) RUL prediction. In order to transform the RUL prediction problem under variable operating conditions into the RUL prediction problem under constant operating condition, two targeted improvements are made: (1) two influencing factors of operating condition on RUL prediction are introduced into model construction, and a two-factor state-space model is established, and (2) the idea of converting varying operating conditions into constant operating condition is introduced. The detailed process of each step is explained as follows.

6.4.2.1 State-Space Model Development

Firstly, without considering the varying of operating conditions, the degradation process of mechanical system is described by the following linear Wiener process:

$$W(t) = W(0) + \eta t + \sigma_B B(t) \tag{6.44}$$

where $W(t)$ is the degradation state at time t, $W(0)$ denotes the initial state, η is the degradation rate, and $\sigma_B B(t)$ is the diffusion term representing the degradation stochasticity, in which $B(t)$ is a standard Browning motion and σ_B is the diffusion term. The degradation rate η is assumed as a constant value if operating condition is but may vary under different conditions.

Without loss of generality, the degradation state is defined as the relative degradation percentile of the mechanical system with reference to complete failure. In another word, the relative degradation is 0. With the propagation of degradation, it gradually increases from 0 to 1 until the mechanical system is completely failed.

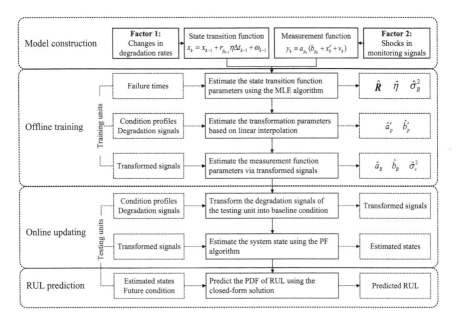

Fig. 6.16 Flowchart of the RUL prediction method considering time-varying operating conditions

With the above definition, the degradation state of mechanical system is transformed into a random variable in [0,1]. The failure time of the above linear Wiener process follows an inverse Gaussian distribution, and the $D = 1$ is correspondingly defined as FT in the following expression.

$$f_W(t) = \frac{D}{\sqrt{2\pi \sigma_B^2 t^3}} \exp\left(-\frac{(t\eta - D)^2}{2\sigma_B^2 t}\right) \tag{6.45}$$

With the consideration of the impacts from time-varying operating conditions, we further formulate the state-space model as follows:

$$x_k = x_{k-1} + r_{p_{k-1}} \eta \Delta t_k + \omega_{k-1} \tag{6.46}$$

$$y_k = a_{p_k}(b_{p_k} + x_k^c + \upsilon_k) \tag{6.47}$$

where x_k and y_k are the degradation state and the monitoring signal at t_k, respectively, $\Delta t_k = t_k - t_{k-1}$ is the monitoring interval, $r_{p_{k-1}}$ is the condition coefficient with p_{k-1} indicating the operating condition at t_{k-1}. To be specific, if there are three different conditions, p_{k-1} then falls within the set $\{1, 2, 3\}$. ω_{k-1} is the state transition noise following the Gaussian distribution $N(0, r_{p_{k-1}} \Delta t_k \sigma_B^2)$, a_{p_k} and b_{p_k} are two operating condition-dependent parameters, c is the exponent of the state representing the nonlinear property of the degradation trend, and υ_k is the measurement

6.4 RUL Prediction with Time-Varying Operational Conditions

noise following the Gaussian distribution $N(0, \sigma^2)$. The above state-space model incorporates time-varying operating conditions by introducing the following two factors:

(1) For the time-varying degradation rate, η is multiplied by a condition coefficient $r_{p_{k-1}}$, which is dependent on operating conditions. A larger $r_{p_{k-1}}$ corresponds to a higher degradation rate, which indicates a heavier operating level.
(2) For the jumps in monitoring signals, a_{p_k} and b_{p_k} are introduced into the measurement function. a_{p_k} and b_{p_k} are assigned with different values in accordance with different operating conditions. Consequently, the measurement function outputs the jumps in the monitoring signal at operating change-points.

For the analytical derivation of RUL PDF, we transform the state transition function in (6.46) as follows according to the independent property of BM.

$$x_k = x_0 + \eta \sum_{i=1}^{k} r_{p_{i-1}} \Delta t_i + \sigma_B B \left(\sum_{i=1}^{k} r_{p_{i-1}} \Delta t_i \right) \tag{6.48}$$

Let $\Delta t_i \to 0$. The above discrete expression converges to its continuous form.

$$x(t) = x_0 + \eta \int_0^t r_{p(v)} dv + \sigma_B B \left(\int_0^t r_{p(v)} dv \right) \tag{6.49}$$

where $r_{p(v)}$ is the condition coefficient under condition $p(v)$ which is a time-dependent continuous function representing the condition index at time v. We regard the integral term $\int_0^t r_{p(v)} dv$ as a new time scale $\tau(t)$, and Let $x_0 = 0$, then (6.49) can be simplified as a new Wiener process (Doksum and Hbyland 1992):

$$x(t) = \eta \tau(t) + \sigma_B B(\tau(t)) = W(\tau(t)) \tag{6.50}$$

Through the above time-scale transformation, the degradation process under time varying degradation rates can be transformed as the degradation process with a fixed baseline degradation rate η. For such degradation process, the cumulative distribution function (CDF) of the failure time holds the following relationship:

$$F(t) = P(x(t) \leq D) = P(W(\tau(t)) \leq D) = F_W(\tau(t)) \tag{6.51}$$

where $F_W(\cdot)$ is the CDF of $f_W(\cdot)$ in Eq. (6.45).

The PDF of the failure time can be obtained by taking the derivative of the CDF with respect to t as

$$f(t) = \frac{dF(t)}{dt} = \frac{dF_W(\tau(t))}{dt} = f_W(\tau(t)) \frac{d\tau(t)}{dt}$$

$$= \frac{r_{p(t)}D}{\sqrt{2\pi\sigma_B^2\tau(t)^3}} \exp\left(-\frac{(\tau(t)\eta - D)^2}{2\sigma_B^2\tau(t)}\right) \tag{6.52}$$

Note that the PDF of the failure time is not continuous at condition change-points. In other words, if the operating condition changes from p_{k-1} to p_k at t_k, the values of the condition coefficient before and after t_k are $r_{p(t_k^-)} = r_{p_{k-1}}$ and $r_{p(t_k^+)} = r_{p_k}$, respectively. As $r_{p(t_k^-)}$ is different from $r_{p(t_k^+)}$, we then have $f(t_k^-) \neq f(t_k^+)$. In conclusion, the PDF of failure time can be calculated by the following equations:

$$\begin{cases} f(t_k^-) = \frac{r_{p_{k-1}}D}{\sqrt{2\pi\sigma_B^2\tau_k^3}} \exp\left(-\frac{(\tau_k\eta - D)^2}{2\sigma_B^2\tau_k}\right) \\ f(t_k^+) = \frac{r_{p_k}D}{\sqrt{2\pi\sigma_B^2\tau_k^3}} \exp\left(-\frac{(\tau_k\eta - D)^2}{2\sigma_B^2\tau_k}\right) \end{cases} \quad \tau_k = \int_0^{t_k} r_{p(v)}\,dv \tag{6.53}$$

In conclusion, Eq. (6.46) and Eq. (6.47) define a two-factor state-space model that incorporates the impacts of time-varying operating conditions. The failure time PDF of such degradation process is calculated by Eq. (6.53). Based on them, we are able to estimate the model parameters based on offline training data in next sections.

6.4.2.2 Model Parameter Estimation Using Offline Data

Assuming there are N available training units for the mechanical system operating under P different operating conditions, the first task is to use the offline data from the training units to estimate the model parameters. For unit n ($n = 1, 2, \cdots, N$), its operating condition series and degradation signal are sampled with same time sequence $(t_{1,n}, t_{2,n}, \cdots, t_{K_n,n})$, where $t_{k,n}$ is the kth sampling time for unit n with $k = 1, 2, \cdots, K_n$, and $t_{K_n,n}$ denotes its corresponding failure time. K_n is the number of samples for unit n. In industrial applications, K_n is determined by both the failure time and the sampling frequency. The sampled sequences of the operating condition series and the condition monitoring signal are denoted as $(p_{1,n}, p_{2,n}, \cdots, p_{K_n,n})$ and $(y_{1,n}, y_{2,n}, \cdots, y_{K_n,n})$, respectively, where $p_{k,n}$ and $y_{k,n}$ are the operating condition series and the measurements of unit n at t_k.

1) Estimation of the State Transition Function Parameters

The state transition function includes three sets of parameters to be estimated, i.e., $\boldsymbol{R} = (r_1, r_2, \cdots, r_P)$, η, and σ_B^2. By using MLE, the likelihood function for these parameters can be formulated, and parameters can correspondingly be estimated. According to Eq. (6.53), the probability that unit n fails at $t_{K_n,n}$ can be given as

$$f_n(t_{K_n}) = \frac{r_{p_{K_n,n}}D}{\sqrt{2\pi\sigma_B^2\tau_{K_n,n}^3}} \exp\left(-\frac{(\tau_{K_n,n}\eta - D)^2}{2\sigma_B^2\tau_{K_n,n}}\right) \tag{6.54}$$

6.4 RUL Prediction with Time-Varying Operational Conditions

It can be seen from the above function that the failure time of a unit is related to its specific operating condition profile. First of all, $r_{p_{K_n,n}}$ records the degradation rate information of unit n at its failure time, which is determined by the condition index $p_{K_n,n}$. Second, the failure time in the transformed time scale $\tau_{K_n,n} = \sum_{k=1}^{K_n} r_{p_{k-1,n}} \Delta t_k$ records all historical information of the condition profile. Based on the above two points, the failure time of a unit is subject to a random distribution highly related to its condition profile. Therefore, we are able to estimate the degradation rates under different condition indexes using the failure time set of training units.

Based on the above, the log-likelihood function of model parameters conditioned on the failure times of training units is:

$$\ell(\boldsymbol{R}, \eta, \sigma_B^2 | T) = \ln\left(\prod_{n=1}^{N} f_n(t_{K_n})\right)$$

$$= -\frac{N}{2}\ln(2\pi\sigma_B^2) + \sum_{n=1}^{N}\ln(r_{p_{K_n,n}}) - \sum_{n=1}^{N}\frac{3}{2}\ln(\tau_{K_n,n}) +$$

$$\sum_{n=1}^{N}\left(-\frac{(\tau_{K_n,n}\eta - D)^2}{2\sigma_B^2 \tau_{K_n,n}}\right) \tag{6.55}$$

where $T = (t_{K_1}, t_{K_2}, \cdots, t_{K_N})$ is the failure time set of the N training units.

Let $\boldsymbol{\Psi}_n = (\psi_{1,n}, \psi_{2,n}, \cdots, \psi_{P,n})$ be the vector of the condition indexes at failure times, of which the element is defined as

$$\psi_{p,n} = \begin{cases} 1, & \text{if } p = \text{the condition index of the } n\text{th unit at } t_{K_n} \\ 0, & \text{otherwise} \end{cases} \tag{6.56}$$

and $\boldsymbol{\Phi}_n = (\phi_{1,n}, \phi_{2,n}, \cdots, \phi_{P,n})$ represents the vector of the time lengths, where $\phi_{p,n}$ is the time length of unit n operating under condition p. Then Eq. (6.55) is equivalent to the following

$$\ell(\boldsymbol{R}, \eta, \sigma_B^2 | T) = -\frac{N}{2}\ln(2\pi\sigma_B^2) + \sum_{n=1}^{N}\ln(\boldsymbol{R}\boldsymbol{\Psi}_n^T) -$$

$$\sum_{n=1}^{N}\frac{3}{2}\ln(\boldsymbol{R}\boldsymbol{\Phi}_n^T) + \sum_{n=1}^{N}\left(-\frac{(\boldsymbol{R}\boldsymbol{\Phi}_n^T\eta - D)^2}{2\sigma_B^2 \boldsymbol{R}\boldsymbol{\Phi}_n^T}\right) \tag{6.57}$$

The derivatives of the log-likelihood function with respect to η and σ_B^2 can be given as

$$\begin{cases} \dfrac{\partial \ell(\boldsymbol{R}, \eta, \sigma_B^2 | \boldsymbol{T})}{\partial \eta} = -\dfrac{\sum_{n=1}^{N} \boldsymbol{R}\boldsymbol{\Phi}_n^{\mathrm{T}} \eta - N}{\sigma_B^2} \\ \dfrac{\partial \ell(\boldsymbol{R}, \eta, \sigma_B^2 | \boldsymbol{T})}{\partial \sigma_B^2} = -\dfrac{N}{2\sigma_B^2} + \dfrac{1}{2\sigma_B^4} \sum_{n=1}^{N} \dfrac{(\boldsymbol{R}\boldsymbol{\Phi}_n^{\mathrm{T}} \eta - D)^2}{\boldsymbol{R}\boldsymbol{\Phi}_n^{\mathrm{T}}} \end{cases} \quad (6.58)$$

Let $\partial \ell(\boldsymbol{R}, \eta, \sigma_B^2 | \boldsymbol{T})/\partial \eta = 0$ and $\partial \ell(\boldsymbol{R}, \eta, \sigma_B^2 | \boldsymbol{T})/\partial \sigma_B^2 = 0$, the MLEs for η and σ_B^2 are

$$\begin{cases} \eta(\boldsymbol{R}) = \dfrac{N}{\sum_{n=1}^{N} \boldsymbol{R}\boldsymbol{\Phi}_n^{\mathrm{T}}} \\ \sigma_B^2(\boldsymbol{R}) = \dfrac{1}{N} \sum_{n=1}^{N} \dfrac{(\boldsymbol{R}\boldsymbol{\Phi}_n^{\mathrm{T}} \eta(\boldsymbol{R}) - D)^2}{\boldsymbol{R}\boldsymbol{\Phi}_n^{\mathrm{T}}} \end{cases} \quad (6.59)$$

The estimation results of η and σ_B^2 are functions of $\boldsymbol{R}$. By submitting Eq. (6.59) into Eq. (6.57), the above log-likelihood function is simplified as a multivariable function dependent on $\boldsymbol{R}$. The estimation of $\hat{\boldsymbol{R}} = (\hat{r}_1, \hat{r}_2, \cdots, \hat{r}_P)$ can be obtained by maximizing the log-likelihood function via multi-dimensional optimization techniques (Lagarias et al. 1998). After that, the estimation results $\hat{\eta}$ and $\hat{\sigma}_B^2$ are calculated by inputting $\hat{\boldsymbol{R}}$ into Eq. (6.59). So far, model parameters of the state transition function are estimated by submitting the failure information and historical condition profiles of N training units into the log-likelihood function. Theoretically speaking, as long as the number of training units is large enough, estimation results are unbiased and reliable.

2) Estimation of the Signal Transformation Parameters

According to the measurement function Eq. (6.48), the measurement signal under condition p can be described as

$$y_{k,n}^p = a_p(b_p + x_{k,n}^c + \upsilon_{k,n}) \quad (6.60)$$

Assuming that the unit is operating under the baseline condition instead of another condition p. Its measurement signal will be

$$y_{k,n}^{\mathrm{B}} = a_{\mathrm{B}}(b_{\mathrm{B}} + x_{k,n}^c + \upsilon_{k,n}) \quad (6.61)$$

where the index B represents the baseline condition that can be predetermined to be any condition. We hereafter define the baseline condition uniformly as the condition with the longest operating period. a_{B} and b_{B} are the model parameters under the baseline condition. From Eq. (6.60) and Eq. (6.61), it is known that the degradation signal under condition p shares the relationship with the signal under the baseline condition as

6.4 RUL Prediction with Time-Varying Operational Conditions

$$y_{k,n}^{\text{B}} = a_{\text{B}}(b_{\text{B}} + \frac{y_{k,n}^p}{a_p} - b_p) = \frac{a_{\text{B}}}{a_p} y_{k,n}^p + a_{\text{B}}(b_{\text{B}} - b_p) \qquad (6.62)$$

Let $a'_p = a_{\text{B}}/a_p$ and $b'_p = a_{\text{B}}(b_{\text{B}} - b_p)$, Eq. (6.62) can be simplified as

$$y_{k,n}^{\text{B}} = a'_p y_{k,n}^p + b'_p \qquad (6.63)$$

With the above relationship, the degradation signals obtained under different conditions can be transformed into the degradation signals under baseline conditions. As the transformed signals can be regarded as recorded in a constant operating condition, the jumps existed in degraded signals caused by condition change will be eliminated during the signal transformation procedure. The only task is to estimate the two transformation parameters a'_p and b'_p according to available degradation signals. The estimation procedure can be elaborated as follows.

(1) Find the samplings under condition p. Obtain their linear interpolation with reference to the range of the signals under the baseline condition. The interpolation of unit n at t_k is then denoted as $\vec{y}_{k,n}^p$. The above interpolations can be considered as signal estimation for baseline condition.
(2) Compute the square error sum between the interpolated signal and the transformed signal

$$J(a'_p, b'_p) = \sum_{n=1}^{N} \sum_{k \in \Omega_{p,n}} (\vec{y}_{k,n}^p - y_{k,n}^{\text{B}})^2 = \sum_{n=1}^{N} \sum_{k \in \Omega_{p,n}} (\vec{y}_{k,n}^p - a'_p y_{k,n}^p - b'_p)^2 \qquad (6.64)$$

where $\Omega_{p,n}$ is the set of the time indexes when the n th unit is operating under condition p.
(3) Let $\partial J(a'_p, b'_p)/\partial b'_p = 0$. The estimation of b'_p can be given as a function conditioned on a'_p.

$$b'_p(a'_p) = \frac{\sum_{n=1}^{N} \sum_{k \in \Omega_{p,n}} (\vec{y}_{k,n}^p - a'_p y_{k,n}^p)}{\sum_{n=1}^{N} |\Omega_{p,n}|} \qquad (6.65)$$

where $|\Omega_{p,n}|$ is the vector length of $\Omega_{p,n}$.
(4) Submit $b'_p(a'_p)$ into Eq. (6.64). Obtain the estimated $\hat{a}'_p = \arg\min(J(a'_p))$ via one-dimension optimization. Then $\hat{b}'_p$ can be computed by submitting $\hat{a}'_p$ into Eq. (6.65).
(5) Repeat step (1) to step (4) for the $P-1$ conditions except the baseline one, and their transformation parameters can correspondingly be obtained.

3) Estimation of the Measurement Function Parameters

So far, all signals have been transformed from different operating conditions to the baseline condition, our next task is to estimate the model parameters under the baseline condition, i.e., a_B, b_B, c and σ^2. The parameter estimation procedure is elaborated as follows.

First of all, the converted signal is smoothed by applying the locally weighted scatter smoothing method (LOESS) (Cleveland and Devlin 1988), which is a common technique in the field of RUL prediction for measurement noise reduction. The smoothed signal is expected to exhibit the overall trend of degraded signal, and its approximate expression is given as

$$\tilde{y}_{k,n}^B \cong a_B(b_B + x_{k,n}^c) \tag{6.66}$$

By submitting $x_{1,n}=0$ and $x_{K_n,n}=1$ into Eq. (6.66), we have $a_B \cong \tilde{y}_{K_n,n}^B - \tilde{y}_{1,n}^B$, $b_B \cong \tilde{y}_{1,n}^B/a_B$ and $\upsilon_k \cong (y_{k,n}^B - \tilde{y}_{k,n}^B)/a_B$. Then the model parameters a_B, b_B and σ^2 can be estimated by computing the mean value of training units estimates.

$$\hat{a}_B \cong \frac{1}{N}\sum_{n=1}^{N}(\tilde{y}_{K_n,n}^B - \tilde{y}_{1,n}^B) \hat{b}_B \cong \frac{1}{N}\sum_{n=1}^{N}\frac{\tilde{y}_{1,n}^B}{\hat{a}_B} \hat{\sigma}^2 \cong \frac{1}{\hat{a}_B^2 N}\sum_{n=1}^{N}\left(\frac{1}{K_n}\sum_{k=1}^{K_n}(y_{k,n}^B - \tilde{y}_{k,n}^B)^2\right) \tag{6.67}$$

According to Eq. (6.66), the degradation state can be approximately given as a function conditioned on C as

$$\hat{x}_{k,n} \cong \left(\frac{\tilde{y}_{k,n}^B - \tilde{y}_{1,n}^B}{\tilde{y}_{K_n,n}^B - \tilde{y}_{1,n}^B}\right)^{\frac{1}{c}} \tag{6.68}$$

It can be known from Eq. (6.48) that the degradation state possesses a mean trend $\hat{\eta}\,\boldsymbol{\tau}_n$, with $\boldsymbol{\tau}_n = (\tau_{1,n}, \tau_{2,n}, ..., \tau_{K_n,n})^T$. Let $\boldsymbol{X}_n(c) = (\hat{x}_{1,n}, \hat{x}_{2,n}, ..., \hat{x}_{K_n,n})^T$. The model parameter c can be estimated by minimizing the sum of squared errors between the approximated curve $\boldsymbol{X}_n$ and its mean trend $\hat{\eta}\,\boldsymbol{\tau}_n$.

$$\hat{c} = \min_c \sum_{n=1}^{N}\left((\boldsymbol{X}_n(c) - \hat{\eta}\,\boldsymbol{\tau}_n)^T(\boldsymbol{X}_n(c) - \hat{\eta}\,\boldsymbol{\tau}_n)\right) \tag{6.69}$$

From Eq. (6.67) it can be noticed that, a_B and b_B are related to only the first and last signal points of N training units. σ^2 can be estimated by computing the mean of the estimated variances of N training units. c can be estimated via the minimization of the sum of squared errors. In a word, a larger N can boost the accuracy of estimation results.

6.4.2.3 State Estimation Using Online Data

After parameter estimation, the next task is the state estimation of the testing unit according to online data. We hereafter adopt the PF algorithm (Jouin et al. 2016) for fulfilling this task.

(1) **Initialization:** At initial time t_0, generate N_s state particles $\{x_0^i = 0\}_{i=1:N_s}$ and assign the weights as $\{w_0^i = 1/N_s\}_{i=1:N_s}$.
(2) **Prediction:** One-step predict the degradation state for each particle via

$$x_k^i = x_{k-1}^i + \hat{r}_{p_{k-1}} \hat{\eta} \Delta t_k + \omega_{k-1}^i \quad (6.70)$$

(3) **Update:** Once the degradation signal y_k and its corresponding condition index p_k are obtained at t_k. Transform signals into the baseline condition if the unit operates under other conditions as

$$y_k^B = \hat{a}'_{p_k} y_k + \hat{b}'_{p_k} \quad (6.71)$$

Then the particle weights can be updated and normalized according to the transformed signal:

$$\tilde{w}_k^i = \frac{w_{k-1}^i}{\sqrt{2\pi \hat{a}_B^2 \hat{\sigma}^2}} \exp\left(-\frac{\left(y_k^B - \hat{a}_B(\hat{b}_B + (x_k^i)^{\hat{c}})\right)^2}{2\hat{a}_B^2 \hat{\sigma}^2}\right) \quad w_k^i = \frac{\tilde{w}_k^i}{\sum_{i=1}^{N_s} \tilde{w}_k^i} \quad (6.72)$$

(4) **Resampling:** Resampling N_s times from the previous particle set and generate a new particle set $\{x_k^j\}_{j=1:N_s}$ (Arulampalam et al. 2002). During resampling, particles are resampled according to $p\left(x_k^j = x_k^i\right) = w_k^i$. Then reset particle weights as $w_k^j = 1/N_s$. The state estimation result is obtained as the medium of the resampled particles $\bar{x}_k$.

Through the above procedure, the degradation state can be dynamically updated using online monitoring data. With the estimated degradation states, the RUL can further be predicted.

6.4.2.4 RUL Prediction

Assuming that the incoming operating conditions can be predetermined with a specific profile. Based on that, the future degradation state $x(l + t_k)$ can be expressed as

$$x(l + t_k) = \bar{x}_k + \eta \int_{t_k}^{l+t_k} r_{p(t)} dt + \sigma_B B\left(\int_{t_k}^{l+t_k} r_{p(t)} dt\right) \quad (6.73)$$

The RUL of the testing unit at t_k is equivalent to the remaining operating period of the above degradation process when state degrades from $\bar{x}_k$ to D, which can be regarded as the failure time of the degradation process $x(l + t_k) - \bar{x}_k$ reaching the FT $D - \bar{x}_k$. Based on the failure PDF in Eq. (6.52), the analytical solution of the RUL PDF can be obtained as

$$f(l) = \frac{r_{p(l+t_k)}(D - \bar{x}_k)}{\sqrt{2\pi \sigma_B^2(\tau(l + t_k) - \tau_k)^3}} \exp\left(-\frac{((\tau(l + t_k) - \tau_k)\eta - D + \bar{x}_k)^2}{2\sigma_B^2(\tau(l + t_k) - \tau_k)}\right) \quad (6.74)$$

6.4.3 RUL Prediction Case of Accelerated Degradation Experiments of Thrusting Bearings

6.4.3.1 Experimental Data Introduction

In this section, the accelerated degradation test (ADT) dataset of rolling element thrust bearings (Bian et al. 2015) is used for effectiveness demonstration of the above introduced method. A schematic diagram of the bearing test rig is shown in Fig. 6.17.

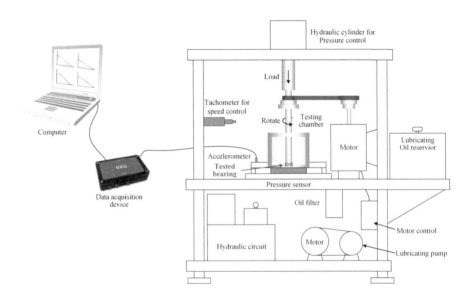

Fig. 6.17 Schematic diagram of the rolling element thrust bearing test rig

In accelerated testing, the tested bearings are subjected to loading conditions that are beyond their design specifications in order to accelerate the failure. During each test, a tested bearing is placed in the testing chamber. The lower race of the bearing is fixed to a stationary housing and the upper race of the bearing is fixed with a rotating shaft. A hydraulic cylinder is on the top of the rotating shaft to load pressure on the tested bearing. A tachometer is installed beside the shaft for measuring and controlling the rotating speed. An accelerometer is mounted in the experimental test rig to obtain bearing vibration signals.

The ADTs were carried out under two rotational speeds: 2200 r/min and 2600 r/min. According to the industrial standards for machine vibration ISO 2372, each ADT was stopped once the RMS of the vibration signals exceeds 2.2g (Bian and Gebraeel 2013). The degradation signals and condition profiles of four bearings are shown in Fig. 6.18. Among those four bearings, bearings 1 and 2 operated under constant speeds 2200 r/min and 2600 r/min, respectively. Bearings 3 and 4 operated under different time-varying speed profiles. As can be seen in Fig. 6.18, the changes of rotational speeds incurred obvious fluctuations in the degradation signal amplitudes. To reduce the jumps caused by time-varying rotational speeds, degradation signals are transformed into the baseline condition 2200 r/min according to the previously described transformation algorithm. Moreover, to reduce the influence of random noise, the transformed signals are further smoothed. The failure times of bearings are determined as the moments once the smoothed signal curves equal or exceed a pre-determined FT. In order to ensure that all bearings are failed at the end of the ADTs, the FT is set as the minimum smoothed signal amplitude of the four bearings at the last sampling points, i.e., 0.0617 Vrms. Note that at the beginning of the operation, the signal fluctuates randomly and does not show any degradation pattern, and is useless for performing RUL prediction. With the increase of operation time, the degradation signal shows an obvious increasing trend, which indicates that the bearing has gradually degraded. The initial time of degradation is selected as the FPT, and RUL prediction is performed since then.

6.4.3.2 RUL Prediction of the Rolling Element Bearings

In this section, four comparison methods are introduced for verifying the effectiveness of the introduced method, which are noted as M1-M4 respectively. Among these four methods, M1-M3 are the reduced versions of the introduced method. To be specific, M1 ignores both two factors addressed in the introduced method, M2 only takes degradation signal jumps into consideration, and M3 only incorporates the degradation changes into the RUL prediction. Additionally, another benchmark method (Bian et al. 2015), which considers both factors but mixes them during the degradation modeling, is also selected for comparison.

During the experimental study, when one bearing is chosen for test, the rest three are utilized as the training units for parameter estimation. From the RUL prediction results provided in Fig. 6.19, it can obviously be seen that F2S2 outputs the most accurate and consistent prediction results among all five methods. To further quantitatively compare the performance of all five methods, the mean and variance of their absolute relative errors (ARE) are computed and provided in Fig. 6.20. It can still be obviously seen that F2S2 always performs the best during the entire lifetime. As for other methods, we can see a distinctive fall-down in their performance. During the

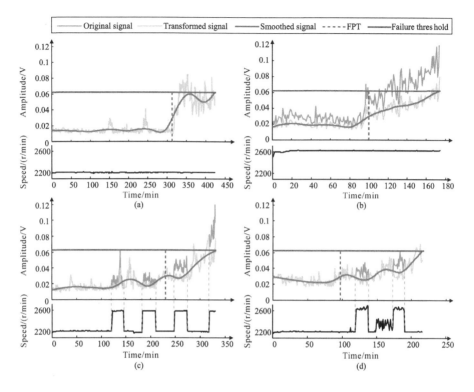

Fig. 6.18 Degradation signals and rotational speeds of the four bearings: (a) Bearing 1, (b) Bearing 2, (c) Bearing 3, and (d) Bearing 4

first half of their lifetime, M1 and M2 perform better than M3 and M4. While during the second half of their lifetime, M3 and M4 gradually outperform M1 and M2. This phenomenon is likely to be caused by the variation of the prominent role of these two factors during the lifetime. To be specific, during the first half of lifetime, degradation signal jumps have a more significant impact on the RUL prediction accuracy. In this context, M3 and M4 are more volatile as they are likely to be disturbed by degradation signal jumps, thus outputting unsatisfactory results. After half of the lifetime, time-varying degradation rates play an increasingly important role rather than degradation signal jumps. In this case, M1 and M2 ignore the changes in degradation rates, and thus obtain larger prediction errors than other methods. Compared with the other four methods, the introduced F2S2 method distinguishes the contributions of the two different factors to the RUL prediction, and thus obtains the best performance during the entire lifetime.

6.4 RUL Prediction with Time-Varying Operational Conditions

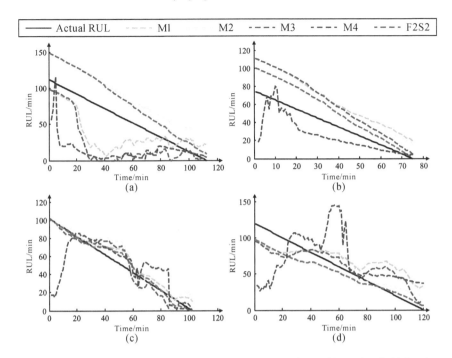

Fig. 6.19 RUL prediction results of the four bearings: (a) Bearing 1, (b) Bearing 2, (c) Bearing 3, and (d) Bearing 4

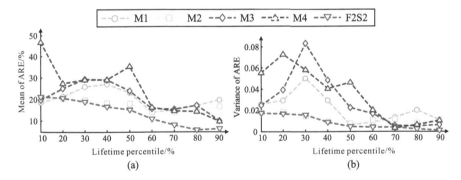

Fig. 6.20 Prediction errors of the four bearings: (a) mean of the AREs, and (b) variance of AREs

6.4.4 Epilog

This section focuses on the RUL prediction of mechanical system under time-varying operating conditions. Firstly, the influencing mechanism of time-varying operating conditions on the degradation process of mechanical system is exhaustively analyzed and summarized into two influencing factors: the degradation rate change of internal degradation state and the amplitude change of external monitoring signal. These two influencing factors are further introduced into the state transfer function and the measurement function, respectively. A two-factor state-space model is correspondingly formulated. Based on the formulated model, a condition transformation method is developed to eliminate the impact of changing operating conditions on the degradation rate through time scale conversion of state transfer function. As for the influence of time-varying operating conditions on degradation signal amplitude, it is eliminated through the transformation of monitoring signal amplitude. With the above two steps, the RUL prediction problem under time-varying operating conditions is transformed into the RUL prediction problem under constant working condition. Finally, the effectiveness of the introduced method is demonstrated via the experimental data of the thrust ball bearing ADTs.

6.5 RUL Prediction with Dependent Competing Failure Processes

6.5.1 Motivation

As elaborated in previous subsections, the operation of mechanical systems is often influenced by multiple internal and external factors, which would in turn lead to the existence of multiple failure processes competing against each other to damage the mechanical system (Hao et al. 2017; Rafiee et al. 2014). Moreover, different failure processes are often interacting with each other, which further substantially increases the complexity of mechanical system degradation. Take rolling element bearings for example, during the lifecycle of a bearing, both wear and surface fatigue are able to occur and eventually lead to the malfunction of it. In the same time, wear process is also able to change the raceway profile of the bearing, which eventually increases the possibility of surface fatigue (Olver 2005). In existing literature(Jiang et al. 2012), such degradation phenomenon is generally called dependent competing failure processes (DCFPs), and an illustration of DCFPs can be seen in Fig. 6.21.

The illustrative example provided in Fig. 6.21 is based on the cumulative damage mechanism of commonly used mechanical components. In this example, two types of failures are involved, i.e., soft failure and hard failure. Soft failure indicates the failure incurred once the cumulative gradual damage surpasses a pre-specified FT, while hard failure indicates that when the cumulative shock damage exceeds a critical threshold, the degradation will accelerate drastically and causing the degradation trend exceeds FT instantaneously. For the complicated degradation process described above, one of the most prominent difficulties in RUL prediction is to obtain the RUL PDF under the first hitting time (FHT) analytically. Specifically speaking, most mechanical and structural systems in practical applications suffer from first hitting failure (Casciati and Roberts 1991), i.e., they are considered failed once their health indicators equal or surpass predefined FTs. In this context, if the RUL prediction

6.5 RUL Prediction with Dependent Competing Failure Processes

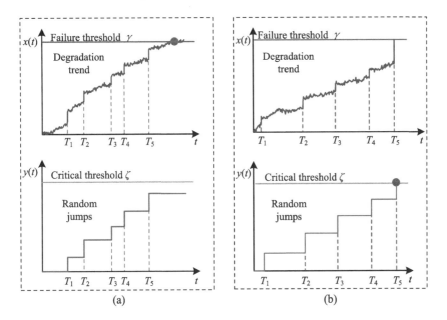

Fig. 6.21 Illustrative example of the competing failure processes: (a) failed by soft failure, and (b) failed by hard failure

method is not based on the FHT concept, an over-estimation of RUL is inevitably caused. As for the importance of analytical solution, it is driven by the fact that simulation-based methods are sensitive to simulation parameters and request huge computational resource, which make them not practically convenient.

Another difficulty incurred by DCFPs is the estimation and update of model parameters. As has been justified in literature (Zhang et al. 2018), the accuracy of model parameters is pivotal in ensuring the effectiveness of RUL prediction. For a specific mechanical system, it is a common situation in industrial cases that historical data collected from similar mechanical systems are available, and the online monitoring data for the considered mechanical system can be dynamically obtained. Based on these data, it is reasonable to perform model parameter estimation based on historical data via methods like maximum likelihood estimation (Ye et al. 2011) for the initialization of the RUL prediction. Moreover, since historical data-based estimation results are merely statistical inferences, which is unable to characterize the individual operating condition and unit-to-unit variability, online monitoring data are further utilized to update the estimated parameters so as to improve the RUL prediction accuracy. However, in the context of DCFPs, the necessity of utilizing multiple models for degradation modeling and the existence of random jumps in degradation trends make it difficult to formulate a likelihood function, not to mention establish an applicable online update framework.

To sum up, the RUL prediction for mechanical system that owns the DCFPs degradation phenomenon suffer from two aspects: (1) the absence of RUL PDF expression under the FHT concept, and (2) the difficulty in estimating and updating model parameters. Targeted at these two aspects, this section will introduce a RUL prediction method for DCFPs (Yan et al. 2021). First, an analytical expression-based RUL PDF solution will be derived, which provides an alternative for obtaining more accurate and time-consuming RUL prediction results. Then, the model parameter offline estimation and online update tasks will both be addressed. The method presented in this section is able to improve the effectiveness of RUL prediction for mechanical system subject to DCFPs, which is promising in providing maintenance scheduling technicians an opportunity to establish maintenance schedules that are more reliable and economical.

6.5.2 RUL Prediction Model Considering Dependent Competing Failure Processes

As provided in Fig. 6.22, four indispensable steps are carried out to accurately predict the RUL of DCFPs. Among these steps, degradation modeling provides a mathematical framework to empirically describe the considered DCPs. Then according to the formulated models, the derivation for the analytical solution of the RUL PDF based on the FHT concept is conducted. By utilizing historical and online data, the estimation and update of model parameters can be fulfilled. Finally, with the analytical solution and the updated model parameters, the RUL prediction results are acquired. This section aims to firstly develop degradation models, and further derive the RUL PDF under the FHT concept. Then, according to the degradation characteristics of DCFPs, the model parameter estimation and update are addressed jointly. In the module of offline estimation, a sequential estimation scheme is presented with the help of EM algorithm. While in the module of online update, a TVMMPF is designed to accurately identify random jumps in the degradation trends and update parameters correspondingly.

6.5.2.1 Model Formulation and RUL PDF Derivation

1) Model Formulation

As described above, two degradation modes are involved in the considered DCFPs, i.e., gradual degradation as well as random shocks (Peng et al. 2010). As for the failure of DCFPs, it can also be divided into two types, i.e., soft failure and hard failure. Based on related literature (Fan et al. 2018; Ranjkesh et al. 2019), it can be considered that soft failure processes consist of gradual degradation trend and nonfatal random shocks, and the eventual failure is considered to occur when the cumulative damage $x(t)$ equals or surpasses the pre-defined FT γ. As for hard failure processes, they are considered to consist of the same random shocks, but the occurrence of hard failure is characterized as an avalanche result of accumulated random shocks. Put otherwise, when the accumulated random shocks $y(t)$ equals or exceeds a specific intrinsic

6.5 RUL Prediction with Dependent Competing Failure Processes

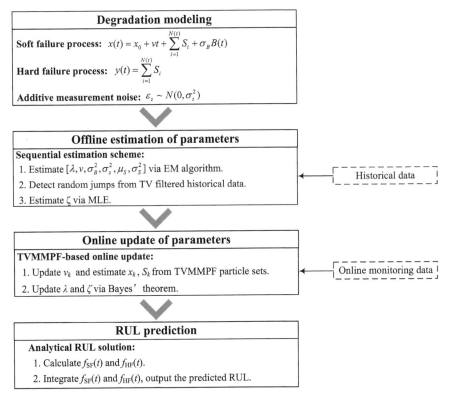

Fig. 6.22 Flowchart of the RUL prediction method considering DCFPs

threshold ζ, the degradation of mechanical system will be suddenly accelerated, and the overall degradation trend will reach the FT γ instantaneously.

With the above characterization, we can describe soft failure processes with the following equation

$$x(t) = x_0 + vt + \sum_{i=1}^{N(t)} S_i + \sigma_B B(t) \qquad (6.75)$$

where x_0 is the initial damage. v is the degradation rate of the gradual degradation component. $N(t)$ is the number of random shock arrival till time t that is assumed to follow a homogeneous Poisson process (HPP) (Fan et al. 2018) Poisson(λ). S_i denotes the shock amplitude of the i th arrived random shock, which is considered as a random variable following $S_i \sim N(\mu_S, \sigma_S^2)$. $\sigma_B B(t)$ describes the randomness of the gradual degradation component, where σ_B is the diffusion parameter, and $B(t)$ denotes a standard Brownian motion.

Similar to soft failure processes, hard failure processes can be described with the following equation

$$y(t) = \sum_{i=1}^{N(t)} S_i. \tag{6.76}$$

Note that although γ and ζ are both the determining criteria of failure, they are not correlated to each other. To be specific, γ is generally a pre-specified value. While ζ, as an intrinsic index, is more reasonable to be estimated from measurement data.

Denote T_{SF} and T_{HF} as the failure times for soft failure and hard failure, respectively. According to the FPT concept, we have

$$\begin{cases} T_{\text{SF}} = \inf\{t; x(t) \geq \gamma\} \\ T_{\text{HF}} = \inf\{t; y(t) \geq \zeta\} \end{cases} \tag{6.77}$$

Correspondingly, the failure time for DCFPs can be obtained as

$$T_{\text{CF}} = \min\{T_{\text{SF}}, T_{\text{HF}}\} \tag{6.78}$$

Therefore, the RUL at time t_k is in the form of

$$L_{\text{CF},k} = T_{\text{CF}} - t_k \tag{6.79}$$

Considering that the actual damage of mechanical system is usually unobservable in practical cases, and measurement data should be utilized to estimate the damage (Tang et al. 2014), which contains measurement noise. Therefore, we hereby denote the historical and online data as z and $z(t)$, respectively, and assume that they are made up as the summation of actual damage and additive measurement noises $\varepsilon_z \sim N(0, \sigma_z^2)$.

2) RUL PDF Derivation

For the above described DCFPs, a feasible derivation framework (Fan et al. 2018; Peng et al. 2010) is to derive the RUL PDF for only the soft failure processes and only the hard failure processes separately. Then based on the definition of DCFP failure time, the above derived separate results can further be integrated to obtain the DCFPs RUL PDF. Following the same procedure, we hereafter also derive the RUL PDF following the separation-integration procedure. The main difference in our derivation procedure is the adoption of FHT concept.

(1) Soft Failure Time PDF Derivation

From the above description, it can be known that soft failure processes are essentially a typical jump-diffusion process. Based on the first hitting law (Coutin and Dorobantu 2011), the soft failure occurrence probability within $(t, t + h)$ can be split as

6.5 RUL Prediction with Dependent Competing Failure Processes

$$\begin{aligned}P_{\text{SF}}(t<T_{\text{SF}}\leq t+h)&=P_{\text{SF}}(t<T_{\text{SF}}\leq t+h,N(t+h)-N(t)=0)+\\ &\quad P_{\text{SF}}(t<T_{\text{SF}}\leq t+h,N(t+h)-N(t)=1)+\\ &\quad P_{\text{SF}}(t<T_{\text{SF}}\leq t+h,N(t+h)-N(t)\geq 2)\end{aligned} \quad (6.80)$$

For convenience, denote

$$\begin{cases}A_0(h)=P_{\text{SF}}(t<T_{\text{SF}}\leq t+h,N(t+h)-N(t)=0)\\ A_1(h)=P_{\text{SF}}(t<T_{\text{SF}}\leq t+h,N(t+h)-N(t)=1)\\ A_2(h)=P_{\text{SF}}(t<T_{\text{SF}}\leq t+h,N(t+h)-N(t)\geq 2)\end{cases} \quad (6.81)$$

Then the PDF of T_{SF} can be obtained by summing up the limits of $A_0(h)$, $A_1(h)$, and $A_2(h)$ with respect to h as

$$f_{\text{SF}}(t)=\lim_{h\to 0}\frac{A_0(h)+A_1(h)+A_2(h)}{h} \quad (6.82)$$

For $A_0(h)$, it indicates the probability conditioned on the event that soft failure occurs during $(t, t+h)$ and there is no positive jump happening within $(t, t+h)$. In (Coutin and Dorobantu 2011), it is well proved that this term can be analytically given with the help of the strong Markov property as

$$A_0(h)=e^{-\lambda h}\int_t^{t+h}E(\tilde{f}(u-T_{N(t)},\gamma-X_{T_{N(t)}}))du \quad (6.83)$$

where $T_{N(t)}$ is the arrival time of the $N(t)$ th random shock, $X_{T_{N(t)}}$ is the cumulative damage at time $T_{N(t)}$. $\tilde{f}(u-T_{N(t)},\gamma-X_{T_{N(t)}})$ denotes the PDF of FPT for the linear Wiener process with the FT of $\gamma-X_{T_{N(t)}}$ at time $u-T_{N(t)}$.

Based on this conclusion, the limit of $A_0(h)$ with respect to h is

$$\lim_{h\to 0}\frac{A_0(h)}{h}=\frac{\int_t^{t+h}E(\tilde{f}(u-T_{N(t)},\gamma-X_{T_{N(t)}}))du}{e^{\lambda h}h} \quad (6.84)$$

Utilizing the L'Hospital's rule and considering that $\tilde{f}(u-T_{N(t)},\gamma-X_{T_{N(t)}})$ is a continuous function, its limit can be obtained as

$$\begin{aligned}\lim_{h\to 0}\frac{A_0(h)}{h}&=\frac{(\int_t^{t+h}E(\tilde{f}(u-T_{N(t)},\gamma-X_{T_{N(t)}}))du)'}{(e^{\lambda h}h)'}\\ &=\frac{E(\tilde{f}(t+h-T_{N(t)},\gamma-X_{T_{N(t)}}))}{e^{\lambda h}+\lambda h e^{\lambda h}}\\ &=E(\tilde{f}(t-T_{N(t)},\gamma-X_{T_{N(t)}}))\end{aligned} \quad (6.85)$$

For $A_1(h)$, it denotes the probability conditioned on the event that soft failure occurs within $(t, t+h)$ and there exists only one positive jump during $(t, t+h)$. By using the well proved conclusion in (Roynette et al. 2008), it can be concluded that when $h \to 0$, we have the following relationship

$$\lim_{h \to 0} \frac{A_1(h)}{h} = \lambda E(1 - F_S(\gamma - x_t)) \tag{6.86}$$

where $F_S(\cdot)$ denotes the CDF of random shocks S_i.

As for $A_2(h)$, it denotes the conditional probability the first passage failure of soft failure process occurs within $(t, t+h)$ along with more than one positive jumps during $(t, t+h)$. For this term, the following relationship holds that

$$\begin{aligned}\lim_{h \to 0} \frac{A_2(h)}{h} &= \lim_{h \to 0} \frac{P_{SF}(t < T_{SF} \leqslant t+h, N(t+h) - N(t) \geqslant 2)}{h} \\ &\leqslant \lim_{h \to 0} \frac{P(N(t+h) - N(t) \geqslant 2)}{h} \\ &\leqslant \lim_{h \to 0} \frac{1 - e^{-\lambda h} - \lambda h e^{-\lambda h}}{h}\end{aligned} \tag{6.87}$$

By applying the L' Hospital's rule again, it can be concluded that

$$\lim_{h \to 0} \frac{A_2(h)}{h} = 0 \tag{6.88}$$

With the above results, the RUL PDF of soft failure processes is expected to be in the form of

$$f_{SF}(t) = \lambda E\{(1 - F_S)(\gamma - x_t)\} + E\left\{\tilde{f}(t - T_{N(t)}, \gamma - x_{T_{N(t)}})\right\} \tag{6.89}$$

So far, we have obtained the general form of RUL PDF expression for soft failure processes, the next step is to explicitly obtain the expressions for each term. As described above, gradual degradation component is made up by a linear Wiener process and random shocks following a compound Poisson process, we then can rewrite Eq. (6.89) as

$$f_{SF}(t) = \lambda \int_{x_t} \left[1 - \Phi\left(\frac{\gamma - x_t - \mu_S}{\sigma_S}\right)\right] f(x_t) dx_t + \int_{x_{T_{N(t)}}} \int_{T_{N(t)}} \frac{\gamma - x_{T_{N(t)}}}{\sqrt{2\pi \sigma_B^2 (t - T_{N(t)})^3}} \times$$

$$\exp\left[-\frac{(\gamma - x_{T_{N(t)}} - \upsilon(t - T_{N(t)}))^2}{2\sigma_B^2(t - T_{N(t)})}\right] f(T_{N(t)}) f(x_{T_{N(t)}}) dT_{N(t)} dx_{T_{N(t)}}$$

$$\tag{6.90}$$

6.5 RUL Prediction with Dependent Competing Failure Processes

In Eq.(6.90), x_t is the cumulative damage given as the summation of both gradual degradation and random shocks. Therefore, the PDF of x_t can be given as

$$f(x_t) = \sum_{n=0}^{\infty} f(x_t|N(t) = n) \cdot P(N(t) = n)$$

$$= \sum_{n=0}^{\infty} N(vt + n\mu_S, \sigma_B^2 t + n\sigma_S^2) \cdot \frac{(\lambda t)^n}{n!} e^{-\lambda t} \qquad (6.91)$$

As $T_{N(t)}$ denotes the time when the $N(t)$ th random shock arrives, which is assumed to follow a homogeneous Poisson process HPP(λ). Based on its intrinsic properties (Ross et al. 1996), the PDF of $T_{N(t)}$ can be given as

$$f_{T_{N(t)}}(l) = \sum_{n=0}^{\infty} f_n(l) \cdot P(N(t) = n) = \sum_{n=0}^{\infty} \frac{\lambda^n l^{n-1}}{(n-1)!} \exp(-\lambda l) \cdot \frac{(\lambda t)^n}{n!} \exp(-\lambda t)$$

$$(6.92)$$

where $f_n(l)$ denotes the PDF of time when the nth random shock arrives.

As a consequence, the analytical expression of $f(x_{T_{N(t)}})$ is expanded as

$$f(x_{T_{N(t)}}) = \sum_{n=0}^{\infty} f(x_{T_{N(t)}}|N(t) = n) \cdot P(N(t) = n)$$

$$= \sum_{n=0}^{\infty} N(vT_{N(t)} + n\mu_S, \sigma_B^2 T_{N(t)} + n\sigma_S^2) \cdot \frac{(\lambda t)^n}{n!} \exp(-\lambda t) \qquad (6.93)$$

$$= \int_l \sum_{n=0}^{\infty} N(vl + n\mu_S, \sigma_B^2 l + n\sigma_S^2) \times \frac{(\lambda t)^n}{n!} \exp(-\lambda t) f_{T_{N(t)}}(l) dl$$

Substitute Eqs. (6.91)–(6.93) into Eq. (6.90), the soft failure time PDF can be analytically obtained. In the integration procedure, the above obtained results will be utilized as intermediate results. For the convenience of reading, the tedious mathematical substitution operation will not be elaborated here.

(2) Hard Failure Time PDF Derivation

Mathematically speaking, the above-described hard failure can be regarded as a special scenario of soft failure when $v \to 0$ and $\sigma_B \to 0$. Based on this premise, one intuitive derivation methodology for the hard failure time PDF is to find the limit of Eq.(6.90) with respect to v and σ_B. However, this derivation methodology involves massive tedious and repetitive mathematical computations. In order to obtain the solution in a more simplified and convenient way, we hereafter perform the hard failure time PDF derivation from a new perspective in which only fundamental stochasticity process properties (Ross et al. 1996) and straightforward mathematical operations will be used. Detailed derivation are as follows.

Denote the hard failure time PDF as $f_{\text{HF}}(t)$, and it is naturally defined as

$$f_{\text{HF}}(t) = -\frac{\mathrm{d}}{\mathrm{d}t}P(T_{\text{HF}} > t) = -\sum_{n=0}^{\infty} F_S^{(n)}(\zeta)\frac{\mathrm{d}}{\mathrm{d}t}P(N(t) = n) \quad (6.94)$$

where $F_S^{(n)}(\cdot)$ represents the n-fold convolution operation of $F_S(\cdot)$.

It should be noted that when $n = 0$, the term $F_S^{(n)}(\cdot)$ is the CDF of Dirac Delta distribution that equals 1.

Based on this fact, we can rewrite Eq. (6.94) as

$$\begin{aligned}
f_{\text{HF}}(t) &= \lambda e^{-\lambda t} - \sum_{n=1}^{\infty} F_S^{(n)}(\zeta)\frac{\mathrm{d}}{\mathrm{d}t}P(N(t) = n) \\
&= \lambda e^{-\lambda t} - \sum_{n=1}^{\infty} F_S^{(n)}(\zeta)\left[\lambda\frac{(\lambda t)^{n-1}}{(n-1)!}e^{-\lambda t} - \lambda\frac{(\lambda t)^n}{n!}e^{-\lambda t}\right] \\
&= \lambda\left\{e^{-\lambda t} + \sum_{n=1}^{\infty} F_S^{(n)}(\zeta)[P(N(t) = n) - P(N(t) = n-1)]\right\} \\
&= \lambda\sum_{n=0}^{\infty} F_S^{(n)}(\zeta)P(N(t) = n) - \lambda\sum_{n=0}^{\infty} F_S^{(n+1)}(\zeta)P(N(t) = n) \quad (6.95)
\end{aligned}$$

Denote $\sum_{n=1}^{\infty} F_S^{(n)}(\zeta)$ as $G(\zeta)$, $\sum_{n=1}^{\infty} F_S^{(n+1)}(\zeta)$ as $H(\zeta)$, we can rewrite Eq. (6.95) as

$$f_{\text{HF}}(t) = \lambda e^{-\lambda t}(1 - F_S(\zeta)) + \lambda(G(\zeta) - H(\zeta)) \quad (6.96)$$

This ends the hard failure time PDF derivation.

(3) Competing Failure Time PDF Derivation

Through the derivations provided in the previous two subsections, we have obtained the separate expressions for the PDFs of T_{SF} and T_{HF}. We hereafter integrate these results to finally obtain the RUL PDF expression for the considered DCFPs.

Based on the definition of T_{CF} in Eq. (6.78), we can obtain that

$$\begin{aligned}
F_{\text{CF}}(t) &= P\{\min[T_{\text{SF}}, T_{\text{HF}}] < t\} = 1 - [1 - F_{\text{SF}}(t)] \cdot [1 - F_{\text{HF}}(t)] \\
&= F_{\text{SF}}(t) + F_{\text{HF}}(t) - F_{\text{SF}}(t)F_{\text{HF}}(t)
\end{aligned} \quad (6.97)$$

6.5 RUL Prediction with Dependent Competing Failure Processes

According to the above equation, we can further obtain the failure time PDF through taking the derivative of Eq. (6.97) with respect to t, i.e.,

$$f_{CF}(t) = f_{SF}(t) + f_{HF}(t) - f_{SF}(t) \cdot F_{HF}(t) - F_{SF}(t) \cdot f_{HF}(t) \tag{6.98}$$

Substitute Eq.(6.90) and Eq.(6.96) into Eq.(6.98), the failure time PDF for DCFPs can be analytically obtained. For the convenience of reading, the substitution operation will not be elaborated here either. With the failure time PDF, it is easy to further obtain the RUL as shown in Eq.(6.79).

6.5.2.2 Parameter Estimation and Update

1) Sequential Estimation Scheme

From the previous subsection, it can be known that $[\lambda, v, \sigma_B^2, \sigma_z^2, \mu_S, \sigma_S^2, \zeta]$ are the model parameters that should be estimated. Among those 7 parameters, it should be noted that the critical threshold parameter ζ denotes the intrinsic level that would initiate a sudden failure of mechanical system. Therefore, instead of specifying it in advance as has been done to γ, it would make more sense to estimate it from measurement data. For model parameter estimation, the most commonly adopted straightforward idea is to formulate and maximize the likelihood function with respect to all parameters. Nevertheless, it should be noted that ζ, as the intrinsic threshold determining the occurrence of fatal shocks, is supposed to be estimated after the estimation of the rest parameters, which in turn makes it unable to establish the likelihood function with respect to all parameters. To tackle this problem, we develop a sequential parameter estimation scheme, in which $[\lambda, v, \sigma_B^2, \sigma_z^2, \mu_S, \sigma_S^2]$ are firstly estimated using the EM algorithm. Then, by using the total variation filter, the historical data can be processed for the identification of random shocks. With the estimated parameters and identified random shocks, we can further formulate and optimize the likelihood function for ζ, which finally ends the parameter estimation task.

Denote $\Delta z_{i,k}$ as the differential sequence of the historical data for the i th failed unit at time t_k, where $k \in \{0, \cdots, L_i\}$, and L_i is the differential data sequence length for unit i. As units are failed either by soft or by hard failure, we also denote I_{SF} to be the soft failure unit set, and I_{HF} being the hard failure unit set. Accordingly, it can be defined that

$$\begin{cases} L_{i,NF} = L_i, i \in I_{SF} \\ L_{i,NF} = L_i - 1, i \in I_{HF} \end{cases} \tag{6.99}$$

Then the log-likelihood function for $[\lambda, v, \sigma_B^2, \sigma_z^2, \mu_S, \sigma_S^2]$ can be formulated as

$$\ln L(\psi | \Delta z) = \sum_{i=1}^{N} \sum_{j=1}^{L_{i,NF}} \ln \sum_{n=0}^{1} f\left(\Delta z_{i,j} | N(\Delta t) = n\right) \times P(N(\Delta t) = n) \tag{6.100}$$

where $f\left(\Delta z_{i,j}|N(\Delta t)=n\right)$ is the PDF of a normal distribution $N(v\Delta t, \sigma_B^2\Delta t+2\sigma_z^2)$ when $n=0$, and the PDF of $N(v\Delta t+\mu_S, \sigma_B^2\Delta t+\sigma_S^2+2\sigma_z^2)$ when $n=1$. Δz is the sequence consisting of all $\Delta z_{i,k,\cdot}$.

Note that in Eq. (6.100), n is considered to be either 0 or 1, which indicates that the interval of the online condition monitoring is short enough that the occurrence number of random shock is at most one during the interval between two consecutive sampling. Thanks to the fast development of instrumentation and measurement technologies, such approximation is not difficult to be satisfied in practical applications (Fan et al. 2018) and is thus also adopted here.

Denote $\gamma(N(\Delta t)|\Delta z)$ as the probability of random jump occurrence conditioned within an interval. According to the Bayes' theorem, it is known that

$$\begin{cases} \gamma(N(\Delta t)=0|\Delta z) = \frac{\pi_0 N(\mu_0,\sigma_0^2)}{\pi_0 N(\mu_0,\sigma_0^2)+\pi_1 N(\mu_1,\sigma_1^2)} \\ \gamma(N(\Delta t)=1|\Delta z) = \frac{\pi_1 N(\mu_1,\sigma_1^2)}{\pi_0 N(\mu_0,\sigma_0^2)+\pi_1 N(\mu_1,\sigma_1^2)} \end{cases} \quad (6.101)$$

where $\mu_0 = v\Delta t$, $\sigma_0^2 = \sigma_B^2\Delta t+2\sigma_z^2$, $\mu_1 = v\Delta t+\mu_S$ and $\sigma_1^2 = \sigma_B^2\Delta t+\sigma_S^2+2\sigma_z^2$. π_0 and π_1 are the prior probabilities of $N(\Delta t_{i,j})=0$ and $N(\Delta t_{i,j})=1$, respectively.

According to the EM algorithm, the estimation of the above six parameters can be obtained via:

(1) **Initialization**: Initialize the parameters as $\mu_{0,1}$, $\sigma_{0,1}^2$, $\mu_{1,1}$, $\sigma_{1,1}^2$, $\pi_{0,1}$ and $\pi_{1,1}$.

(2) **E step**: Use current parameters to evaluate Eq. (6.100).

(3) **M step**: Re-estimate the parameters as:

$$\mu_{0,k+1} = \frac{\sum_{i=1}^{N}\sum_{j=1}^{L_{i,NF}}\gamma(N(\Delta t)=0|\Delta z)\Delta z_{i,j}}{\sum_{i=1}^{N}\sum_{j=1}^{L_{i,NF}}\gamma(N(\Delta t)=0|\Delta z)} \quad (6.102)$$

$$\mu_{1,k+1} = \frac{\sum_{i=1}^{N}\sum_{j=1}^{L_{i,NF}}\gamma(N(\Delta t)=1|\Delta z)\Delta z_{i,j}}{\sum_{i=1}^{N}\sum_{j=1}^{L_{i,NF}}\gamma(N(\Delta t)=1|\Delta z)} \quad (6.103)$$

$$\sigma_{0,k+1}^2 = \frac{\sum_{i=1}^{N}\sum_{j=1}^{L_{i,NF}}\gamma(N(\Delta t)=0|\Delta z)\left(\Delta z_{i,j}-\mu_{0,k+1}\right)^2}{\sum_{i=1}^{N}\sum_{j=1}^{L_{i,NF}}\gamma(N(\Delta t)=0|\Delta z)} \quad (6.104)$$

6.5 RUL Prediction with Dependent Competing Failure Processes

$$\sigma_{1,k+1}^2 = \frac{\sum_{i=1}^{N} \sum_{j=1}^{L_{i,\text{NF}}} \gamma(N(\Delta t) = 1|\Delta z)(\Delta z_{i,j} - \mu_{1,k+1})^2}{\sum_{i=1}^{N} \sum_{j=1}^{L_{i,\text{NF}}} \gamma(N(\Delta t) = 1|\Delta z)} \quad (6.105)$$

$$\pi_{0,k+1} = \frac{\sum_{i=1}^{N} \sum_{j=1}^{L_{i,\text{NF}}} \gamma(N(\Delta t) = 0|\Delta z)}{\sum_{i=1}^{N} L_{i,\text{NF}}} \quad (6.106)$$

$$\pi_{1,k+1} = 1 - \pi_{0,k+1} \quad (6.107)$$

(4) **Evaluation & output**: According to Eq. (6.100), perform the log-likelihood function evaluation. If the convergence of the log-likelihood function does not meet the stopping criteria, return to (2). Otherwise, obtain the estimation results as $\hat{\mu}_0, \hat{\mu}_1, \hat{\sigma}_0^2, \hat{\sigma}_1^2, \hat{\pi}_0$ and $\hat{\pi}_1$.

Based on the estimation results, we have

$$\begin{cases} \hat{\lambda} = -\frac{\ln \hat{\pi}_0}{\Delta t} \\ \hat{v} = \frac{\hat{\mu}_0}{\Delta t} \\ \hat{\mu}_S = \hat{\mu}_1 - \hat{\mu}_0 \\ \hat{\sigma}_S^2 = \hat{\sigma}_1^2 - \hat{\sigma}_0^2 \\ \hat{\sigma}_B^2 = \frac{\hat{\sigma}_0^2 - 2\sigma_z^2}{\Delta t} \end{cases} \quad (6.108)$$

Substitute Eq.(6.108) into Eq.(6.100) and utilize one dimensional optimization technique to find the maxima of Eq. (6.100) with respect to σ_z^2, the rest parameters can be correspondingly be estimated.

After the EM estimation, we can further estimate the parameter ζ. Since the estimation of ζ is based on the accurate knowledge of the exact number of random shocks for failed unit, we firstly utilize total variation filter to preprocess the historical data by removing the measurement noises and random fluctuations, but preserving random shocks trends that exhibit step-wise phenomenon (Ke et al. 2017):

$$\bar{z}_{i:L_i} = \arg\min_{z_{i:L_i}^*} \left\{ \left\| z_{i:L_i} - z_{i:L_i}^* \right\|_2^2 - \rho \left\| Dz_{i:L_i}^* \right\| \right\} \quad (6.109)$$

where $z_{i:L_i}$ and $\bar{z}_{i:L_i}$ are the historical and filtered data collected from the ith unit, respectively. ρ is the regularization parameter, and

$$D = \begin{bmatrix} -1 & 1 & \\ & -1 & 1 \\ & & \ddots \end{bmatrix} \quad (6.110)$$

Based on the widely adopted 3-σ criterion (Bian et al. 2015), we can easily locate the numbers of random shocks for all failed units and will thus not be elaborated here. After the acquisition of the detected random shocks, the log-likelihood function for ζ can be formulated as

$$\ln L_\zeta = \sum_{i=1}^N \ln L_{\zeta,i} \quad (6.111)$$

where

$$\ln L_{\zeta,i} = \begin{cases} \ln\left\{\Phi\left[\frac{\zeta-(m_i\hat{\mu}_S+m_i\hat{v}\Delta t)}{m_i\hat{\sigma}_S^2+m_i\hat{\sigma}_B^2\Delta t+2m_i\hat{\sigma}_y^2}\right]\right\}, & i \in I_{\text{SF}} \\ \ln\left\{1-\Phi\left[\frac{\zeta-(m_i\hat{\mu}_S+m_i\hat{v}\Delta t)}{m_i\hat{\sigma}_S^2+m_i\hat{\sigma}_B^2\Delta t+2m_i\hat{\sigma}_y^2}\right]\right\}, & i \in I_{\text{HF}} \end{cases} \quad (6.112)$$

Maximizing Eq.(6.112) with respect to ζ, the estimation result for parameter $\hat{\zeta}$ can eventually be obtained. Till now, all model parameters are successfully estimated.

2) Online Update based on TVMMPF

We hereafter elaborate on a specially designed TVMMPF on parameter update and damage estimation according to online monitoring data. It should be noticed that although there are 7 parameters in total characterizing degradation models, jointly updating all parameters puts a high requirement on computational resource, which may compromise the online update module in practical applications. In most related works (Lei et al. 2016), it is generally the case that only the degradation rate parameters are updated. However, for the DCFPs considered in this chapter, if their critical threshold ζ is not accurately obtained, the RUL prediction accuracy would be seriously compromised. In conclusion, to make a tradeoff between computational feasibility and prediction accuracy, v, λ, and ζ are chosen to be online updated. The fundamental idea of online update is to filter the monitoring data for random shock identification, then use the raw measurement data for damage estimation and parameter update with the help of particle filtering framework. The designed TVMMPF can be elaborated as follows.

(1) **Initialization:** At initial time t_0, generate $2 \times N_P$ state particles $\left\{x_0^{1,p}\right\}_{p=1:N_p}$, $\left\{x_0^{2,p}\right\}_{p=1:N_p}$. According to the offline parameter estimation results, sample N_P gradual degradation rate particles $\left\{v_0^p\right\}_{p=1:N_p}$ and N_P random jump particles $\left\{S_0^p\right\}_{p=1:N_p}$. All particles are set with the weight of $1/N_P$.

(2) **Prediction:** At time t_k, perform one-step state prediction via the state transition function

$$\begin{cases} \text{Model 1}: x_k^{1,p} = x_{k-1}^p + v_{k-1}^p \Delta t + \hat{\sigma}_B B(\Delta t) \\ \text{Model 2}: x_k^{2,p} = x_{k-1}^p + v_{k-1}^p \Delta t + S_{k-1}^p + \hat{\sigma}_B B(\Delta t) \end{cases} \quad (6.113)$$

(3) **Weight update & model assignment:** When the online condition monitoring data z_k is obtained, filter $z_{1:k}$ using Eq. (6.109), and update the particle weights as

$$\overline{w}_k^{1,p} = w_{k-1}^{1,p} \frac{1}{\sqrt{2\pi \hat{\sigma}_z^2 \Delta t}} \exp\left[-\frac{(\bar{z}_k - x_k^{1,p})^2}{2\hat{\sigma}_z^2}\right] \quad (6.114)$$

$$\overline{w}_k^{2,p} = w_{k-1}^{2,p} \frac{1}{\sqrt{2\pi \hat{\sigma}_z^2 \Delta t}} \exp\left[-\frac{(\bar{z}_k - x_k^{2,p})^2}{2\hat{\sigma}_z^2}\right] \quad (6.115)$$

where $\bar{z}_k$ is the filtered data at time t_k. $\overline{w}_k^{1,p}$ and $\overline{w}_k^{2,p}$ are the updated particle weights.

Denote $P_1 = \sum_{p=1}^{N_p} \tilde{w}_k^{1,p}$ and $P_2 = \sum_{p=1}^{N_p} \tilde{w}_k^{2,p}$. Re-update and normalize the particle weights as

$$\tilde{w}_k^{d,p} = w_{k-1}^{d,p} \frac{1}{\sqrt{2\pi \hat{\sigma}_B^2 \Delta t}} \exp\left[-\frac{(z_k - x_k^{1,p})^2}{2\hat{\sigma}_z^2}\right] \quad (6.116)$$

$$\begin{cases} w_k^{d,p} = \tilde{w}_k^{d,p} \Big/ \sum_{p=1}^{N_p} \tilde{w}_k^{d,p} \\ w_k^{\bar{d},p} = 1/N_P \end{cases} \quad (6.117)$$

where $d = 1$ and $\bar{d} = 2$ if $P_1 \geq P_2$, otherwise $d = 2$ and $\bar{d} = 1$.

(4) **Resampling and parameter update:** Resample particles and reset their weights to be $1/N_P$. Denote the resampled particles as $\{x_k^p\}_{p=1:N_p}$, $\{v_k^p\}_{p=1:N_p}$ and $\{S_k^p\}_{p=1:N_p}$, then the damage estimation and gradual degradation rate update results at time t_k can be obtained as the particle medians.

For λ and ζ, they are updated once a random shock is detected in(3). Specifically speaking, when $d = 2$ in c, a random shock is detected at t_k. Then according to the Bayes' theorem, the posterior probabilities of λ and ζ can be given as

$$P(\lambda | z_{1:k}) \propto P(\Delta_k | \lambda) \times \pi(\lambda) \quad (6.118)$$

$$P(\zeta|z_{1:k}) \propto P(z_k|\zeta, z_{k-1}) \times \pi(\zeta) \tag{6.119}$$

where Δ_k is the random shock occurrence time interval at time t_k. $\pi(\lambda)$ and $\pi(\zeta)$ are the prior probabilities of λ and ζ, respectively. According to the damage estimation result, $P(z_k|\zeta, z_{k-1})$ can be given as

$$P(z_k|\zeta, z_{k-1}) = \frac{P(z_k|z_{k-1})P(x_k < \zeta|z_k)}{\int_{x_k} P(z_k|z_{k-1})P(x_k < \zeta|z_k)} \tag{6.120}$$

Till now, through the utilization of online monitoring data, the above 3 critical parameters can be updated, and cumulative damages are estimated as well, ensuring the effective application of the RUL prediction method.

6.5.3 RUL Prediction Case of Accelerated Degradation Experiments of Rolling Element Bearings

6.5.3.1 Experimental Data Introduction

In this part, the vibration data of rolling element bearings collected from ADTs are used for demonstration (Wang et al. 2018). The testbed is described in Sect. 5.2.4. During each ADT, the radial force is applied in the horizontal direction. Therefore, the vibration data collected horizontally are expected to contain more sensitive degradation information. We thus only use the horizontal vibration signals for analysis. The detailed information of the 20 tested bearings is tabulated in Table 6.3.

In this section, RMS is extracted from vibration signals as the health indicator. Four typical tested bearings under four operating conditions and their corresponding vibration signals as well as RMS values are displayed in Fig. 6.23. From Fig. 6.23, it can be seen that these tested bearings exhibit competing degradation pattern before failure. In addition, there exists also a normal operating stage before degradation occurs, during which the tested bearings show no obvious degradation trend. To avoid unnecessary computation and increase the RUL prediction accuracy, the FPT

Table 6.3 Operating conditions of tested bearings

Operating condition	Radial force /kN	Rotating speed /(r/min)	Sampling interval	Bearing No.
No. 1	12	2100	1 min	B 1_1 ~ B 1_5
No. 2	11	2250	1 min	B 2_1 ~ B 2_5
No. 3	11	2400	16 s	B 3_1 ~ B 3_5
No. 4	10	2700	16 s	B 4_1 ~ B 4_5

6.5 RUL Prediction with Dependent Competing Failure Processes

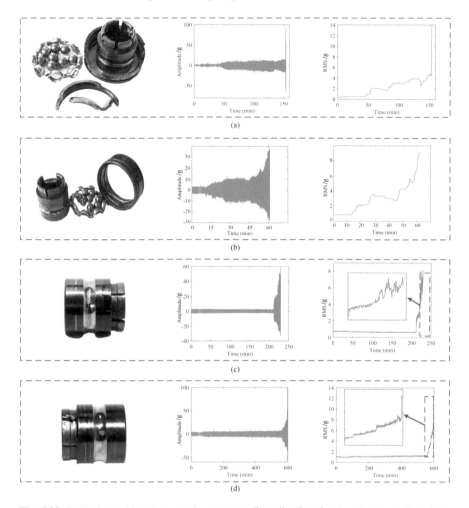

Fig. 6.23 Typical tested bearings and the corresponding vibration signals: (a) B 1_1, (b) B 2_1, (c) B 3_1, and (d) B 4_1

selection as introduced in Sect. 6.2 is also utilized here to determine the start of bearing degradation.

6.5.3.2 RUL Prediction

In this part, the presented method is applied on all of the 20 tested bearings. Moreover, 3 state-of-the-art methods are selected for comparison. For convenience, we denote the method introduced in this section as PM, and the comparison methods are denoted as CM 1 to CM 3, respectively. As can be seen in Table 6.4, CM 1 is presented by Ke et al. in (Ke et al. 2017) for the RUL prediction of soft failure processes

Table 6.4 Differences of RUL prediction methods

	CM 1	CM 2	CM 3	PM
Considering DCFPs?	No	Yes	Yes	**Yes**
Under the FPT concept?	Yes	No	Yes	**Yes**
Based on analytical solution?	Yes	Yes	No	**Yes**

made up by gradual degradation and nonfatal random shocks, so hard failure is neglected in CM 1. CM 2 based on (Peng et al. 2010), which represents a typical competing failure RUL prediction methodology that neglects the FHT concept. CM 3 is a derivative method based on (Fan et al. 2018), which utilizes the FHT concept but is based on simulation algorithm. By comparing with these three methods, we aim to demonstrate the effectiveness of the introduced method via addressing (1) the importance of adopting the FHT concept, (2) the necessity of jointly considering soft and hard failure processes and (3) the computational efficiency of the analytical expression.

In Fig. 6.24, the RUL prediction results for the above 4 bearings are presented. It can be seen that PM yields the most accurate RUL prediction results compared with the rest 3 methods. Moreover, to quantitively demonstrate the method performance, we also provide the relative RMSE, Score, and computation time results as comparison metrics. The terms relative RMSE and relative Score are defined as follows to avoid scale difference among different units:

$$\text{RMSE}_{\text{relative}}(m) = \frac{\text{RMSE}(m)}{\text{RMSE(PM)}} \tag{6.121}$$

$$\text{Score}_{\text{relative}}(m) = \frac{\text{Score}(m)}{\text{Score(PM)}} \tag{6.122}$$

According to the difference of sampling interval, the results for B 1_1 ~ B 1_5 and B 2_1 ~ B 2_5 are provided in Fig. 6.25, and the results for B 3_1 ~ B 3_5 and B 4_1 ~ B 4_5 are provided in Fig. 6.26, respectively. From them, it can be seen that in most cases, PM generally outputs an overall better regarding both accuracy and computational efficiency.

It can be observed that in Fig. 6.25, there are some cases where CM 2 and CM 3 slightly outperform PM. While in Fig. 6.26, CM 2 generally performs worse compared with that in Fig. 6.25 and there is only one case in which CM 3 performs slightly better than PM. The less accurate results of PM in those few cases can be explained as: (1) for the better performance of CM 2 in Fig. 6.25, one possible reason is the inappropriately set long sampling interval in Condition 1 and Condition 2. Specifically speaking, if the sampling interval is set too long, it may lead to the situation that the tested bearing has already exceeded the FT but not been recorded, which causes the delayed determination of failure time. In this situation, the over-estimation drawback of CM 2 is not clearly exposed. While in Fig. 6.26, the shorter sampling interval makes this drawback exposed clearly, and eventually

6.5 RUL Prediction with Dependent Competing Failure Processes

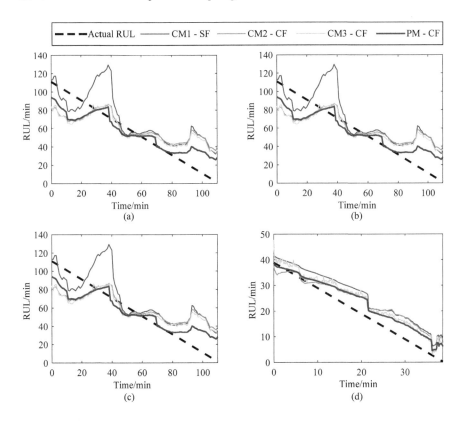

Fig. 6.24 RUL prediction results for (a) B 1_1, (b) B 2_1, (c) B 3_1, and (d) B 4_1

leads to the unsatisfactory prediction accuracy. In the era of Industrial Big Data, the advancement of instrumentation and measurement encourages the sampling interval to be set very short to capture in-time degradation information (Lei et al. 2018). In this case, PM possess more practical meaning for practical applications. (2) for the better performance of CM 3, the premise that should be neglected is that CM 3 is based on simulation. As elaborated in (Zhang et al. 2018), the accuracy of simulation-based methods has been well acknowledged, and is thus utilized as the benchmark of evaluating new RUL prediction methods. Nevertheless, CM 3 still has two intrinsic shortcomings, i.e., sensitive to parameter configuration (Li et al. 2017a), and high requirement of computational resource. The first shortcoming is directly related to the inaccuracy of CM 3 in most cases shown in Figs. 6.25 and 6.26: As described in (Zhang et al. 2018), the prediction accuracy of simulation-based methods will converge to the actual results given the number of trajectories tends to infinity and the step length tends to infinitesimal, which are impossible to be met in practical cases. Therefore, the prediction accuracy of CM 3 will inevitably be compromised. The second shortcoming can be observed in the computation time provided in both Figs. 6.25 and 6.26: given the computation platform on a PC @ 6-Core 3.6 GHz

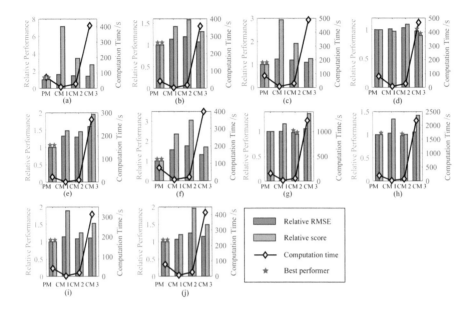

Fig. 6.25 Relative performance and computation time: (a) B 1_1, (b) B 1_2, (c) B 1_3, (d) B 1_4, (e) B1_5, (f) B 2_1, (g) B 2_2, (h) B 2_3, (i) B 2_4, and (j) B 2_5

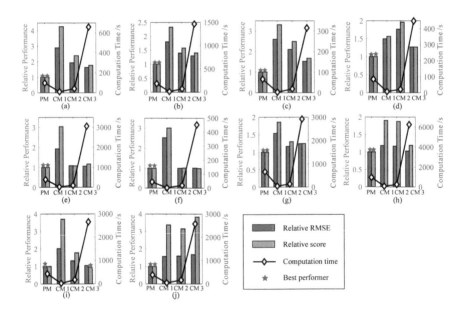

Fig. 6.26 Relative performance and computation time: (a) B 3_1, (b) B 3_2, (c) B 3_3, (d) B 3_4, (e) B3_5, (f) B 4_1, (g) B 4_2, (h) B 4_3, (i) B 4_4, and (j) B 4_5

CPU and 16 G RAM, CM 3 requires much more time to obtain the predicted results compared with the rest methods. From a maintenance management viewpoint, such high computational resource requirement would jeopardize the timeliness of maintenance planning, leading to a higher total cost and lower operational safety (Zio 2013). In a word, compared with CM 1 ~ 3, PM is able to predict the RUL for DCFPs with consistently high accuracy and with tractable computational load.

6.5.4 Epilog

Owing to the existence of multiple internal and external factors influencing the degradation of mechanical system, dependent competing failure phenomenon are commonly seen in practical cases. In view of this, a RUL prediction for mechanical system subject to DCFPs. An analytical expression of RUL PDF is firstly given under the FHT concept. Then considering the existence of random jumps in degradation trends, an integrated offline estimation and online update framework of model parameters is detailed. Finally, accelerated degradation tests of rolling element bearings are employed for verification. The results show that the introduced method owns the capability of obtaining more accurate RUL prediction results with acceptable computational load, which make it promising for maintenance management in industrial applications.

6.6 Conclusions

This chapter concentrates again on the RUL prediction of mechanical system but with a specific emphasis on data-model fusion RUL prediction methods. In industries, it is often feasible to obtain both the prior knowledge of mechanical system degradation mechanisms and condition monitoring data, and the fusion of them would yield more promising RUL prediction results. However, due to the existence of both internal and external variabilities that influence the degradation of mechanical systems, the effectiveness of data-model fusion methods is consequently compromised. Driven by this situation, this chapter focuses on four major types of variabilities that influence the RUL prediction accuracy, i.e., random fluctuation variabilities, unit-to-unit variability, time-varying operation conditions, and competing failure processes, and develops corresponding methods and algorithms to further promote the effectiveness of data-model fusion RUL prediction.

For the random fluctuation variabilities, an FPT selection method based on consecutive triggering logic is firstly developed to identify the appropriate initial time of RUL prediction from randomly fluctuated condition monitoring data. Then a RUL prediction method combining both EM algorithm and particle filtering is developed to reduce the impact of random fluctuation on RUL prediction, so as to output accurate and consistent prediction results. Through experimental analyses of FEMTO-ST accelerated degradation tests of rolling element bearings, the effectiveness of the presented method is eventually verified.

For the unit-to-unit variability, an WPM is firstly formulated for degradation modeling. Then an UMLE algorithm and a fuzzy resampling PF algorithm are jointly developed for parameter estimation and update. The experimental analyses of turbofan engine degradation dataset verify that the presented method is effective in addressing the unit-to-unit variability of degradation process, and is able to achieve satisfying RUL prediction accuracy.

For time-varying operation conditions, the impact of varying conditions on RUL prediction are firstly analyzed. Then an F2S2 method involving degradation modeling, model parameter estimation, system state estimation, and RUL prediction, is systematically developed. The developed F2S2 method is verified with accelerated degradation experiments of thrusting bearings, and the experimental results show that the presented method owns the capability of obtaining consistently accurate RUL prediction results for mechanical systems operating in varying conditions.

As for competing failure processes, the failure process competing mechanism is firstly analyzed. According to the analyzed mechanism, the degradation models for both soft and hard failure processes are formulated, and the analytical RUL PDF under FHT concept is correspondingly derived. Furthermore, a sequential estimation scheme and a TVMMPF are designed for parameter estimation and update. By utilizing the accelerated Degradation experiments of rolling element bearings, the effectiveness of the presented method in competing failure scenario is verified.

References

An D, Choi J-H, Kim NH (2013) Prognostics 101: a tutorial for particle filter-based prognostics algorithm using Matlab. Reliab Eng Syst Saf 115:161–169

Arulampalam MS, Maskell S, Gordon N, Clapp T (2002) A tutorial on particle filters for online nonlinear/non-Gaussian Bayesian tracking. IEEE Trans Signal Process 50(2):174–188

Bian L, Gebraeel N (2013) Stochastic methodology for prognostics under continuously varying environmental profiles. Stat Anal Data Min: ASA Data Sci J 6(3):260–270

Bian L, Gebraeel N, Kharoufeh JP (2015) Degradation modeling for real-time estimation of residual lifetimes in dynamic environments. IIE Trans 47(5):471–486

Casciati F, Roberts JB (1991) Reliability problems: general principles and applications in mechanics of solids and structures, vol 317. Springer

Chen C, Zhang B, Vachtsevanos G, Orchard M (2010) Machine condition prediction based on adaptive neuro–fuzzy and high-order particle filtering. IEEE Trans Industr Electron 58(9):4353–4364

Cleveland WS, Devlin SJ (1988) Locally weighted regression: an approach to regression analysis by local fitting. J Am Stat Assoc 83(403):596–610

Coutin L, Dorobantu D (2011) First passage time law for some Lévy processes with compound poisson: existence of a density. Bernoulli 17(4):1127–1135

Doksum KA, Hbyland A (1992) Models for variable-stress accelerated life testing experiments based on wener processes and the inverse gaussian distribution. Technometrics 34(1):74–82

Elattar HM, Elminir HK, Riad A (2016) Prognostics: a literature review. Complex Intell Syst 2(2):125–154

Fan M, Zeng Z, Zio E, Kang R, Chen Y (2018) A sequential Bayesian approach for remaining useful life prediction of dependent competing failure processes. IEEE Trans Reliab 68(1):317–329

Gebraeel NZ, Lawley MA, Li R, Ryan JK (2005) Residual-life distributions from component degradation signals: a Bayesian approach. IIE Trans 37(6):543–557

Ginart A, Barlas I, Goldin J, Dorrity JL (2006) Automated feature selection for embeddable prognostic and health monitoring (PHM) architectures. In: 2006 IEEE Autotestcon. IEEE, pp 195–201

Hanachi H, Yu W, Kim IY, Liu J, Mechefske CK (2019) Hybrid data-driven physics-based model fusion framework for tool wear prediction. Int J Adv Manuf Technol 101(9):2861–2872

Hao S, Yang J, Ma X, Zhao Y (2017) Reliability modeling for mutually dependent competing failure processes due to degradation and random shocks. Appl Math Modell 51:232–249

Jiang L, Feng Q, Coit DW (2012) Reliability and maintenance modeling for dependent competing failure processes with shifting failure thresholds. IEEE Trans Reliab 61(4):932–948

Jouin M, Gouriveau R, Hissel D, Péra M-C, Zerhouni N (2016) Particle filter-based prognostics: review, discussion and perspectives. Mech Syst Sig Proc 72:2–31

Ke X, Xu Z, Wang W, Sun Y (2017) Remaining useful life prediction for non-stationary degradation processes with shocks. Proceedings of the Institution of Mechanical Engineers, Part O. J Risk Reliab 231(5):469–480

Lagarias JC, Reeds JA, Wright MH, Wright PE (1998) Convergence properties of the Nelder-Mead simplex method in low dimensions. SIAM J Optim 9(1):112–147

Le Son K, Fouladirad M, Barros A, Levrat E, Iung B (2013) Remaining useful life estimation based on stochastic deterioration models: a comparative study. Reliab Eng Syst Saf 112:165–175

Lee J, Lapira E, Bagheri B, Kao H-a (2013) Recent advances and trends in predictive manufacturing systems in big data environment. Manuf Lett 1(1):38–41

Lei Y, Li N, Guo L, Li N, Yan T, Lin J (2018) Machinery health prognostics: a systematic review from data acquisition to RUL prediction. Mech Syst Signal Proc 104:799–834

Lei Y, Li N, Lin J (2016) A new method based on stochastic process models for machine remaining useful life prediction. IEEE Trans Instrum Meas 65(12):2671–2684

Li N, Lei Y, Lin J, Ding SX (2015) An improved exponential model for predicting remaining useful life of rolling element bearings. IEEE Trans Ind Electron 62(12):7762–7773. https://doi.org/10.1109/TIE.2015.2455055

Li N, Lei Y, Guo L, Yan T, Lin J (2017a) Remaining useful life prediction based on a general expression of stochastic process models. IEEE Trans Ind Electron 64(7):5709–5718

Li X, Duan F, Mba D, Bennett I (2017b) Multidimensional prognostics for rotating machinery: a review. Adv Mech Eng 9(2):1687814016685004

Li N, Gebraeel N, Lei Y, Bian L, Si X (2019) Remaining useful life prediction of machinery under time-varying operating conditions based on a two-factor state-space model. Reliab Eng Syst Saf 186:88–100

Liao H, Tian Z (2013) A framework for predicting the remaining useful life of a single unit under time-varying operating conditions. IIE Trans 45(9):964–980

Liu K, Gebraeel NZ, Shi J (2013) A data-level fusion model for developing composite health indices for degradation modeling and prognostic analysis. IEEE Trans Autom Sci Eng 10(3):652–664

Nectoux P, Gouriveau R, Medjaher K, Ramasso E, Chebel-Morello B, Zerhouni N, Varnier C (2012) PRONOSTIA: an experimental platform for bearings accelerated degradation tests. In: IEEE International Conference on Prognostics and Health Management, PHM'12. IEEE Catalog Number: CPF12PHM-CDR, pp 1–8

Olver A (2005) The mechanism of rolling contact fatigue: an update. Proceedings of the Institution of Mechanical Engineers, Part J. J Eng Tribol 219(5):313–330

Orchard ME, Hevia-Koch P, Zhang B, Tang L (2012) Risk measures for particle-filtering-based state-of-charge prognosis in lithium-ion batteries. IEEE Trans Industr Electron 60(11):5260–5269

Patil S, Patil A, Handikherkar V, Desai S, Phalle VM, Kazi FS (2018) Remaining useful life (RUL) prediction of rolling element bearing using random forest and gradient boosting technique.

In: ASME international mechanical engineering congress and exposition. American Society of Mechanical Engineers, p V013T005A019

Peng H, Feng Q, Coit DW (2010) Reliability and maintenance modeling for systems subject to multiple dependent competing failure processes. IIE Trans 43(1):12–22

Rafiee K, Feng Q, Coit DW (2014) Reliability modeling for dependent competing failure processes with changing degradation rate. IIE Trans 46(5):483–496

Ramasso E (2014) Investigating computational geometry for failure prognostics. Int J Prognostics Health Manag 5(1):005

Ranjkesh SH, Hamadani AZ, Mahmoodi S (2019) A new cumulative shock model with damage and inter-arrival time dependency. Reliab Eng Syst Saf 192:106047

Ricciardi LM (1976) On the transformation of diffusion processes into the Wiener process. J Math Anal Appl 54(1):185–199

Ross SM, Kelly JJ, Sullivan RJ, Perry WJ, Mercer D, Davis RM, Washburn TD, Sager EV, Boyce JB, Bristow VL (1996) Stochastic processes, vol 2. Wiley, New York

Roynette B, Vallois P, Volpi A (2008) Asymptotic behavior of the hitting time, overshoot and undershoot for some Lévy processes. ESAIM: probability and statistics 12:58–93

Saxena A, Goebel K (2008) Turbofan engine degradation simulation data set. https://tiarcnasagov/tech/dash/groups/pcoe/prognostic-data-repository/

Saxena A, Celaya J, Saha B, Saha S, Goebel K (2010) Metrics for offline evaluation of prognostic performance. Int J Prognostics Health Manage 1(1):4–23

Si X-S, Wang W, Hu C-H, Zhou D-H (2011) Remaining useful life estimation—a review on the statistical data driven approaches. Eur J Oper Res 213(1):1–14

Si X-S, Wang W, Hu C-H, Zhou D-H, Pecht MG (2012) Remaining useful life estimation based on a nonlinear diffusion degradation process. IEEE Trans Reliab 61(1):50–67

Si X-S, Wang W, Chen M-Y, Hu C-H, Zhou D-H (2013) A degradation path-dependent approach for remaining useful life estimation with an exact and closed-form solution. Eur J Oper Res 226(1):53–66

Si X-S, Hu C-H, Kong X, Zhou D-H (2014) A residual storage life prediction approach for systems with operation state switches. IEEE Trans Ind Electron 61(11):6304–6315

Tang S, Yu C, Wang X, Guo X, Si X (2014) Remaining useful life prediction of lithium-ion batteries based on the wiener process with measurement error. Energies 7(2):520–547

Wang B, Lei Y, Li N, Li N (2018) A hybrid prognostics approach for estimating remaining useful life of rolling element bearings. IEEE Trans Reliab 69(1):401–412

Yan T, Lei Y, Li N, Wang B, Wang W (2021) Degradation modeling and remaining useful life prediction for dependent competing failure processes. Reliability Engineering & System Safety 212:107638. https://doi.org/10.1016/j.ress.2021.107638

Ye ZS, Tang LC, Xu HY (2011) A distribution-based systems reliability model under extreme shocks and natural degradation. IEEE Trans Reliab 60(1):246–256

Zhang B, Sconyers C, Byington C, Patrick R, Orchard ME, Vachtsevanos G (2010) A probabilistic fault detection approach: application to bearing fault detection. IEEE Trans Ind Electron 58(5):2011–2018

Zhang Z, Si X, Hu C, Lei Y (2018) Degradation data analysis and remaining useful life estimation: a review on Wiener-process-based methods. Eur J Oper Res 271(3):775–796

Zheng J-F, Si X-S, Hu C-H, Zhang Z-X, Jiang W (2016) A nonlinear prognostic model for degrading systems with three-source variability. IEEE Trans Reliab 65(2):736–750

Zio E (2013) Prognostics and health management of industrial equipment. Diagnostics and prognostics of engineering systems: methods and techniques, pp 333–356

Glossary

AACO accumulative amplitudes of carrier orders
AC alternating current
ADT accelerated degradation test
AE acoustic emission
AI artificial intelligence
ANN artificial neural network
AR autoregressive
ARE absolute relative error
ARMA autoregressive moving average
AUC area under the receiver operation characteristic curve
BFP bearing fault in the planet gear
BM Brownian motion
BN batch normalization
CaAE capsule auto-encoder
CBM condition-based maintenance
CD contrastive divergence
CDET compensation distance evaluation technique
CDF cumulative distribution function
CLSTM convolutional long short-term memory
CNC computer numerical control
CNN convolutional neural network
CPG crack in the planetary gear
CS crack in the sun gear
CWRU Case Western Reserve University
DAFD domain adaptation for fault diagnosis
DAN deep adaptation network
DAQ data acquisition
DBN deep belief network
DBNCL deep belief network with continual learning
DCFP dependent competing failure process
DCN deep convolutional network

DDC deep domain confusion
DDL dynamic dense layer
DNN deep neural network
DPS degradation process simulation
DPTLN deep partial transfer learning network
DSCN deep separable convolutional network
EEMD ensemble empirical mode decomposition
EM expectation maximization
EMD empirical mode decomposition
ERDS energy ratio based on difference spectrum
ERM empirical risk minimization
FCL fully-connected layer
FFT fast Fourier transform
FHT first hitting time
FPT first predicting time
FT failure threshold
GAN generative adversarial network
GAP global average pooling
GA genetic algorithm
GMP global max pooling
HI health indicator
HPP homogeneous Poisson process
i.i.d. independent identically distributed
IF inner race failure
ILWAL instance-level weighted adversarial learning
IMF intrinsic mode function
IoT Internet of things
KNN *K* nearest neighbor
KKT Karush-Kuhn-Tucker
KL Kullback-Leibler
LOESS locally weighted scatter smoothing method
mAP mean average precision
MCNN multi-scale convolutional neural network
MLAN multi-layer adaptation network
MLP multi-layer perceptron
MMD maximum mean discrepancy
MPE multi-scale permutation entropy
MSCAN multi-scale convolutional attention network
MSE mean square error
MRVM multiclass relevance vector machine
OAA one-against-all
OAO one-against-one
OF outer race failure
OSVM open set support vector machine
PCA principal component analysis

PDF probability density function
PE permutation entropy
PHM prognostics and health management
RBF radial basis function
RBM restricted Boltzmann machine
RCNN recurrent convolutional neural network
RDN residual dense network
ReLU rectified linear unit
ResNet residual network
RF rolling element failure
RKHS reproducing kernel Hilbert space
RMS root mean square
RMSE root mean square error
RUL remaining useful life
RVM relevance vector machine
SD standard deviation
SE squeeze and excitation
SLT statistical learning theory
SNR signal-to-noise ratio
SRM structural risk minimization
SPRO spectrum peak ratio of bearing outer race
SPRI spectrum peak ratio of bearing inner race
SPRR spectrum peak ratio of bearing roller
STFT short-time Fourier transform
SVM support vector machine
SW wear in the sun gear
TCA transfer component analysis
TD temporal dimension
TSA time synchronous average
t-SNE t-distributed stochastic neighbor embedding
UMLE unit maximum likelihood estimation
UtUV unit-to-unit variability
VC Vapnik-Chervonenkis
WKNN weighted K nearest neighbor
WNN wavelet neural network
WPM Wiener process-based method
WPT wavelet packet transform
WPTLS wavelet packet transform with the lifting scheme
WPW wear in the planetary gear
WS wear in the sun gear